MUSHROOMS OF THE SOUTHEASTERN UNITED STATES

MUSHROOMS OF THE SOUTHEASTERN UNITED STATES

ALAN E. BESSETTE · WILLIAM C. ROODY
ARLEEN R. BESSETTE · DAIL L. DUNAWAY

SYRACUSE UNIVERSITY PRESS

Copyright © 2007 by Syracuse University Press
Syracuse, New York, 13244-5160

ALL RIGHTS RESERVED

First Edition 2007

07 08 09 10 11 12   6 5 4 3 2 1

The paper used in this publication meets the minimum requirements of American National Standard for Information Sciences—Permanence of Paper for Printed Library Materials, ANSI Z39.48-1984.∞™

For a listing of books published and distributed by Syracuse University Press, visit our Web site at SyracuseUniversityPress.syr.edu

ISBN-13: 978-0-8156-3112-5
ISBN-10: 0-8156-3112-X

LIBRARY OF CONGRESS CATALOGING-IN-PUBLICATION DATA

Mushrooms of the southeastern United States / Alan E. Bessette . . . [et al.].
— 1st ed.
    p. cm.
    Includes bibliographical references and index.
    ISBN-13: 978-0-8156-3112-5 (cloth : alk. paper)
    ISBN-10: 0-8156-3112-X (cloth : alk. paper)
    1. Mushrooms—Southern States—Identification. I. Bessette, Alan.
QK605.5.S67M87 2007
579.60975—dc22   2006032144

Front cover illustration: *Clavaria zollingeri;* p. ii: *Gomphus floccosus;* p. x: *Bolbitius vitellinus;* p. xii: *Marasmius rotula;* p. xvi: *Hygrophorus marginatus* var. *concolor;* p. 12: *Morchella esculenta* complex.

Color separations by Pre Tech, Hartford, Vermont
Printed and bound by CS Graphics Pte., Ltd., Singapore

> **NOTICE** Though this book includes information regarding the edibility of various mushrooms, it is *not* intended to function as a manual for the safe consumption of wild mushrooms. Readers interested in consuming wild fungi should consult other sources of information—especially experienced mycophagists, mycologists, and literary sources—before eating any wild mushrooms. Neither the authors nor the publisher are responsible for any undesirable outcomes that may occur for those who fail to heed this warning.

To Emily Johnson, Ray Fatto, and Dan Guravich
Gentle smiles, gentle hearts,
Greatly missed

Distribution Range for Mushrooms of the Southeastern United States

Contents

Preface / xi

Acknowledgments / xiii

Introduction to Mycology / 1
    Mushroom Facts and Fallacies 1 · Fungal Anatomy: Parts of a Mushroom 2
    Mycorrhizal Relationships 3 · Equipment for Collecting Mushrooms 4
    When to Collect Mushrooms 4 · Where to Collect Mushrooms 5 · How to Collect
    Mushrooms 5 · Spore Prints 6 · Notes on Descriptions of Illustrated Species 6
    How to Use This Book 7 · Mushroom Identification Procedure 8

Key to the Major Groups of Mushrooms / 9

COLOR PLATES
    Chanterelles and Allies 14 · Split Gill and Ally 15 · Gilled Mushrooms 16
    Tooth Fungi 47 · Boletes 48 · Polypores 58 · Stinkhorns 65 · Morels,
    False Morels, and Allies 69 · Fiber Fans and Vases 71 · Branched and Clustered
    Corals 72 · Earth Tongues, Earth Clubs, and Allies 73 · *Cordyceps, Claviceps,* and
    Allies 75 · Bird's-nest Fungi 76 · Cup and Saucer Fungi 77 · Jelly Fungi 79
    Cauliflower Mushrooms 81 · Puffballs, Earthballs, Earthstars, and Allies 81
    Carbon and Cushion Fungi 87 · Crust and Parchment Fungi and Allies 88
    Hypomyces 90 · Cedar-Apple Rust 90

SPECIES DESCRIPTIONS

Chanterelles and Allies / 93
    Cantharellus 93 · *Craterellus* 96 · *Gomphus* 97

Split Gill and Ally / 99
    Plicaturopsis 99 · *Schizophyllum* 99

Gilled Mushrooms / 101
    Agaricus 101 · *Agrocybe* 103 · *Amanita* 104 · *Anellaria* 114 · *Anthracophyllum* 115
    Armillaria 115 · *Asterophora* 116 · *Baeospora* 117 · *Bolbitius* 117 · *Callistosporium* 118
    Catathelasma 118 · *Chlorophyllum* 119 · *Chroogomphus* 119 · *Claudopus* 120
    Clitocybe 120 · *Clitocybula* 122 · *Collybia* 122 · *Conocybe* 123 · *Coprinus* 123
    Cortinarius 126 · *Crepidotus* 129 · *Crinipellis* 129 · *Cyptotrama* 130 · *Cystoderma* 130
    Entoloma 131 · *Flammulina* 131 · *Galerina* 132 · *Gerronema* 132 · *Gymnopilus* 133

Gymnopus 135 · *Hebeloma* 136 · *Hohenbuehelia* 138 · *Hygrocybe* 138 · *Hygrophorus* 142 *Hypholoma* 144 · *Hypsizygus* 144 · *Inocybe* 145 · *Laccaria* 146 · *Lactarius* 147 *Lentinellus* 163 · *Lentinula* 163 · *Lentinus* 164 · *Lepiota* 166 · *Leucoagaricus* 168 *Leucocoprinus* 169 · *Limacella* 171 · *Macrocybe* 171 · *Macrolepiota* 172 · *Marasmiellus* 173 *Marasmius* 174 · *Melanoleuca* 175 · *Mycena* 176 · *Nolanea* 178 · *Omphalina* 180 *Omphalotus* 180 · *Panaeolus* 180 · *Panellus* 181 · *Paxillus* 181 · *Pholiota* 182 *Phyllotopsis* 183 · *Pleurotus* 184 · *Pluteus* 185 · *Pouzarella* 186 · *Psathyrella* 186 *Psilocybe* 188 · *Rhodocollybia* 188 · *Rhodocybe* 189 · *Ripartitella* 190 · *Russula* 190 *Squamanita* 196 · *Strobilurus* 197 · *Stropharia* 197 · *Tricholoma* 198 · *Tricholomopsis* 200 *Volvariella* 200 · *Xeromphalina* 201

Tooth Fungi / 202
Hericium 202 · *Hydnellum* 203 · *Hydnum* 204 · *Mycorrhaphium* 204 · *Phellodon* 205

Boletes / 206
Austroboletus 206 · *Boletellus* 208 · *Boletus* 208 · *Chalciporus* 230 · *Gyrodon* 230 *Gyroporus* 231 · *Leccinum* 233 · *Phylloporus* 234 · *Pulveroboletus* 235 · *Strobilomyces* 236 *Suillus* 237 · *Tylopilus* 239 · *Xanthoconium* 244

Polypores / 247
Albatrellus 247 · *Antrodia* 249 · *Bjerkandera* 249 · *Bondarzewia* 250 · *Cerrena* 250 *Coltricia* 251 · *Daedaleopsis* 251 · *Fistulina* 252 · *Fomes* 252 · *Fomitopsis* 253 *Ganoderma* 253 · *Gloeophyllum* 255 · *Gloeoporus* 257 · *Hapalopilus* 257 · *Hexagonia* 258 *Hydnopolyporus* 259 · *Inonotus* 259 · *Ischnoderma* 261 · *Laetiporus* 261 · *Lenzites* 263 *Meripilus* 263 · *Microporellus* 264 · *Nigroporus* 265 · *Phaeolus* 265 · *Phellinus* 266 *Polyporus* 266 · *Pseudofavolus* 267 · *Pseudofistulina* 268 · *Pycnoporus* 268 *Schizopora* 269 · *Spongipellis* 269 · *Trametes* 270 · *Trichaptum* 271

Stinkhorns / 273
Aseroe 273 · *Blumenavia* 274 · *Dictyophora* 274 · *Linderia* 274 · *Lysurus* 275 *Mutinus* 276 · *Phallogaster* 277 · *Phallus* 277

Morels, False Morels, and Allies / 279
Gyromitra 279 · *Helvella* 280 · *Morchella* 282

Fiber Fans and Vases / 284
Thelephora 284

Branched and Clustered Corals / 286
Clavaria 286 · *Clavicorona* 287 · *Clavulina* 287 · *Clavulinopsis* 288 · *Ramaria* 288 *Tremellodendropsis* 289

Earth Tongues, Earth Clubs, and Allies / 291
Clavariadelphus 291 · *Cudonia* 292 · *Leotia* 292 · *Microglossum* 293 *Spathularia* 293 · *Trichoglossum* 293 · *Underwoodia* 294 · *Xylocoremium* 294

Cordyceps, *Claviceps*, and Allies / 295
Cordyceps 295 · *Nomuraea* 296

Bird's-nest Fungi / 298
Crucibulum 298 · *Cyathus* 298

Cup and Saucer Fungi / 299
Aleuria 299 · *Ascocoryne* 299 · *Chlorociboria* 300 · *Galiella* 300 · *Helvella* 300
Humaria 301 · *Peziza* 301 · *Plicaria* 301 · *Sphaerosporella* 301 · *Urnula* 302
Wolfina 302

Jelly Fungi / 303
Arrhytidia 303 · *Auricularia* 303 · *Calocera* 304 · *Dacrymyces* 304
Dacryopinax 304 · *Exidia* 305 · *Sebacina* 305 · *Syzygospora* 306 · *Tremella* 306

Cauliflower Mushrooms / 307
Sparassis 307

Puffballs, Earthballs, Earthstars, and Allies / 308
Arachnion 308 · *Astraeus* 309 · *Bovista* 309 · *Bovistella* 309 · *Calostoma* 310
Calvatia 311 · *Chorioactis* 312 · *Disciseda* 312 · *Elaphomyces* 313 · *Endoptychum* 313
Geastrum 314 · *Lycoperdon* 316 · *Macowanites* 317 · *Pisolithus* 318 · *Rhizopogon* 318
Rhopalogaster 319 · *Scleroderma* 320 · *Vascellum* 321 · *Zelleromyces* 322

Carbon and Cushion Fungi / 323
Daldinia 323 · *Diatrype* 323 · *Hypoxylon* 324 · *Peridoxylon* 324 · *Ustulina* 325
Xylaria 325

Crust and Parchment Fungi and Allies / 327
Aleurodiscus 327 · *Cotilydia* 327 · *Hydnochaete* 328 · *Phlebia* 328
Pulcherricium 329 · *Stereum* 329 · *Xylobolus* 330

Hypomyces / 331
Hypomyces 331

Cedar-Apple Rust / 333
Gymnosporangium 333

Appendixes / 337
A. Microscopic Examination of Mushrooms / 337
B. Chemical Reagents and Mushroom Identification / 338
C. Classification / 339
D. Mycophagy / 343

Glossary / 353

Recommended Reading / 359

Index to Common Names / 363

Index to Scientific Names / 367

Preface

The objective of this book is to provide a relatively comprehensive guide to the mushrooms of the southeastern United States. Although the book will stand on its own, it is intended to compliment and serve as a companion to *Mushrooms of Northeastern North America*. These volumes together form a foundation and reference for identifying the mushrooms found in eastern North America from Canada to the subtropics of Florida and Texas. The current volume features more than 450 species that are fully described and illustrated with photographs, many for the first time in color. The photographs were selected for high-quality color fidelity and documentary merit. We hope they also reflect our subject's aesthetic appeal. The number of species described and illustrated in color is substantially higher than has previously appeared in any other single work devoted to the mushrooms of the southeastern United States. We provide cross-references to additional species occurring in the region that are illustrated in *Mushrooms of Northeastern North America* (abbreviated *MNE* in the species descriptions). If you find a mushroom that is not described in this book, it may be described and illustrated in *MNE*.

Although we have endeavored to supply the necessary detail required by advanced students and professional mycologists, this book emphasizes identification based primarily on macroscopic field characters for easier use by a general audience. Each species illustration is accompanied by a detailed description of macroscopic and microscopic features based on the concepts of the authors who first described them.

From the interior mountain habitats to the Piedmont and the coastal lowlands, the southeastern United States is virtually unrivaled in its fungal diversity. It is estimated that up to five thousand or more mushroom species occur here. Because of the many habitats and the extensive biodiversity, it is impossible to feature every species, variety, or form of mushroom that may be encountered in this vast and diverse region. Many species are not yet scientifically described, and others remain to be discovered. This work is by no means a complete accounting of the mycoflora found in the Southeast.

We offer this guide as a step in the collective process of better comprehending the mushrooms of the southeastern United States. It is our hope that whatever the reason for one's interest in mushrooms—whether it be scientific study, the search for edible species, or sheer appreciation of their beauty—this book will serve as a trustworthy guide to the mushrooms of the southeastern United States.

Acknowledgments

Many individuals have contributed to this work. Our thanks to Brian Akers, Meredith Blackwell, Ernst Both, Bill Burk, Bob Gilbertson, Jim Kimbrough, Steve Miller, Donna Mitchell, Barrie Overton, Ron Petersen, Gary Samuels, and Walt Sundberg for their assistance with mycological notes, technical information, and species identification. For contributing their valuable photographs that greatly enhanced the beauty and functionality of this book, our thanks to Brian Akers, Ulla Benny, Meredith Blackwell, Robert Chapman, Tim Geho, Don Gray, David Lewis, Joe Liggio, Owen McConnell, Van Metzler, Jim Murray, Robert Williams, and David Work. We are especially grateful for Marcia Guravich and Seanna Annis's kindness in allowing us to use photography of the late Dan Guravich and Richard Homola, respectively. Ike Forester and the North American Mycological Association were generous in sharing several illustrations from the Dan Guravich Slide Collection. Thanks also to Tina Dunaway and Billie Both for collecting specimens for photography and study, to Sam Norris for the beautiful mushroom illustration included in the introduction, and to all the members of the many mushroom clubs and mycological societies who have included us in their activities and have been willing to share their fungi and their knowledge of them over the years. We greatly appreciate the efforts and contributions of Annie Barva, who copyedited the manuscript, and of Christopher Kuntze, who designed the book. Finally, this book would not have been possible without the support of Peter Webber and the staff at Syracuse University Press.

MUSHROOMS OF THE SOUTHEASTERN UNITED STATES

Introduction to Mycology

MUSHROOMS FACTS AND FALLACIES

Mushrooms, being neither plant nor animal, belong to their own kingdom, the kingdom Fungi. Fungi lack chlorophyll and cannot produce food for themselves. They obtain nutrients through a process of external digestion and absorption. Some, as decomposers, extract what they need from dead and decaying materials and are called *saprobes*. Those that attack live plants, animals, or other fungi are called *parasites*. The third group exists in a mutually beneficial relationship with living trees or other plants. This relationship is called *mycorrhizal,* one in which both partners obtain what they need, in part, from the other. Learning which food source, or *substrate*, a particular kind of mushroom requires greatly improves the likelihood of successfully finding it.

Still, what exactly is a mushroom? If you go by names, you will surely be confused. Many mushrooms have at least two names: a common name and a scientific name. The common name may be widely used or regional. It may reflect a particular feature of the mushroom or honor an individual after whom the mushroom was named. For example, *Boletus frostii* has two common names: Frost's Bolete (in honor of Charles Christopher Frost) and the Apple Bolete (describing its appearance). *Boletus edulis* has no fewer than six common names! Some species lack common names altogether, but each mushroom has a scientific name. Scientific names always have two parts. The first part is the genus, or "generic name," the first letter of which is always capitalized. The second part is the species, or "specific name," with all letters in lower case. Owing to disagreement among taxonomists, there are discrepancies in what are considered some mushrooms' "correct" scientific names. Therefore, more than one scientific name may be assigned to a single mushroom, depending on the reference used.

Imagine a vast underground network of fine, interconnected, and interwoven filaments (called *hyphae*). When conditions are correct (temperature, moisture, nutrients, pH, daylight length), this living mat, called a *mycelium*, periodically sends forth an above-ground fruiting body, a mushroom. Mushrooms produce seed-like microscopic reproductive structures known as *spores*. Picking a mushroom, if carefully done, has no more ecological impact than picking a piece of fruit from a tree.

FUNGAL ANATOMY: PARTS OF A MUSHROOM

When first starting out as a mushroom collector, you would be wise to become acquainted with mushroom anatomy because it differs from that of other organisms and will likely be unfamiliar to you. Refer to the accompanying illustrations as you read about the basic macroscopic features described.

As previously stated, a mushroom is the fruiting body that arises from the larger fungal organism, the mycelium, which is typically concealed beneath the soil or within decaying wood or other substrate. A mushroom begins as an immature form called a *button*. Depending on the species, the button may initially be entirely surrounded by a membranous structure known as the *universal veil*. As the mushroom within expands, it stretches and tears the universal veil, often leaving remnants on the cap. These remnants are referred to as *patches* or *warts*. There may also be a cup-like remnant of the universal veil called a *volva* around the base of the mushroom stalk. Most mushrooms lack a universal veil and therefore have neither patches nor warts nor a volva.

The typical mature mushroom has a *cap* and a *stalk*. The stalk may be attached to the cap's center, off its center, or at its side. On the cap's underside, there may be *gills*, which are radiating, blade-like structures upon which spores are produced. In place of gills, there may be *teeth* or *spines* or *tubes* (the open end of each tube is called a *pore*). The tubes are packed closely together; their collective pores are known as the *pore*

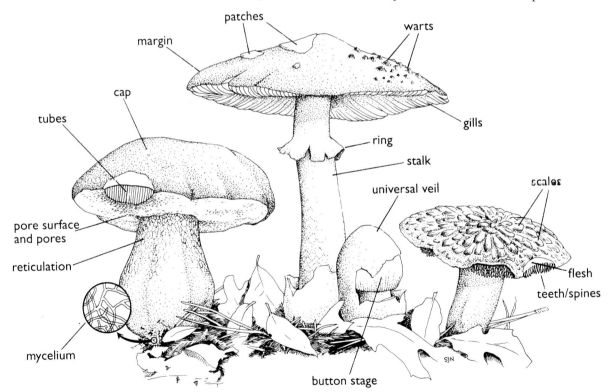

Parts of a Mushroom. Drawing by Sam Norris.

surface. Both teeth or spines and tubes serve the same basic reproductive function as gills. In some species, the underside of the immature cap is covered by a piece of tissue stretching from the cap's edge, or *margin,* to the stalk. This tissue, the *partial veil* (not shown), covers and protects the developing gills or tubes. As the mushroom cap expands, the partial veil tears, often leaving remnants on the cap margin or adhering to the stalk, where it forms a *skirt* or *ring.*

Refer to the glossary for more precise definitions of these structures and for other mycological terms that you will encounter in this book.

MYCORRHIZAL RELATIONSHIPS

Mycorrhiza is a mutually beneficial symbiotic association between fungal hyphae and the roots of living trees or other type of host plant. Host trees engaged in a mycorrhizal relationship tend to grow more rapidly, better tolerate changes in the soil pH, require fewer nutrients, and be more resistant to many soil-borne diseases caused by pathogenic fungi and nematodes. The fungal partner in a mycorrhizal relationship enhances the host plant's ability to obtain water and nutrients, including phosphorus and nitrogen. In exchange for providing these benefits to the host plant, the fungal partner receives nutrients such as carbohydrates and vitamins essential for its growth.

Two major types of mycorrhizae are recognized: *endotrophic mycorrhizae (endomycorrhizae)* and *ectotrophic mycorrhizae (ectomycorrhizae).* The fungal partner in an endotrophic mycorrhizal relationship enters the root hairs of the host plant and penetrates the cells of the root cortex, forming extensive fine branches called *arbuscules.* These arbuscular fungi do not form the typical mushroom fruiting bodies with which we are familiar. Although commonly observed microscopically in many plants such as American Beach Grass, Beach Plum, some orchids, and Rugosa Rose, this type of mycorrhiza is not formed with most tree species. Instead, trees such as pine and most other conifers, beech, birch, and oak form ectotrophic mycorrhizae. In this relationship, the hyphae of the fungal partner surround the root tip, forming an extensive layer called a mantle. The hyphae enter the first few layers of the cortex, growing between the cortical cells but not penetrating them. This intercortical fungal growth is called a Hartig net.

More than two thousand ectotrophic mycorrhizal fungi have been identified to date. Most of these fungi are members of the class Holobasidiomycetes, which includes boletes, earthballs and their relatives, fiber fans, and gilled mushrooms. Some of the more commonly encountered genera that have ectotrophic mycorrhizal members include *Amanita, Boletus, Gomphidius, Hebeloma, Laccaria, Lactarius, Pisolithus, Russula, Scleroderma, Suillus, Thelephora,* and *Tricholoma.*

The mutually beneficial relationship between fungi and plants can be described as one of nature's perfect partnerships. Without their fungal partners, many trees would grow much more slowly and in some cases not at all. Impaired growth of tree species would threaten forest habitats and may have devastating effects on the biodiversity found there.

EQUIPMENT FOR COLLECTING MUSHROOMS

One of the advantages of collecting mushrooms is that the required equipment is both minimal and inexpensive. The following list includes suggested items that we have found to be useful:

1. A basket or sturdy-sided container for transporting your finds (plastic buckets or bags are not recommended because specimens tend to become crushed in them)
2. A sturdy knife for digging, cutting, and cleaning
3. Waxed paper, brown lunch bags, or waxed paper sandwich bags for wrapping specimens and keeping them separate
4. Insect repellant
5. Compass, map, and whistle
6. Pencil and paper for field notes and spore prints
7. Camera equipment and film
8. Walking stick

WHEN TO COLLECT MUSHROOMS

Mushroom fruiting patterns are affected by various conditions, including humidity, temperature, daylight length, and precipitation. Although it is impossible to predict exactly when mushrooms will fruit, following some basic guidelines will help ensure a successful foray.

Some mushrooms have only one fruiting season, whereas others have split or multiple fruiting periods. These periods must be discovered for the particular geographic area in which you are collecting. In general, the collecting season in the southeastern United States begins in March and extends through November in the upper range or through January or even year-round in the southern range.

The best time to collect is typically from two to five days after a significant soaking rainfall or sooner if rains have been falling at frequent intervals. Sunny, windy, dry days may assist with spore dispersal, but they reduce fruiting by minimizing the moisture essential to it. Some mushrooms fruit optimally during hot, humid weather, whereas others prefer cooler temperatures. Summer and fall are usually the most productive seasons for collecting; there are exceptions, however.

Take the time to learn weather and fruiting patterns for your own area. Keep notes of when and where you collect species so that you can refer to them in the future. In this way, you have the best chance of keeping your basket filled.

WHERE TO COLLECT MUSHROOMS

Once you know when to collect mushrooms, the next question is, Where? Again, the answer depends on several factors: local weather conditions, the type of mushroom being sought, time of year, and geographic location.

Because most fungi require moisture to fruit, the best locations to search for mushrooms during times of extended dry weather are naturally moist areas: along the shorelines of ponds and lakes; along the banks of streams, creeks, and rivers; and in cool, sheltered ravines. Fallen trees and stumps often retain moisture longer than the surrounding soil and are good places to explore. However, after several days of rain, these same locations may be too wet, and you might do better to search in drier locations: hillsides, meadows, and sandy areas.

Because all mushrooms require rather specific substrates, and because some exist in a mycorrhizal relationship with specific trees or other plants, the kind of mushroom you are hunting will affect where you should look: on the ground or on trees and decaying wood; beneath conifers or hardwoods; in meadows; or in bogs. It is extremely helpful to learn what a mushroom requires in order to know where to look for it.

Fruiting patterns are usually seasonal, so some mushrooms, such as morels, are found only in the spring. Others, such as boletes, are most abundant during summer and fall. Still others—*Tricholoma* species, for example—are most abundant during fall. Fruiting patterns vary from one geographic location to another throughout North America.

HOW TO COLLECT MUSHROOMS

While you are collecting, your primary concern should be to keep the specimens in as good condition as possible so that key identifying features are not lost or damaged by storage and transport. It is also important to keep species from becoming mixed together. Therefore, we recommend wrapping collections of a single species together in either waxed paper or brown paper sandwich bags. These materials allow moisture to escape, unlike plastic bags and wraps, which cause moisture to build up, hastening decay and the development of harmful bacteria.

For identification purposes, it is also important to collect whole specimens, including parts that may be partially buried. Carefully dig up specimens rather than pulling them up or cutting them off at the base. Gather specimens in various stages of development whenever possible to help ensure accurate identification. Note the substrate, location, and nearby tree types. Place your notes with your collections. You might use an extra sheet of paper to make an in-the-field spore print, as described in the next section.

SPORE PRINTS

Prominent field features alone will not guarantee accurate identification. Spore prints often bridge the gap between macroscopic and microscopic identification, making it possible to differentiate between similar and not so easily identifiable fungi.

Spore prints are formed when the mushroom spores are allowed to drop undisturbed onto a surface. They are simple to make. Cut the cap from the stalk, leaving about a ½-inch stub to serve as a pedestal. Place the cap with the gill or pore side down on a piece of clean white paper and cover it with a cup or bowl to prevent dispersal by drafts. Allow several hours for a good, thick spore deposit. The color of the spore deposit can be essential in separating similar species, especially at the genus level. While you are collecting in the field, you might also wrap one or two mature caps, with the fertile surface facing down, in a piece of white paper. In this way, you might have an adequate spore print ready for the next step in identification once you arrive home.

A spore print is also useful if you intend to do microscopic work because it is the best source of mature spores.

NOTES ON DESCRIPTIONS OF ILLUSTRATED SPECIES

SCIENTIFIC NAME: A Latin scientific name is provided for each species. The name used may not be the same one commonly found in other field guides; it may reflect a recent taxonomic change. We often provide an alternative name in the Comments section of the species description. We have also provided something called "author citations" for the convenience of advanced mycologists who may find this information useful. In some instances, the author citation is simple—for example, *Cantharellus ignicolor* Petersen was first described and named by Ronald Petersen. Sometimes, however, the names of the original authors are enclosed within parentheses and followed by the names of those who later reclassified the fungi. *Gyrodon merulioides* (Schweinitz) Singer was first described as *Daedalea merulioides* by Lewis Schweinitz, but Rolf Singer later concluded that the species should be placed in the genus *Gyrodon*.

COMMON NAME: Common names are provided whenever they are known to exist.

MACROSCOPIC FEATURES: The appearance of the fruiting body is described, including size, shape, color, staining reactions, odor, and taste. The morphological features of some mushrooms, such as puffballs and jelly fungi, are described under the single heading "Fruiting." Others have morphological features described under separate headings such as "Cap," "Gills," "Pore Surface," and "Stalk." Many mushrooms have distinctive odors, which is noted if useful. The flesh of some species has a distinctive taste, which is also indicated if known. *If you choose to taste the tissues of a mushroom, be advised that some mushrooms taste hot and peppery and may irritate, burn, or numb your mouth if chewed for an extended period.* Note also that there is no significant risk in properly tasting mushrooms *so long as you don't swal-*

low the tissue! To taste mushrooms safely, place a small piece in your mouth, chew it for only a few seconds and spit it out. If the taste is mild (not bitter or peppery), wait a minute, then chew a second small piece for fifteen to thirty seconds, and again spit it out; some mushrooms' bitter or acrid tastes are subtle or tardy.

SPORE PRINT COLOR: Spore print color is a valuable characteristic for mushroom identification, especially for gilled mushrooms and boletes.

MACROCHEMICAL TESTS: Various chemicals, including KOH, NH_4OH, and $FeSO_4$, may be applied to mushroom tissues and the combination observed for any color reaction produced. Macrochemical testing is best done on fresh specimens at the time of collection or soon thereafter. Additional information is provided in appendix B, "Chemical Reagents and Mushroom Identification."

MICROSCOPIC FEATURES: Information about spore size, shape, surface features, and microscopic color is presented in the descriptions. Additional information such as length of *asci*, shape of *paraphyses*, presence of *setae*, and other useful microscopic characters is included where appropriate.

FRUITING: The habit (solitary, scattered, in groups, or in fused clusters), the substrate, the habitat, the fruiting period, and the frequency of mushroom growth are described. The fruiting period is stated as a month-to-month range and describes the time during which the mushroom is likely to occur. On occasion, mushrooms will appear outside of their expected fruiting period owing to atypical weather conditions. Frequency is estimated for the entire region; species listed as "occasional" may be locally abundant in some areas or rare to absent in others.

EDIBILITY: Species known to be edible are listed so. The term *inedible* is used for mushrooms that are too acrid or bitter or too fibrous or woody for consumption. Any species described as poisonous, not recommended, or edibility unknown *should not be eaten!*

COMMENTS: This section includes brief descriptions of similar species, alternate names, warnings, explanations, and other useful or interesting information about a species.

HOW TO USE THIS BOOK

The mushroom species illustrated in this book are arranged in major groups based on similarities in their appearance. The sequence in which those species occur corresponds to the sequence in which they appear in the Key to the Major Groups of Mushrooms. They are not intentionally arranged by order, family, or genus. Representatives of each of the twenty major groups are illustrated in the key. This key and the accompanying brief descriptions in it constitute the foundation upon which this entire work is based.

If you know the identification of a species and want to read about it, consult the index. If, however, you wish to identify an unknown mushroom, follow the steps presented in the next section. Before attempting the identification procedure, be sure to collect as many different stages of growth as possible (in as good condition as possible), make notes about the habitat and substrate, and obtain a spore print if possible. Identifying mushrooms can sometimes be a very difficult task, and every bit of information is useful.

MUSHROOM IDENTIFICATION PROCEDURE

1. Always start at the beginning of the Key to the Major Groups of Mushrooms and determine which major group best describes the mushroom you are attempting to identify by reading the information provided and by examining the color illustrations.
2. Once you have determined the major group, turn to the page indicated for that group and read the introductory information presented.
3. Compare the features of your unknown mushroom with the descriptions of the species within the major group. Determine which species description most closely matches your specimen. Be sure to read the Comments section when provided.

When reading descriptions, pay particular attention to qualifiers such as *usually, sometimes, frequently, typically, when young, at maturity,* and others. *Frequently* does not mean "always"; it typically means "usually, but not always." We have endeavored to avoid such qualifiers, but in many cases they are unavoidable.

The language used in this book is nontechnical, except in instances where simpler definitions become cumbersome or less precise and in the Microscopic Features sections, as needed. The technical terms used are defined in the glossary.

Key to the Major Groups of Mushrooms

1A   Fruit body with gills, vein-like ridges, teeth (spines), or pores present on the undersurface → 2

1B   Fruit body lacking gills, vein-like ridges, teeth (spines), or pores → 7

    2A   Fruit body with gills or gill- to vein-like ridges on the undersurface → 3

    2B   Fruit body lacking gills or vein-like ridges → 5

3A   Fruit body with cap and stalk or with funnel-like shape; undersurface with blunt, gill- to vein-like ridges that are often forked and crossveined or nearly smooth; usually growing on the ground → **Chanterelles and Allies** (p. 14)

3B   Fruit body with normal, split, or crimped gills present on the undersurface → 4

    4A   Fruit body small, stalkless; undersurface gill-like but longitudinally split or distinctly crimped, often forked and vein-like; growing on wood → **Split Gill and Ally** (p. 15)

    4B   Fruit body small to large; undersurface with knifeblade-like gills radiating from a stalk or, on stalkless species, from point of attachment to the substrate; growing on a variety of substrates → **Gilled Mushrooms** (p. 16)

5A   Fruit body fleshy, corky, or leathery with downward-oriented spines or teeth; shape varies from cap-and-stalk to branched and icicle-like, fan-shaped, or shelf-like; growing on the ground, on wood, or on fallen pine cones → **Tooth Fungi/Spine Fungi** (p. 47). *Note:* Also check the section on polypores (p. 58), which includes some species that become tooth-like on the undersurface in age.

5B   Fruit body with pores on the undersurface → 6

    6A   Fruit body fleshy, with cap and a typically central stalk; cap undersurface with a sponge-like layer of vertically arranged tubes, each terminating in a pore; the sponge-like layer usually separates easily from the cap tissue; growing on the ground or infrequently on wood → **Boletes** (p. 48)

    6B   Fruit body woody or fibrous-fleshy to leathery, with pores on its undersurface (the pores are sometimes minute; use a hand lens); the pore layer typically does not separate easily from the cap tissue; shape varies from cap-and-stalk to stalkless and shelf-like or rather complex clusters; growing on wood or arising from buried wood → **Polypores** (p. 58)

7A   Fruit body erect, with a stalk or stalk-like base, sometimes fan- or vase-shaped or coral-like; interior tissue *never* powdery at maturity and *not* hard and black unless growing on the inflorescences of grasses → 8

7B   Fruit body not as in 7A → 13

    8A   Fruit body erect and phallus-like with a stalk and head, or pear-shaped to nearly round and stalkless, or squid-like with arched and tapered arms; fertile surface usually coated with a foul-smelling slimy layer; growing on the ground, mulch, wood chips, or decaying wood → **Stinkhorns** (p. 65)

    8B   Fruit body not as in 8A → 9

9A Fruit body with a conic to bell-shaped cap that has pits and ridges; or cap brain-like, saddle-shaped, or irregularly lobed; stalk typically hollow or multichambered, indistinct to massive; growing on the ground or sometimes on decaying wood → **Morels, False Morels, and Allies** (p. 69)

9B Fruit body not as in 9A → 10

   10A Fruit body leathery or fibrous tough, fan- to vase-shaped, often with a split or torn margin; typically some shade of brown at maturity, with or without whitish tips or margins; fertile surface smooth or wrinkled or warty, but lacking pores (use a hand lens); growing on the ground or enveloping roots, branches, seedlings, or mosses → **Fiber Fans and Vases** (p. 71)

   10B Fruit body not as in 10A → 11

11A Fruit body erect, worm-like, typically unbranched, usually arranged in clusters or colonies, often fused at their bases; or erect, coral-like, and repeatedly branched; growing on the ground or on wood → **Branched and Clustered Corals** (p. 72)

11B Fruit body not as in 11A → 12

   12A Fruit body erect, resembling tongues or clubs; includes species with a clearly defined head and stalk as well as those with club-like fruit bodies that lack a clearly differentiated head; fertile surface not roughened like sandpaper; growing on the ground or on decaying wood → **Earth Tongues, Earth Clubs, and Allies** (p. 73)

   12B Fruit body curved and spindle-shaped to cylindric, hard, purplish to brownish black, growing on the inflorescences of grasses; or cylindric to oval or club to spindle-shaped and attached to buried insects or to buried false truffles; or cylindric to club-shaped, whitish to yellowish or brownish orange, fertile surfaces roughened like sandpaper, growing on the ground or on decaying wood → ***Cordyceps, Claviceps,* and Allies** (p. 75)

13A Fruit body small, cylindric to vase-shaped, containing numerous egg-like peridioles (tiny spore sacs); growing on wood chips, mulch, branches, dung, and other organic debris → **Bird's-nest Fungi** (p. 76)

13B Fruit body not as in 13A → 14

   14A Fruit body resembling a small cup or saucer; flesh thin and brittle; with or without a stalk; growing on the ground or on decaying wood → **Cup and Saucer Fungi** (p. 77)

   14B Fruit body not as in 14A → 15

15A Fruit body distinctly gelatinous, usually rubbery but sometimes soft, with considerable variation in shape and color; growing on the ground or on wood → **Jelly Fungi** (p. 79)

15B Fruit body not as in 15A → 16

   16A Fruit body rather large, a more or less globose, cauliflower- or lettuce-like cluster; branches leaf-like, lacking pores on the undersurface (use a hand lens); usually growing on the ground at the base of trees or near decaying stumps → **Cauliflower Mushrooms** (p. 81)

   16B Fruit body not as in 16A → 17

17A Fruit body round, oval, pear- to turban-shaped, irregularly rounded, or star-shaped, with a powdery interior at maturity; usually stalkless but occasionally stalked; growing on the ground or on decaying wood, sometimes partially or completely buried → **Puffballs, Earthballs, Earthstars, and Allies** (p. 81)

17B Fruit body not as in 17A → 18

   18A Fruit body extremely variable, mostly cushion-shaped to round, sometimes crust-like and spreading; or erect and cylindric to club-shaped or antler-like; if erect, fruit body also hard and black, at least on the lower half; fertile surfaces roughened like sandpaper (use a hand lens), wrinkled, or furrowed; fibrous tough to woody or hard and carbonaceous or sometimes gelatinous; creamy white, yellow, green, brick red, rusty brown, or black; usually growing on decaying wood or leaves → **Carbon and Cushion Fungi** (p. 87)

18B Fruit body not as in 18A → 19

19A Fruit body typically hard, thin, spreading, crust-like to leathery or papery; some nearly flat, but others small, with projecting shelf-like caps; fertile surface warted, wrinkled, cracked, toothed or smooth, but lacking pores and not roughened like sandpaper (use a hand lens); usually growing on decaying wood → **Crust and Parchment Fungi and Allies** (p. 88)

19B Fruit body not as in 19A → 20

20A Fruit body a parasitic fungus that covers and usually disfigures gilled mushrooms, boletes, polypores, and some Ascomycetes; each has a roughened, sandpaper-like, moldy, feathery, or powdery appearance and texture → ***Hypomyces*** (p. 90)

20B Fruit body a swollen woody gall up to 4" (10 cm) in diameter, at first a small greenish brown swelling on the upper surface of a cedar needle, enlarging rapidly to form a reddish brown to dark brown gall that periodically produces conspicuous orange to orange-brown jelly-like horns that arise from small circular depressions → **Cedar-Apple Rust** (p. 90)

COLOR PLATES

Chanterelles and Allies

Cantharellus cibarius

Cantharellus cinnabarinus

Cantharellus confluens

Cantharellus ignicolor

Cantharellus lateritius

Cantharellus persicinus

Cantharellus tabernensis

Cantharellus tubaeformis

Craterellus fallax

Craterellus odoratus

Gomphus clavatus

Gomphus floccosus

Split Gill and Ally

Plicaturopsis crispa

Schizophyllum commune

Gilled Mushrooms

Agaricus abruptibulbus

Agaricus campestris

Agaricus pocillator

Agaricus porphyrocephalus

Agaricus subrutilescens

Amanita abrupta

Agrocybe semiorbicularis

Gilled Mushrooms / 17

Amanita atkinsoniana

Amanita chlorinosma

Amanita citrina f. *citrina*

Amanita cokeri

Amanita daucipes

Amanita flavorubescens

18 / *Gilled Mushrooms*

Amanita frostiana

Amanita hesleri

Amanita komarekensis

Amanita longipes

Amanita muscaria var. *flavivolvata*

Amanita mutabilis (A) maturing specimens

Amanita mutabilis (B) section showing staining

Gilled Mushrooms / 19

Amanita parcivolvata

Amanita polypyramis

Amanita ravenelii

Amanita rhopalopus f. *rhopalopus*

Amanita rubescens

Amanita spreta

Amanita virosa

Amanita volvata

Anellaria sepulchralis (A) immature specimens

Anellaria sepulchralis (B) mature specimens

Anthracophyllum lateritium

Armillaria mellea

Armillaria ostoyae

Armillaria tabescens

Asterophora lycoperdoides

Baeospora myosura

Bolbitius vitellinus

Callistosporium luteo-olivaceum

Catathelasma ventricosa

Chlorophyllum molybdites (A) immature specimens

Chlorophyllum molybdites (B) mature specimens

Chroogomphus rutilus

Clitocybe clavipes

Clitocybe nuda

Clitocybe subconnexa

Collybia cookei

Copelandia westii

Clitocybula familia

Coprinus americanus

Conocybe lactea

Coprinus atramentarius

Coprinus comatus

Coprinus disseminatus

Coprinus floridanus

Coprinus laniger

Coprinus micaceus

Coprinus plicatilis

Cortinarius delibutus

Cortinarius iodes

Cortinarius lewisii

Cortinarius marylandensis

Cortinarius semisanguineus

Crepidotus mollis

Gilled Mushrooms / 25

Cyptotrama asprata

Cystoderma amianthinum var. *rugosoreticulatum*

Entoloma abortivum

Flammulina velutipes

Galerina marginata

Gerronema strombodes

Gymnopilus fulvosquamulosus

Gymnopilus liquiritiae

Gymnopilus palmicola

Gymnopilus penetrans

Gymnopilus spectabilis

Gymnopus confluens

Gymnopus dryophilus

Gymnopus iocephalus

Hebeloma crustuliniforme

Hebeloma sinapizans

Hebeloma syriense

Hohenbuehelia petaloides

Hygrocybe acutoconica

Hygrocybe andersonii

Hygrocybe chlorophana

Hygrocybe conica

Hygrocybe conicoides

Hygrocybe ovina

Hygrocybe nitida

Hygrophorus hypothejus

Hygrophorus marginatus var. *marginatus*

Hygrophorus pratensis

Gilled Mushrooms / 29

Hygrophorus roseibrunneus

Hypholoma fasciculare (A) immature specimens

Hypholoma fasciculare (B) mature specimens

Hypholoma sublateritium

Hypsizygus tessulatus

Inocybe geophylla var. *lilacina*

Inocybe rimosa

30 / *Gilled Mushrooms*

Laccaria laccata

Laccaria ochropurpurea

Laccaria trullisata

Lactarius agglutinatus

Lactarius alachuanus var. *alachuanus*

Lactarius allardii

Lactarius atroviridis

Lactarius chelidonium var. *chelidonium*

Lactarius chrysorheus

Lactarius corrugis

Lactarius croceus

Lactarius deceptivus

Lactarius floridanus

Lactarius glaucescens

Lactarius hygrophoroides var. *rugatus*

Lactarius imperceptus

Lactarius indigo var. *diminutivus*

Lactarius indigo var. *indigo*

Lactarius luteolus

Lactarius maculatipes

Lactarius paradoxus

Lactarius peckii var. *peckii*

Lactarius proximellus

Lactarius psammicola f. *psammicola*

Gilled Mushrooms / 33

Lactarius pseudodeliciosus var. *pseudodeliciosus*

Lactarius quietus var. *incanus*

Lactarius salmoneus var. *salmoneus*

Lactarius subplinthogalus

Lactarius tomentoso-marginatus

Lactarius volemus var. *flavus*

Lactarius volemus var. *volemus*

Lactarius xanthydrorheus

Lactarius yazooensis

Lentinellus ursinus

Lentinula raphanica

Lentinus crinitis (A) pale form

Lentinus crinitis (B) dark form

Lentinus levis

Lentinus strigosus

Lentinus tephroleucus

Leucocoprinus birnbaumii (A) immature specimens

Leucocoprinus birnbaumii (B) mature specimens

Leucocoprinus cepaestipes

Leucocoprinus fragilissimus

Leucocoprinus lilacinogranulosus

Limacella illinita var. *illinita*

Macrocybe titans (A) immature specimens

Macrocybe titans (B) more mature specimens

Macrolepiota procera

Macrolepiota subrachodes

Marasmiellus albuscortiscis

Marasmius cohaerens

Marasmius fulvoferrugineus

Marasmius rotula

Melanoleuca melaleuca

Mycena epipterygia var. *epipterygioides*

Mycena epipterygia var. *viscosa*

Mycena haematopus

Gilled Mushrooms / 39

Mycena luteopallens

Mycena pura

Nolanea murraii

Nolanea quadrata

Nolanea strictia

Nolanea verna

Omphalina ectypoides

40 / *Gilled Mushrooms*

Omphalotus olearius (A) immature specimens

Omphalotus olearius (B) mature specimens

Panaeolus semiovatus

Panellus stipticus

Paxillus atrotomentosus

Paxillus involutus

Gilled Mushrooms / 41

Pholiota highlandensis

Pholiota polychroa (A) variety with purplish tones

Pholiota polychroa (B) variety with orange tones

Phyllotopsis nidulans

Pleurotus dryinus

Pleurotus ostreatus

Pluteus cervinus

Pluteus petasatus

Pouzarella nodospora

Psathyrella candolleana

Psathyrella hydrophila

Psathyrella umbonata

Psilocybe cubensis

Rhodocollybia butyracea

Gilled Mushrooms / 43

Rhodocollybia maculata

Rhodocybe roseoavellanea

Ripartitella brasiliensis (A) immature specimens

Ripartitella brasiliensis (B) mature specimens

Russula amoenolens

Russula ballouii

Russula compacta

Russula crustosa

Russula densifolia

Russula perlactea

Russula foetentula

Russula sanguinea

Russula sericeonitans

Russula silvicola

Gilled Mushrooms / 45

Russula subalbidula

Russula vinacea

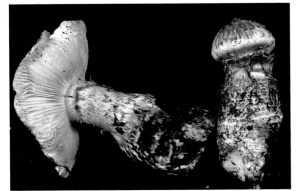
Squamanita umbonata

Strobilurus conigenoides

Stropharia bilamellata

Tricholoma caligatum

Tricholoma flavovirens

46 / Gilled Mushrooms

Tricholoma odorum

Tricholoma sulphureum

Tricholomopsis decora

Tricholomopsis formosa

Volvariella gloiocephala

Xeromphalina campanella

Tooth Fungi

Hericium coralloides

Hericium erinaceus

Hydnellum aurantiacum

Hydnellum spongiosipes

Hydnum repandum

Hydnum repandum var. *album*

Mycorrhaphium adustum

Phellodon niger

Boletes

Austroboletus betula

Austroboletus gracilis var. *gracilis*

Austroboletus subflavidus

Boletellus ananas

Boletes / 49

Boletus abruptibulbus

Boletus albisulphureus

Boletus auriflammeus

Boletus auriporus

Boletus bicolor var. *bicolor*

Boletus carminiporus

Boletus curtisii

Boletus dupainii

Boletus edulis

Boletus firmus

Boletus flammans

Boletus floridanus

Boletus griseus

Boletus hortonii

Boletus hypocarycinus

Boletus innixus

Boletus longicurvipes

Boletus luridellus

Boletus luridiformis

Boletus luridus

Boletus mahagonicolor

Boletus nobilis

Boletus ornatipes

Boletus pallidus

Boletus patrioticus

Boletus pulverulentus

Boletus purpureorubellus

Boletus roseolateritius

Boletus roseopurpureus

Boletus rubricitrinus (A) immature specimens

Boletus rubricitrinus (B) mature specimens

Boletus rufomaculatus

Boletus sensibilis

Boletus subglabripes

Boletus subluridellus

Boletus subvelutipes

Boletus tenax

Boletus variipes

Boletus weberi

Chalciporus pseudorubinellus

Gyrodon merulioides

Gyroporus castaneus

Gyroporus cyanescens var. *cyanescens*

Gyroporus subalbellus

Leccinum albellum

Leccinum nigrescens

Leccinum snellii

Boletes / 55

Phylloporus boletinoides

Phylloporus leucomycelinus

Pulveroboletus ravenelii

Strobilomyces dryophilus

Suillus brevipes

Suillus decipiens

Suillus hirtellus

Suillus salmonicolor (A) immature specimens

Suillus salmonicolor (B) mature specimens

Suillus subalutaceus

Tylopilus alboater

Tylopilus ballouii

Tylopilus conicus var. *conicus*

Tylopilus plumbeoviolaceus

Tylopilus rhoadsiae

Tylopilus tabacinus var. *amarus*

Tylopilus tabacinus var. *tabacinus*

Tylopilus variobrunneus

Tylopilus violatinctus

Xanthoconium affine var. *maculosus*

Xanthoconium separans (A) immature specimens

Xanthoconium separans (B) mature specimens

Xanthoconium stramineum

Polypores

Albatrellus cristatus

Albatrellus pes-caprae

Antrodia albida

Albatrellus subrubescens

Bondarzewia berkeleyi

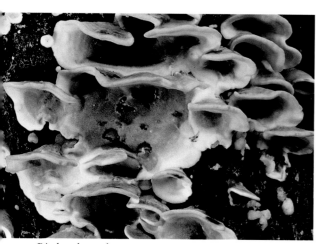

Bjerkandera adusta

Cerrena unicolor

Coltricia cinnamomea

Daedaleopsis confragosa

Fistulina hepatica

Fomes fasciatus

Fomitopsis nivosa

Ganoderma applanatum

Ganoderma lucidum

Ganoderma tsugae

Ganoderma zonatum

Gloeoporus dichrous

Hapalopilus croceus

Gloeophyllum sepiarium

Gloeophyllum striatum

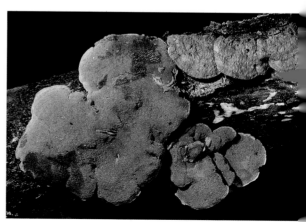

Hapalopilus nidulans

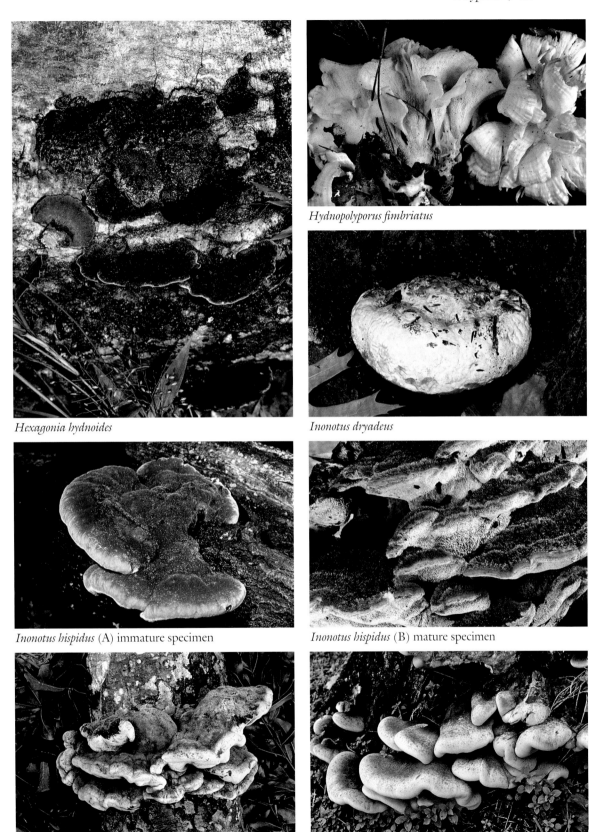

Hexagonia hydnoides

Hydnopolyporus fimbriatus

Inonotus dryadeus

Inonotus hispidus (A) immature specimen

Inonotus hispidus (B) mature specimen

Inonotus quercustris

Ischnoderma resinosum

Laetiporus cincinnatus

Laetiporus persicinus

Laetiporus sulphureus

Lenzites betulina

Meripilus sumstinei (A) immature specimen

Meripilus sumstinei (B) mature specimen

Microporellus dealbatus

Microporellus obovatus

Nigroporus vinosus

Phellinus chrysoloma

Polyporus elegans

Polyporus tenuiculus

Phaeolus schweinitzii

Phellinus gilvus

Polyporus varius

Pseudofavolus cucullatus

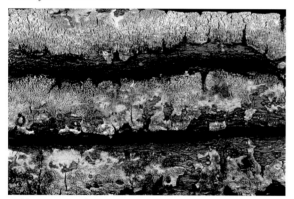
Schizopora paradoxa

Spongipellis pachyodon

Pseudofistulina radicata

Pycnoporus sanguineus

Trametes elegans

Polypores – Stinkhorns / 65

Trametes menziesii

Trametes versicolor

Trichaptum biforme

Stinkhorns

Aseroe rubra (A) entire fruit body

Aseroe rubra (B) close-up of arms and spore mass

Blumenavia angolensis

Dictyophora duplicata

Linderia columnata (A) entire fruit bodies

Linderia columnata (B) close-up of arms and spore mass

Lysurus gardneri

Lysurus periphragmoides (A) immature and developing stages

Lysurus periphragmoides (B) mature specimens

Mutinus elegans (A) immature specimens

Mutinus elegans (B) mature specimen

68 / Stinkhorns

Mutinus ravenelii

Phallogaster saccatus

Phallus hadriani

Phallus ravenelii

Morels, False Morels, and Allies

Gyromitra caroliniana (A) mature specimen

Gyromitra caroliniana (B) section showing chambers

Gyromitra infula

Gyromitra esculenta

Helvella albella

70 / *Morels, False Morels, and Allies*

Helvella crispa

Helvella elastica

Helvella macropus

Morchella elata

Morchella esculenta

Morchella semilibera

Fiber Fans and Vases

Thelephora anthocephala var. *americana*

Thelephora palmata

Thelephora terrestris f. *concrescens*

Thelephora vialis

Branched and Clustered Corals

Clavaria vermicularis

Clavaria zollingeri

Clavicorona pyxidata

Clavulina cristata

Clavulinopsis aurantio-cinnabarina

Branched and Clustered Corals – Earth Tongues, Earth Clubs, and Allies

Clavulinopsis fusiformis

Ramaria murrillii

Ramaria subbotrytis

Tremellodendropsis semivestitum

Earth Tongues, Earth Clubs, and Allies

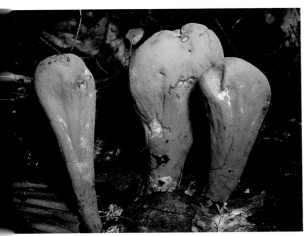
Clavariadelphus pistillaris

Clavariadelphus truncatus

Underwoodia columnaris

Xylocoremium flabelliforme

Cordyceps, Claviceps, and Allies

Cordyceps capitata

Cordyceps melolonthae

Cordyceps olivascens

Cordyceps sphecocephala

Nomuraea atypicola

Bird's-nest Fungi

Crucibulum laeve

Cyathus striatus

Cup and Saucer Fungi

Aleuria aurantia

Ascocoryne sarcoides

Chlorociboria aeruginascens

Galiella rufa

Helvella acetabulum

Humaria hemisphaerica

Peziza ammophila

Peziza repanda

78 / *Cup and Saucer Fungi*

Plicaria anthracina

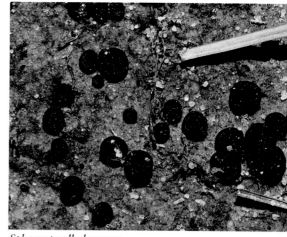
Sphaerosporella brunnea

Urnula craterium

Wolfina aurantiopsis

Jelly Fungi

Auricularia auricula

Auricularia polytricha

Calocera cornea

Dacrymyces palmatus

Dacryopinax elegans

Dacryopinax spathularia

Exidia glandulosa

80 / Jelly Fungi

Exidia recisa

Syzygospora mycetophila

Tremella foliacea

Sebacina incrustans

Tremella concrescens

Tremella fuciformis

Cauliflower Mushrooms

Sparassis crispa

Sparassis herbstii

Puffballs, Earthballs, Earthstars, and Allies

Arachnion album

Astraeus hygrometricus

Bovista plumbea, immature specimens

Bovistella radicata

Calostoma cinnabarina

Calostoma lutescens

Calostoma ravenelii

Calvatia cyathiformis

Calvatia rubroflava

Chorioactis geaster (A) immature specimens

Puffballs, Earthballs, Earthstars, and Allies / 83

Chorioactis geaster (B) intermediate specimens

Chorioactis geaster (C) mature specimens

Disciseda candida

Elaphomyces granulatus

Endoptychum agaricoides

Geastrum fornicatum

Geastrum mirabilis

Geastrum quadrifidum

Geastrum saccatum

Geastrum vulgatum

Lycoperdon acuminatum

Lycoperdon marginatum

Lycoperdon pulcherrimum

Puffballs, Earthballs, Earthstars, and Allies / 85

Lycoperdon pyriforme

Macowanites arenicola

Pisolithus tinctorius

Rhizopogon atlanticus

Rhizopogon nigrescens

Rhopalogaster transversarium (A) immature specimens

Rhopalogaster transversarium (B) mature specimens

Scleroderma bovista

Scleroderma floridanum

Scleroderma meridionale

Scleroderma polyrhizon

Scleroderma texense

Scleroderma verrucosum

Vascellum pratense

Zelleromyces cinnabarinus

Carbon and Cushion Fungi

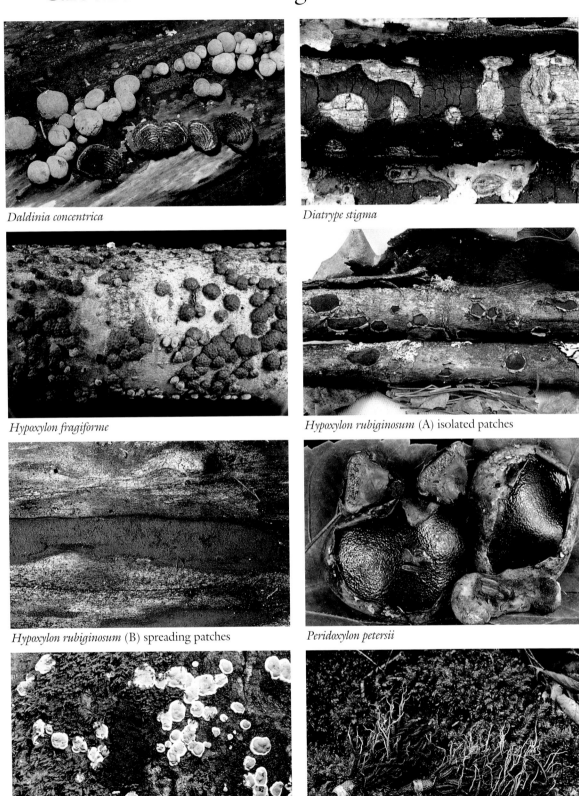

Daldinia concentrica

Diatrype stigma

Hypoxylon fragiforme

Hypoxylon rubiginosum (A) isolated patches

Hypoxylon rubiginosum (B) spreading patches

Peridoxylon petersii

Ustulina deusta

Xylaria magnoliae

88 / *Carbon and Cushion Fungi – Crust and Parchment Fungi and Allies*

Xylaria oxyacanthae

Xylaria persicaria

Xylaria polymorpha

Xylaria tentaculata

Crust and Parchment Fungi and Allies

Aleurodiscus oakesii

Cotilydia diaphana

Crust and Parchment Fungi and Allies / 89

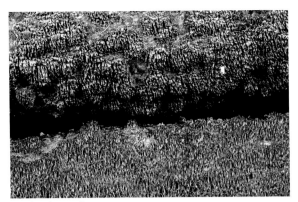
Hydnochaete olivaceum

Phlebia incarnata

Phlebia tremellosa

Pulcherricium caeruleum

Stereum complicatum

Xylobolus frustulatus

Stereum ostrea

Hypomyces

Hypomyces chrysospermus

Hypomyces lactifluorum

Hypomyces luteovirens

Hypomyces hyalinus

Cedar-Apple Rust

Gymnosporangium juniperi-virginiana

SPECIES DESCRIPTIONS

Chanterelles and Allies

Members of this small group produce fruit bodies that are often funnel- to vase-shaped at maturity. Many resemble gilled mushrooms, but their spores are not produced on true gills. Their fertile surfaces are typically blunt, with gill- to vein-like ridges that are often forked or joined together by crossveins, and a few have nearly smooth fertile surfaces. Chanterelles usually grow on the ground, and several are believed to be mycorrhizal with various trees. Many are popular edibles. Although a few species produce gastrointestinal upset when eaten, none is known to be dangerously poisonous.

Cantharellus cibarius Fries Illus. p. 14
  COMMON NAME: Chanterelle, Golden Chanterelle.
  CAP: ⅝–5½" (1.6–14 cm) wide, convex to nearly plane, sometimes with a depressed center (fruit body at times more or less funnel-shaped); surface dry, nearly smooth, orange-yellow to yellow; margin thin, incurved to inrolled when young, often remaining so for a long time, becoming uplifted and wavy in age, sometimes crimped or lobed.
  FLESH: thick, firm, white; odor fragrant like apricots or not distinctive; taste peppery or not distinctive.
  FERTILE SURFACE: decurrent, with forked, blunt, gill-like ridges, with or without crossveins, pale yellow to yellow or pale orange.
  STALK: ⅝–2¾" (1.6–7 cm) long, up to 1" (2.5 cm) thick, equal or enlarged at either end, smooth, pale yellow to orange-yellow.
  SPORE PRINT COLOR: pinkish cream to pale buff.
  MICROSCOPIC FEATURES: spores 8–11 × 4.5–6 μm, elliptic, smooth, hyaline.
  FRUITING: solitary, scattered, in groups, or sometimes clustered on the ground in woods; June–October; widely distributed in the Southeast; common.
  EDIBILITY: edible, choice.
  COMMENTS: Compare with the Jack O'Lantern, *Omphalotus olearius* (p. 180), poisonous, which has true gills with sharp edges and grows on wood or buried wood, typically in large overlapping clusters.

Cantharellus cinnabarinus Schweinitz Illus. p. 14
  COMMON NAME: Cinnabar-red Chanterelle.
  CAP: ⅜–1¾" (1–4.5 cm) wide, convex, becoming broadly convex and depressed, then funnel-shaped; surface smooth or nearly so, reddish orange, fading to pale pinkish orange in age; margin incurved when young, becoming uplifted and wavy in age.
  FLESH: whitish, tinged reddish orange near the cap surface; odor not distinctive; taste not distinctive or slightly acrid.

FERTILE SURFACE: decurrent, with distant, forking, blunt, gill-like ridges with crossveins; pale reddish orange to salmon orange.

STALK: ¾–2" (2–5 cm) long, ⅛–⅜" (3–10 mm) wide, nearly equal or tapered downward, often curved, reddish orange, fading to pale pinkish orange.

SPORE PRINT COLOR: pinkish cream.

MICROSCOPIC FEATURES: spores 6–10.5 × 4–6 μm, elliptic to oblong, smooth, hyaline.

FRUITING: scattered or in groups on soil or among mosses; June–October; widely distributed in the Southeast; fairly common.

EDIBILITY: edible.

Cantharellus confluens (Berk. and Curt.) Petersen — Illus. p. 14

COMMON NAME: none.

CAP: ¾–3¾" (2–9.5 cm) wide, convex to nearly plane, sometimes somewhat depressed at the center (entire fruit body may be funnel-shaped); surface dry, nearly smooth, yellow-orange to yellow; margin thin, incurved when young, becoming uplifted and wavy in age, often crimped or lobed.

FLESH: thick, firm, white; odor somewhat fragrant or not distinctive; taste not distinctive.

FERTILE SURFACE: decurrent, smooth or with blunt gill-like ridges near the margin, pale yellow to orange buff or ochraceous buff.

STALK: 1–3" (2.5–7.5 cm) long, ¼–1" (6–25 mm) thick, tapered downward to a somewhat pointed base, often curved, dry, hollow in age, pale yellow above, whitish toward the base, staining rusty orange where bruised, confluent.

SPORE PRINT COLOR: pinkish yellow.

MICROSCOPIC FEATURES: spores 6–10 × 5–6.5 μm, elliptic, smooth, hyaline.

FRUITING: scattered or in groups of two or more caps arising from a common base attached to soil in woods, usually associated with oak or other broadleaf trees; June–September; widely distributed along the Gulf region; occasional to fairly common.

EDIBILITY: edible.

COMMENTS: *Cantharellus lateritius* (p. 95), edible, is nearly identical but tends to form individual, not confluent, fruit bodies. Some authors consider *Cantharellus confluens* and *Cantharellus lateritius* to be synonyms.

Cantharellus ignicolor Petersen — Illus. p. 14

COMMON NAME: Flame-colored Chanterelle.

CAP: ⅜–3⅛" (1–8 cm) wide, convex with a shallow central depression, soon perforated; surface scurfy with tiny erect fibrous scales, especially along the margin, sometimes nearly smooth at maturity, orange to yellow-orange, becoming pale yellow-orange to brownish orange or pale yellow-brown in age; margin incurved when young, becoming uplifted, wavy, and crimped to lobed at maturity.

FLESH: thin, flexible, orange to yellow-orange, with brownish tints in age; odor of citronella or not distinctive; taste not distinctive.

FERTILE SURFACE: slightly decurrent, with gill- to vein-like blunt ridges, often forking, sometimes with crossveins, orange-yellow, developing a pinkish tinge in age.

STALK: ¾–2⅜" (2–6 cm) long, ⅛–⅜" (3–10 mm) thick, nearly equal, smooth, orange to yellow-orange or yellow, sometimes with white to yellowish matted fibers at the base.

SPORE PRINT COLOR: pale ochraceous salmon.

MICROSCOPIC FEATURES: spores 9–13 × 6–9 μm, elliptic to broadly elliptic, smooth, hyaline.

FRUITING: scattered or in groups or clusters on the ground in mixed woods or among

sphagnum mosses in bogs and wet woodlands; July–October; widely distributed in the Southeast; fairly common.
EDIBILITY: unknown.
COMMENTS: Commonly misidentified; use a hand lens and look for the tiny erect fibrous scales, especially along the cap margin.

Cantharellus lateritius (Berkeley) Singer   Illus. p. 14
COMMON NAME: Smooth Chanterelle.
CAP: ⅝–4¾" (1.6–12 cm) wide, convex to nearly plane, sometimes with a low, broad umbo or slightly depressed (entire fruit body is often funnel-shaped); surface dry, nearly smooth, yellow-orange to yellow; margin thin, incurved when young, becoming uplifted and wavy in age, often crimped or lobed.
FLESH: thick, firm, white; odor fragrant like apricots or not distinctive; taste not distinctive.
FERTILE SURFACE: decurrent, with forked, blunt, gill-like ridges and crossveins, or sometimes nearly smooth, pale yellow to yellow or orange.
STALK: ⅝–3⅛" (1.6–8 cm) long, up to 1" (2.5 cm) thick at the apex, tapering downward, smooth, pale yellow to orange-yellow.
SPORE PRINT COLOR: pinkish yellow to pale yellow-orange.
MICROSCOPIC FEATURES: spores 7.5–12 × 4.5–6.5 μm, elliptic, smooth, hyaline.
FRUITING: scattered or in groups or clusters on the ground under broadleaf trees, especially oaks; June–October; widely distributed in the Southeast; occasional to fairly common.
EDIBILITY: edible, choice.
COMMENTS: Compare with the Jack O'Lantern, *Omphalotus olearius* (p. 180), poisonous, which has true gills with sharp edges and grows on wood or buried wood, typically in large overlapping clusters.

Cantharellus persicinus Petersen   Illus. p. 14
COMMON NAME: none.
CAP: ⅜–1¾" (1–4.5 cm) wide, hemispheric when young, becoming convex and finally nearly plane in age; surface smooth, moist or dry, salmon to peach color, paler along the margin, hygrophanous; margin incurved when young, becoming uplifted at maturity.
FLESH: whitish; odor faintly fruity; taste not distinctive.
FERTILE SURFACE: decurrent, with distant, blunt, gill-like ridges and crossveins, pale salmon orange to ochraceous salmon.
STALK: 1–2⅜" (2.5–6 cm) long, up to ⅜" (1 cm) wide, tapered downward, dry, solid, colored like the cap, with a white base.
SPORE PRINT COLOR: whitish to pale salmon.
MICROSCOPIC FEATURES: spores 10–12 × 6–7.5 μm, narrowly ovate to ellipsoid, smooth, hyaline.
FRUITING: scattered or in groups on the ground in mixed woodlands, usually with oak or hemlock; June–September; reported from Tennessee and West Virginia, distribution limits yet to be determined; occasional.
EDIBILITY: edible.
COMMENTS: *Cantharellus cinnabarinus* (p. 93), edible, is similar but has a more reddish orange fruit body and smaller spores.

Cantharellus tabernensis Feibelman and Cibula   Illus. p. 14
COMMON NAME: none.
CAP: ⅜–2" (1–5 cm) wide, convex to nearly plane; disc depressed, often umbilicate, sometimes perforate; surface dry, glabrous; disc and young specimens light yellowish brown to strong yellowish brown or brown, becoming strong orange-yellow at maturity; margin inrolled at first, becoming decurved and remaining so well into maturity, orange yellow, even or minutely crenulate.
FLESH: moderately thick, firm, yellowish white; odor fragrant, fruity, like apricots; taste mild to moderately acrid.
FERTILE SURFACE: decurrent, with forked, blunt, gill-like ridges and crossveins, vivid orange-yellow to yellow.
STALK: ⅝–1¾" (1.5–4.5 cm) long, ⅛–⅜" (2–8 mm) thick, tapered downward, smooth, dry, hollow, usually longitudinally furrowed, yellow to orange-yellow, paler toward the base.
SPORE PRINT COLOR: yellowish white to pale orange-yellow.
MICROSCOPIC FEATURES: spores 6–9 × 4–6 μm, elliptical to subglobose, smooth, hyaline, inamyloid.
FRUITING: scattered or in groups on the ground in well-drained mixed pine and broadleaf forests, usually near mature pine; July–September; reported from Louisiana and Mississippi; occasional.
EDIBILITY: edible.
COMMENTS: The name *tabernensis* is a reference to a gathering place, "tavern," not far from the Stennis Space Center, near which the species was first encountered.

Cantharellus tubaeformis Fries   Illus. p. 14
COMMON NAME: Trumpet Chanterelle.
CAP: ⅜–3" (1–7.5 cm) wide, convex to nearly plane or funnel-shaped, with a depressed center, sometimes perforated; surface moist or dry, smooth or roughened with small fibrous scales, especially along the margin; reddish brown, fading to pale yellow-brown; margin incurved to inrolled when young, becoming even and wavy at maturity.
FLESH: thin, firm, grayish to yellowish brown; odor and taste not distinctive.
FERTILE SURFACE: decurrent, with blunt, often forked, widely spaced, gill-like ridges and crossveins, yellowish when young, becoming violaceous or violaceous gray.
STALK: 1⅛–3¾" (3–9.5 cm) long, ⅛–⅜" (3–10 mm) thick, nearly equal, smooth, moist, hollow at maturity, yellow-orange to brownish orange, sometimes whitish at the base, with or without a whitish mycelium.
SPORE PRINT COLOR: cream to pale yellow.
MICROSCOPIC FEATURES: spores 8–12 × 6–10 μm, granular, broadly elliptic, smooth, hyaline.
FRUITING: scattered or in groups among sphagnum mosses in bogs and on soil in wet areas under conifers, sometimes growing on decaying wood; July–January; widely distributed in the Southeast; occasional to locally common.
EDIBILITY: edible.
COMMENTS: *Cantharellus infundibuliformis*, edible, is very similar, but it has a blackish brown to dark olive brown cap with a brownish yellow margin (see *MNE*, p. 37).

Craterellus fallax Smith   Illus. p. 15
COMMON NAME: Black Trumpet.
CAP: ⅜–3⅛" (1–8 cm) wide, funnel-shaped and deeply depressed; upper surface grayish

brown to dark brown or blackish, with darker radiating fibers or tiny fibrous scales; margin inrolled at first, becoming arched, wavy, and irregular.

FLESH: thin, brittle to fibrous, colored like the surface; odor pleasant, often somewhat fruity; taste not distinctive.

FERTILE SURFACE: decurrent, smooth or with shallow, blunt, vein-like ridges, gray to brown or blackish, often with ochre-orange tints, bruising blackish.

STALK: indistinct, a short extension below the fertile surface, often hollow, dark brown to blackish.

SPORE PRINT COLOR: ochraceous orange to ochraceous buff.

MICROSCOPIC FEATURES: spores 11–18 × 7–11 μm, broadly elliptic, smooth, hyaline.

FRUITING: scattered or in groups or clusters on the ground in broadleaf or mixed woods; June–December; widely distributed in the Southeast; occasional to fairly common.

EDIBILITY: edible and very popular.

COMMENTS: Although this mushroom is widely considered an excellent edible, bitter-tasting specimens are occasionally encountered. *Craterellus cornucopioides* (not illustrated), edible, is nearly identical but has a whitish spore print.

Craterellus odoratus (Schw.) Fries   Illus p. 15

COMMON NAME: Fragrant Chanterelle.

FRUIT BODY: up to 6" (15.5 cm) wide and high, a large orange cluster composed of several funnel-shaped, hollow, trumpet-like extensions.

CAP: ½–2" (1.3–5 cm) wide, funnel-shaped; surface moist or dry, smooth, bright orange at the center, paler toward the margin, several arising from a common stalk; margin wavy and lobed.

FLESH: thin, orange; odor variously described as pleasant, apricot-like, fragrant, fruity, or resembling violets; taste not distinctive.

FERTILE SURFACE: smooth or slightly wrinkled, pale orange-yellow to creamy yellow.

STALK: up to 2" (5 cm) high, not clearly demarcated from the fertile surface, up to ¾" (2 cm) thick, smooth, dry, hollow.

SPORE PRINT COLOR: pale apricot.

MICROSCOPIC FEATURES: spores 8–12 × 4–6.5 μm, elliptic to narrowly oval, smooth, hyaline.

FRUITING: in clusters on the ground in woodlands; April–October; widely distributed in the Southeast; fairly common.

EDIBILITY: edible when young and fresh, becoming unpleasant in age.

COMMENTS: Also known as *Cantharellus odoratus*.

Gomphus clavatus (Fries) S. F. Gray   Illus. p. 15

COMMON NAME: Pig's Ear Gomphus.

CAP: 1⅛–4" (3–10 cm) wide, cylindric and truncate with a depressed center when young, becoming funnel-shaped to somewhat flattened in age, often perforated at the center; surface smooth, lacking prominent scales, violet when young, becoming tan to brownish yellow in age; margin uplifted, wavy, often lobed at maturity.

FLESH: thick, soft, brittle to fibrous, whitish to pale buff; odor and taste not distinctive.

FERTILE SURFACE: decurrent, with blunt, vein-like ridges and crossveins, violet to grayish violet, becoming dull ochre to tan in age.

STALK: up to 2" (5 cm) long below the fertile surface, up to ¾" (2 cm) thick at the apex, tapering downward, often fused with adjacent stalks, solid, colored like the fertile surface.

SPORE PRINT COLOR: ochraceous.

MICROSCOPIC FEATURES: spores 10–13 × 4–6.5 μm, narrowly elliptic to spindle-shaped, minutely warted, hyaline.

FRUITING: scattered or in groups or clusters on the ground under conifers; August–November; North Carolina west to Tennessee and northward; occasional.

EDIBILITY: edible and rated as choice by some.

COMMENTS: The violet color of the young cap and gills, which becomes tan to brownish yellow in age, is the distinctive feature.

Gomphus floccosus (Schweinitz) Singer   Illus. p. 15

COMMON NAME: Scaly Vase Chanterelle.

CAP: 1½–6¼" (4–16 cm) wide, funnel- to vase-shaped and deeply depressed; upper surface moist, with numerous coarse orange to reddish orange scales over a paler orange ground color, becoming orange-brown in age; margin thin, wavy, typically lobed at maturity.

FLESH: moderately thick, fibrous, whitish; odor and taste not distinctive.

FERTILE SURFACE: strongly decurrent, with forked, blunt, vein-like ridges and crossveins; pale yellow to creamy white, becoming ochre-tinged in age.

STALK: up to 4" (10 cm) long below the fertile surface, up to 2" (5 cm) thick at the apex, tapering downward, solid, becoming hollow in age, nearly smooth, pale yellow to creamy white, browning in age.

SPORE PRINT COLOR: ochraceous.

MICROSCOPIC FEATURES: spores 11.5–14.5 × 7–8 μm, elliptic, minutely warted, hyaline.

FRUITING: solitary, scattered, or in groups on the ground under conifers or in mixed conifer and broadleaf forests; widely distributed in the Southeast; June–October; fairly common.

EDIBILITY: not recommended; although enjoyed by some, a common cause of gastrointestinal upset.

Split Gill and Ally

Fruiting bodies of the split gill and ally grow on wood and are stalkless or have only a short lateral stalk. Their fertile surfaces are gill-like and split longitudinally or are distinctly crimped, often forked, and vein-like. We have included the two most common species that are likely to be encountered.

Plicaturopsis crispa (Fries) Reid   Illus. p. 15
    COMMON NAME: Crimped Gill.
    CAP: ⅜–⅞" (1–2.3 cm) wide, fan-shaped to shell-shaped, concentrically zoned, yellow-orange to reddish brown or yellow-brown, finely tomentose, dry; margin undulating, lobed, scalloped, decurved to inrolled, whitish to pale yellow.
    FLESH: membranous, thin, flexible when moist, hard and brittle when dry; odor and taste not distinctive.
    GILL-LIKE FOLDS: crimped, often forked and vein-like, frequently anastomosing; moderately distant, narrow, whitish to grayish.
    STALK: very short, a narrow central extension of the cap, sometimes absent.
    SPORE PRINT COLOR: white.
    MICROSCOPIC FEATURES: spores $3-4 \times 1-1.5$ μm, sausage-shaped to elliptic, smooth, hyaline, often containing two oil drops.
    FRUITING: in overlapping clusters or groups on branches and trunks of broadleaf trees, especially beech and birch; year-round; distribution limits yet to be determined; infrequent.
    EDIBILITY: inedible.
    COMMENTS: Formerly known as *Trogia crispa*. *Stereum* species lack gills and have a smooth undersurface. Although not as common, widely distributed, or well known as the Common Split Gill, *Schizophyllum commune* (p. 99), inedible, the Crimped Gill has distinctive markings that make it easy to identify.

Schizophyllum commune Fries   Illus. p. 15
    COMMON NAME: Common Split Gill.
    CAP: ⅜–1¾" (1–4.5 cm) wide, fan- to shell-shaped, white to grayish white, covered by a dense layer of fine hairs, dry; margin incurved to inrolled, becoming nearly flat when moist, wavy and usually torn in age.
    FLESH: thin, leathery, flexible, whitish or grayish, often with brownish tones in age.
    GILL-LIKE FOLDS: split lengthwise along their free edge, often serrated or torn, subdistant, narrow, white to gray or pinkish gray.
    STALK: absent or a short, narrow extension of the cap.
    SPORE PRINT COLOR: white.

MICROSCOPIC FEATURES: spores 5–7.5 × 2–3 μm, cylindric, smooth, hyaline.

FRUITING: solitary, scattered, or often in overlapping clusters on fallen branches and decaying wood of broadleaf trees; year-round; widely distributed in the Southeast; common.

EDIBILITY: inedible.

COMMENTS: Easily recognized by its densely hairy, fan-shaped cap and split gills. The Common Split Gill is one of the most widely distributed mushrooms in the world.

Gilled Mushrooms

Gilled mushrooms, also known as agarics, belong to a very large group of fungi that have caps with knifeblade-like gills on the undersurface. Many have a central to eccentric stalk, whereas others are laterally stalked or stalkless. The cap diameter of some species rarely exceeds ⅛" (3 mm) at maturity, but the diameter of others can be 32" (80 cm) or more. A universal veil surrounds the button stage of some gilled mushrooms, often leaving warts and patches on the cap and a volva encompassing the stalk base. Many species produce a partial veil that encloses the immature gills and leaves a ring on the stalk at maturity.

Gilled mushrooms occur in a seemingly endless array of colors, and some change color as they mature. They grow on a wide variety of substrates, including soil, humus, wood, sawdust, straw, cones, fruits, manure, and other mushrooms. Many are excellent edibles; others are poisonous (some even deadly!). The edibility of the vast majority is unknown.

Agaricus abruptibulbus Peck Illus. p. 16

COMMON NAME: Abruptly-bulbous Agaricus.
CAP: 1½–6" (4–15.5 cm) wide, convex to nearly flat in age; surface dry, fibrillose to more or less glabrous, shiny, white, staining yellowish when bruised.
FLESH: white, bruising yellow; odor and taste of anise or not distinctive.
GILLS: free, crowded, narrow, white at first, becoming pink, then dark brown at maturity.
STALK: 2–6" (5–15.5 cm) long, ⅜–⅝" (1–1.5 cm) thick, enlarged slightly downward to a small, abruptly bulbous base; dry, solid then hollow in age, smooth, white; partial veil membranous, white with patches on the lower surface, leaving a persistent, pendant, superior ring.
SPORE PRINT COLOR: purple-brown.
MACROCHEMICAL TESTS: cap surface stains yellow with the application of KOH.
MICROSCOPIC FEATURES: spores 6–8 × 4–5 μm, elliptic, smooth, purple-brown.
FRUITING: solitary, scattered, or in groups on the ground in broadleaf and mixed woodlands; June–November; widely distributed in the Southeast; frequent.
EDIBILITY: edible.
COMMENTS: *Agaricus silvicola*, edible, is very similar and also grows in woodlands, but its stalk lacks the abruptly bulbous base (see *MNE*, p. 271). The cap of both *Agaricus abruptibulbus* and *Agaricus silvicola* may develop pinkish red tints during wet conditions. *Agaricus arvensis*, edible, is similar but more robust, lacks the abruptly bulbous stalk base, and grows in grassy open areas (see *MNE*, p. 271).

Agaricus campestris Linnaeus : Fries Illus. p. 16
  COMMON NAME: Meadow Mushroom, Pink Bottom.
  CAP: 1–4" (2.5–10 cm) wide, convex to nearly flat in age; surface fibrillose or nearly smooth, sometimes fibrillose-scaly, dry, white to grayish or grayish brown.
  FLESH: white; odor and taste not distinctive.
  GILLS: free, crowded, white at first, becoming pink, then dark brown at maturity.
  STALK: 1–2⅜" (2.5–6 cm) long, ⅜–⅝" (1–1.6 cm) thick, nearly equal or enlarged or tapered at the base, smooth, white; partial veil membranous, white, leaving a sparse, persistent, or evanescent superior ring.
  SPORE PRINT COLOR: dark brown.
  MICROSCOPIC FEATURES: spores 6–9 × 4–6 μm, elliptic, smooth, pale brown.
  FRUITING: scattered, in groups, or sometimes in clusters, arcs, or fairy rings on lawns, pastures, golf courses, and other grassy areas; June–October; widely distributed in the Southeast; common.
  EDIBILITY: edible and often rated choice.
  COMMENTS: *Agaricus arvensis*, edible, is similar, but it has a larger white cap, up to 7" (18 cm) wide, that bruises yellow; white flesh that bruises yellow; an anise or almond-like odor; and a white stalk that stains yellowish when bruised (see *MNE*, p. 271).

Agaricus pocillator Murrill Illus. p. 16
  COMMON NAME: none.
  CAP: 1⅛–4" (3–10 cm) wide, convex, becoming nearly plane, sometimes with a flattened disc; surface dry, covered with minute grayish to grayish brown scales on a white ground color that darkens in age; disc grayish brown.
  FLESH: thin, white; odor unpleasant, resembling phenol, or not distinctive; taste unpleasant or sometimes not distinctive.
  GILLS: attached then free at maturity, crowded, white at first, becoming pink and finally dark brown in age.
  STALK: 1½–3½" (4–9 cm) long, ¼–½" (6–12 mm) thick, nearly equal down to a bulbous or sometimes abruptly bulbous base, often bent near the base, dry, hollow at maturity, smooth, white, slowly staining brown when bruised; partial veil conspicuously membranous, white, with cottony patches, leaving a superior and conspicuously double-layered ring with cottony patches on the undersurface; flesh white, staining yellow at the base when exposed.
  SPORE PRINT COLOR: dark brown.
  MICROSCOPIC FEATURES: spores 4.5–6 × 3–4 μm, ellipsoid, smooth, pale brown.
  FRUITING: scattered or in groups on the ground in broadleaf and mixed woodlands or in grassy areas with trees; June–October; widely distributed in the Southeast; fairly common.
  EDIBILITY: unknown.
  COMMENTS: *Agaricus placomyces*, edibility unknown, is nearly identical but has a single-layered superior ring and large reddish brown scales on the cap (see *MNE*, p. 271).

Agaricus porphyrocephalus Møller Illus. p. 16
  COMMON NAME: none.
  CAP: 1⅛–3" (3–8 cm) wide, convex, becoming broadly convex with a somewhat flattened disc; surface dry, covered with appressed reddish brown to purplish brown fibrillose scales over a whitish ground color.
  FLESH: firm, white or slightly pinkish; odor and taste not distinctive.

GILLS: attached then free at maturity, crowded, pink when young, becoming dark brown in age.
STALK: 1–2" (2.5–5 cm) long, ⅜–¾" (1–2 cm) thick, enlarged downward or sometimes nearly equal, dry, hollow at maturity, fibrillose near the base, smooth near the apex, white; partial veil thin, membranous, white, leaving a thin superior ring.
SPORE PRINT COLOR: dark brown.
MICROSCOPIC FEATURES: spores 5–7 × 3–4 μm, ovoid, smooth, dark brown.
FRUITING: solitary, scattered, or in groups on the ground in lawns and meadows; June–November; widely distributed in the Southeast; occasional.
EDIBILITY: edible and good.
COMMENTS: The Silvery Agaricus, *Agaricus argenteus* (not illustrated), edible, is similar and also grows on lawns and in meadows, but it has silky brownish to grayish brown cap fibrils and larger spores that measure 8–13 × 6–7 μm.

Agaricus subrutilescens (Kauff.) Hot. and Stuntz   Illus. p. 16
COMMON NAME: Wine-colored Agaricus.
CAP: 2–6" (5–15.5 cm) wide, convex with a flattened disc, becoming nearly plane at maturity; surface dry, with fine purplish brown scales on a whitish ground color.
FLESH: thick, whitish; odor and taste not distinctive.
GILLS: free from the stalk, close to crowded, white at first, becoming pinkish then purplish brown to dark brown at maturity.
STALK: 2–8" (5–20 cm) long, ⅜–¾" (1–2 cm) thick, enlarged downward, dry, solid or hollow in age, coated with white cottony patches below the ring, smooth above it; partial veil white, membranous, leaving a large, flaring, superior ring.
SPORE PRINT COLOR: dark brown.
MACROCHEMICAL TESTS: pileipellis stains green with the application of KOH.
MICROSCOPIC FEATURES: spores 5–6 × 3–3.5 μm, elliptic, smooth, brownish.
FRUITING: solitary, scattered, or in groups on the ground in conifer or mixed woodlands; August–December; reported from Florida, Mississippi, and Texas, distribution limits yet to be determined; occasional.
EDIBILITY: edible with caution.
COMMENTS: The green KOH reaction separates this species from other *Agaricus* species. Some persons have reported gastric upset after eating this mushroom.

Agrocybe semiorbicularis (Bulliard) Fayod   Illus. p. 16
COMMON NAME: Hemispheric Agrocybe.
CAP: ⅜–2" (1–5 cm) wide, hemispheric when young, becoming broadly convex in age; surface viscid to slightly glutinous to suede-like when fresh, becoming dry and shiny and often cracked on the disc in age, ochre to rusty ochre at first, becoming ochre-yellow to yellow or buff in age; margin incurved and remaining so well into maturity, even.
FLESH: thin, whitish; odor farinaceous; taste farinaceous or somewhat bitter to disagreeable.
GILLS: attached, close, broad, buff to pale brownish when young, becoming orange-brown to pale rusty brown at maturity.
STALK: ¾–2¾" (2–7 cm) long, 1/16–½" (1.5–12 mm) thick, nearly equal or tapered downward, pale yellow to ochre or pale brown, finely fibrillose, typically slightly bulbous at the base, often with white rhizomorphs.
SPORE PRINT COLOR: dark brown.

MICROSCOPIC FEATURES: spores 10–14 × 7–10 μm, elliptical with a prominent germ pore, smooth, pale brown.

FRUITING: scattered or in dense groups on lawns and fields, on wood mulch, or on sawdust mixed with manure; May–November; widely distributed in the Southeast; common.

EDIBILITY: not worthwhile, edible but of poor quality.

COMMENTS: *Agrocybe pediades* is a synonym. Specimens gathered on sawdust mixed with manure tend to be much more robust than specimens collected on lawns or on wood mulch.

Amanita abrupta Peck   Illus. p. 16

COMMON NAME: none.

CAP: 1½–4" (4–10 cm) wide, convex, becoming nearly plane; surface dry, covered with conical warts, shiny, nonstriate, white; margin appendiculate.

FLESH: moderately thick at the disc and tapered toward the margin, white; odor and taste not distinctive.

GILLS: free, narrow, crowded, white; edges even.

STALK: 2⅜–5" (6–12.5 cm) long, ¼–⅝" (5–15 mm) thick, enlarged downward to a large, globose to subglobose, abrupt basal bulb; solid, fibrillose to glabrous, white; partial veil submembranous, white, leaving a superior, pendant ring that is usually striate above and floccose below; basal bulb sometimes bearing a few small warts or ridges.

SPORE PRINT COLOR: white.

MICROSCOPIC FEATURES: spores 6–10 × 5–9 μm, globose to elliptic, smooth, hyaline, amyloid.

FRUITING: solitary, scattered, or in groups on the ground in mixed conifer and broadleaf forests; September–December; widely distributed in the Southeast; fairly common.

EDIBILITY: unknown.

COMMENTS: This mushroom is one of the more commonly collected *Amanita* species during fall in the Southeast. *Amanita excelsa* var. *alba* (not illustrated), possibly poisonous, is similar but more robust and lacks the abrupt bulbous base. *Amanita pantherina* var. *multisquamosa* (not illustrated), poisonous, is also similar but has a distinct rim on the basal bulb of the stalk.

Amanita atkinsoniana Coker   Illus. p. 17

COMMON NAME: none.

CAP: 2–5⅛" (5–13 cm) wide, convex, becoming nearly plane in age, sometimes with a depressed center; surface dry, whitish to buff or brownish, with reddish brown to grayish brown warts; margin smooth or slightly striate, appendiculate.

FLESH: white, sometimes staining yellowish or pinkish; odor faint, resembling chlorine, or not distinctive.

GILLS: free, moderately broad, crowded, creamy white, sometimes with a pale reddish tint; edges even.

STALK: 2½–8⅜" (6.5–21 cm) long, ⅜–1⅛" (8–28 mm) thick, typically enlarged slightly downward to a ventricose to napiform or occasionally pointed basal bulb, dry, floccose to glabrous, whitish; partial veil submembranous, whitish, leaving a somewhat fragile, whitish superior ring that often collapses against the stalk; basal bulb bearing reddish brown rings or scales that sometimes extend up the stalk for a short distance.

SPORE PRINT COLOR: white.

MICROSCOPIC FEATURES: spores 9–13 × 5–8 μm, elliptic to elongate, smooth, hyaline, amyloid.
FRUITING: solitary, scattered, or in groups on the ground in mixed conifer and broadleaf forests; August–October; widely distributed in the Southeast; fairly common.
EDIBILITY: unknown.
COMMENTS: *Amanita onusta*, edibility unknown, is similar, but it forms somewhat smaller fruiting bodies with dark gray to brownish gray warts and a basal bulb that is usually somewhat rooting (see *MNE*, p. 273).

Amanita chlorinosma (Austin) Lloyd
Illus. p. 17

COMMON NAME: Chlorine Amanita.
CAP: 3–10" (7.5–25.5 cm) wide, convex, becoming nearly flat in age; surface dull white, covered with thick, powdery universal veil material that forms patches or warts on the disc; margin even, not striate.
FLESH: thick, white; odor and taste unpleasant, somewhat reminiscent of chlorine (chloride of lime) or of old ham.
GILLS: narrowly free or barely reaching the stalk, crowded, broad, white to cream; edges minutely fringed.
STALK: 4–12" (10–30 cm) long, ½–1" (1.3–2.5 cm) thick, tapered upward from an enlarged club- to spindle-shaped or elongated base, powdery with floccose patches on the base, dry, solid, white; partial veil whitish, soon disintegrating, rarely leaving an evanescent delicate ring on the stalk.
SPORE PRINT COLOR: white.
MICROSCOPIC FEATURES: spores 8–11 × 5–7 μm, elliptical, smooth, hyaline, inamyloid.
FRUITING: solitary or loosely scattered on the ground in mixed woods, often under oak; June–November; widely distributed in the Southeast; occasional.
EDIBILITY: unknown.
COMMENTS: This mushroom is easily confused with other large, odorous, whitish amanitas, including *Amanita polypyramis* (p. 111), edibility unknown, which has pyramidal warts on the cap disc, a large club-shaped swollen stalk base (like a bowling pin), and larger spores that measure 9–13 × 5.5–9.5 μm. *Amanita longipes* (p. 109), edibility unknown, has a relatively slender rooting stalk base that is often bent "doglegged" beneath the ground. *Amanita daucipes* (p. 106), edibility unknown, has salmon-colored tints on the cap and stalk and a large turnip-shaped rooting stalk base.

Amanita citrina f. citrina (Schaeff) per Roques
Illus. p. 17

COMMON NAME: Citron Amanita.
CAP: 1½–5⅛" (4–13 cm) wide, convex, becoming nearly plane in age; surface glabrous, slightly viscid when moist, pale greenish yellow or sometimes with pale brownish areas, with whitish to pale yellowish scattered patches; margin entire or obscurely striate.
FLESH: moderately thick, white; odor of raw potatoes, sometimes faint or absent; taste not distinctive.
GILLS: free or nearly so, broad, crowded, white to whitish; edges even to slightly fringed.
STALK: 2⅜–5½" (6–14 cm) long, ¼–¾" (6–20 mm) thick, enlarged downward to a dingy white to pale yellowish submarginate to marginate-depressed basal bulb, dry, hollow in age, fibrillose above the ring and glabrous below, white or sometimes with pale brown areas; partial veil membranous, whitish to pale yellow, leaving a pendant superior ring that often collapses.

SPORE PRINT COLOR: white.

MICROSCOPIC FEATURES: spores 6–7 × 6–7 μm, globose to subglobose, smooth, hyaline, amyloid.

FRUITING: solitary, scattered, or in groups on the ground in oak and pine woods; July–March; widely distributed in the Southeast; fairly common.

EDIBILITY: toxicity suspected.

COMMENTS: *Amanita citrina* f. *lavendula,* toxicity suspected, is much more common in the Southeast and is nearly identical, but it has lavender tints, especially on the volva (see *MNE*, p. 272). Both forms are especially common in sandy soil on the coastal plain and barrier islands during the winter months.

Amanita cokeri (Bilb. and Kühner) Gilbert   Illus. p. 17

COMMON NAME: Coker's Amanita.

CAP: 2¾–5¾" (7–14.5 cm) wide, convex, becoming broadly convex to nearly plane in age; surface subviscid when moist, nonstriate, shiny when dry, white to whitish, densely covered with fairly large white to pale brownish pyramidal warts; margin entire.

FLESH: moderately thick, white; odor and taste not distinctive.

GILLS: free or nearly so, broad, crowded, white to pale cream.

STALK: 4–8" (10–20 cm) long, ⅜–¾" (1–2 cm) thick, enlarged downward to a slightly rooting, fusiform to ventricose basal bulb; dry, solid, fibrillose to lacerate-scaly, white; partial veil membranous, white, leaving a large pendant superior ring; ring frequently double-edged, striate above, floccose-lacerate below; basal bulb bearing concentrically arranged, white to pale brownish pyramidal warts and recurved scales that extend upward on the lower portion of the stalk.

SPORE PRINT COLOR: white.

MICROSCOPIC FEATURES: spores 11–14 × 7–9 μm, elliptic to elongate, smooth, hyaline, amyloid.

FRUITING: solitary, scattered, or in groups on the ground in mixed oak and pine woods; July–November; widely distributed in the Southeast; occasional to fairly common.

EDIBILITY: unknown.

COMMENTS: This mushroom was named in honor of American mycologist William Chambers Coker (1872–1953). It is easily recognized by its shiny white cap with conspicuous warts, the double-edged ring, and the recurved scales on the lower stalk and basal bulb.

Amanita daucipes (Mont.) Lloyd   Illus. p. 17

COMMON NAME: Carrot-foot Amanita.

CAP: 2⅜–11¾" (6–30 cm) wide, convex to nearly plane in age, occasionally with a very broad umbo; surface dry to shiny, pale cream to pale pinkish orange with pale cream to pale pinkish orange conical warts; margin nonstriate, appendiculate.

FLESH: thick, whitish; odor variously described as sweetish, meaty, or distinctive but difficult to characterize; taste not recorded in the original description.

GILLS: free, moderately narrow, crowded, white to cream.

STALK: 3⅛–8¼" (8–21 cm) long, ⅜–3⅛" (1–3 cm) thick, enlarged downward to a very large, broadly fusiform to napiform, submarginate basal bulb that often splits longitudinally; dry, solid, floccose, whitish to pale pinkish orange; partial veil submembranous, whitish to cream, typically tearing and not leaving a ring, or sometimes leaving a collapsed superior ring; basal bulb sometimes bearing conical warts.

SPORE PRINT COLOR: white.

MICROSCOPIC FEATURES: spores 8.5–12 × 5–7 μm, elliptic to elongate, smooth, hyaline, amyloid.

FRUITING: solitary, scattered, or in groups on the ground in mixed conifer and broadleaf forests; July–November; widely distributed in the Southeast; occasional to fairly common.

EDIBILITY: unknown.

COMMENTS: The massive size, pinkish orange coloration, and huge basal bulb make this mushroom most distinctive. The basal bulb may attain a length of 6" (15.5 cm) and a diameter of 4¾" (12 cm).

Amanita flavorubescens Atkinson Illus. p. 17

COMMON NAME: Yellow Blusher.

CAP: 1½–4⅜" (4–11 cm) wide, convex, becoming nearly plane; surface smooth, somewhat viscid when moist, golden yellow to brownish yellow, typically with dingy yellowish warts and patches; margin lacking striations.

FLESH: whitish, slowly staining reddish when bruised; odor and taste not distinctive.

GILLS: free or nearly so, close to crowded, creamy white; edges even.

STALK: 2–5½" (5–14 cm) long, ⅜–¾" (1–2 cm) thick, enlarged slightly downward to a club-shaped bulb and often tapered below the bulb, whitish to yellowish, usually with reddish stains near the base, nearly smooth; flesh yellowish, slowly staining reddish when exposed; partial veil whitish to yellowish, leaving a superior, thin, membranous, pendant, yellowish ring; bulb slight, ovoid, sometimes coated with yellowish patches.

SPORE PRINT COLOR: white.

MICROSCOPIC FEATURES: spores 8–10 × 5.5–7 μm, elliptic, smooth, hyaline, amyloid.

FRUITING: solitary, scattered, or in groups on the ground in mixed woods, especially with oak; June–October; widely distributed in the Southeast; occasional.

EDIBILITY: toxicity suspected.

COMMENTS: *Amanita rubescens* (p. 112), edible, is similar but lacks the brighter yellow coloration and has more pronounced reddish stains, especially on the stalk base.

Amanita frostiana (Peck) Sacc. Illus. p. 18

COMMON NAME: Frost's Amanita.

CAP: 1–3" (2.5–7.5 cm) wide, convex, becoming flattened in age; surface covered with cottony yellow warts or patches, smooth beneath, bright orange, paler toward the margin in age; margin striate.

FLESH: thin, white; odor and taste not distinctive.

GILLS: free from the stalk, crowded, narrow, white to yellowish.

STALK: 2–3" (5–7.5 cm) long, ¼–½" (5–13 mm) thick, slightly tapered upward from an ovoid or globose bulbous base that may or may not have a rim, fibrillose to somewhat scurfy, often with yellow universal veil patches on or near the base, stuffed, white to yellowish; partial veil leaving a delicate membranous yellowish superior ring that often falls off in age.

SPORE PRINT COLOR: white.

MICROSCOPIC FEATURES: spores 7–10 μm, globose to subglobose, smooth, hyaline, inamyloid.

FRUITING: solitary to loosely scattered in mixed woods; June–December; widely distributed in the Southeast; occasional.

EDIBILITY: unknown.

COMMENTS: The patches or warts on the cap are easily washed off in wet weather. *Ama-*

nita flavoconia, toxicity suspected, is very similar but has amyloid spores and an even cap margin that lacks striations (see *MNE*, p. 273). *Amanita muscaria* var. *flavivolvata* (p. 109) is much larger overall, has a more persistent stalk ring, paler yellowish to tannish warts or patches on the cap, and broadly elliptical to elongated spores.

Amanita hesleri Bas   Illus. p. 18

COMMON NAME: Hesler's Amanita

CAP: 2¾–4½" (7–11.5 cm) wide, convex to nearly plane in age, with a low, broad umbo, sometimes slightly depressed on the disc; surface dry, becoming slightly viscid in age, fibrillose, white to dingy white with large, brown to brownish gray pyramidal warts and scales; margin nonstriate, appendiculate.

FLESH: moderately thick, whitish; odor and taste not distinctive.

GILLS: attached and notched, moderately broad, crowded, white or sometimes with a pinkish tint; edges even.

STALK: 2½–5½" (6.5–14 cm) long, ¼–⅝" (7–16 mm) thick, enlarged downward to a slender clavate to fusiform basal bulb, dry, solid, floccose to fibrillose, white; partial veil and ring absent; basal bulb bearing scales or warts that extend upward onto the lower stalk.

SPORE PRINT COLOR: white.

MICROSCOPIC FEATURES: spores 10–13 × 5–8 μm, elliptic to elongate, smooth, hyaline, amyloid.

FRUITING: solitary, scattered, or in groups on the ground in mixed conifer and broadleaf woodlands; May–September; widely distributed in the Southeast; occasional to fairly common.

EDIBILITY: unknown.

COMMENTS: This mushroom was named in honor of American mycologist Lexemuel R. Hesler (1888–1977).

Amanita komarekensis Jenkins and Vinopal   Illus. p. 18

COMMON NAME: none

CAP: 1½–3" (4–7.5 cm) wide, convex to nearly plane in age, with a slight umbo; surface slightly viscid when moist, glabrous, creamy white at first, becoming more beige at maturity, with pinkish beige patches; margin striate.

FLESH: white; odor and taste not recorded in the original description.

GILLS: free, crowded, creamy white; edges entire.

STALK: 2⅜–4⅜" (6–11 cm) long, ¼–⅝" (5–15 mm) thick, enlarged slightly downward to a creamy white, subglobose to ovoid basal bulb; dry, hollow, creamy white with pale pinkish pulverulence near the apex; more glabrous toward the base; partial veil membranous, pinkish beige, leaving an apical, moderately persistent ring; basal bulb apex bearing pinkish beige patches or floccose chunks.

SPORE PRINT COLOR: white.

MICROSCOPIC FEATURES: spores 8–10 × 5.5–8 μm, broadly elliptic to elliptic, smooth, hyaline, inamyloid.

FRUITING: solitary, scattered, or in groups on the ground in mixed conifer and broadleaf forests; June–October; widely distributed in the Southeast; occasional.

EDIBILITY: unknown.

COMMENTS: The photographs labeled *Amanita roseitincta* on pages 156 and 157 of *A Field Guide to Southern Mushrooms* (Weber and Smith 1985), are actually *Amanita komarek-*

ensis. *Amanita roseotincta* (not illustrated), edibility unknown, is similar but it has globose, amyloid spores.

Amanita longipes Bas in Tulloss and Jenkins Illus. p. 18
  COMMON NAME: none
  CAP: 1–3" (2.5–8 cm) wide, hemispheric when young, becoming broadly convex; surface dry, coated with a dense layer of fine powder, white, pale grayish buff to pale grayish brown on the disc; margin nonstriate, incurved when young, typically rimmed with flaps of torn partial veil.
  FLESH: white; odor and taste usually not distinctive.
  GILLS: barely to distinctly attached, close, whitish; edges powdery; partial veil fibrous-cottony, white, not forming a ring.
  STALK: 2–5½" (5–14 cm) long, ¼–¾" (6–20 mm) thick, enlarged downward and club-shaped then tapered abruptly and radicating somewhat, typically flattened or dog-legged or both, dry, powdery to scurfy, white, sometimes with reddish stains or spots
  SPORE PRINT COLOR: white.
  MICROSCOPIC FEATURES: spores 8–18 × 4–7 μm, elliptic, smooth, hyaline, amyloid.
  FRUITING: solitary, scattered, or in groups on sandy soil in mixed woods, especially oak-pine; June–October; widely distributed in the Southeast; fairly common.
  EDIBILITY: unknown.
  COMMENTS: The stalk is often deeply buried and coated with sand.

Amanita muscaria var. **flavivolvata** (Singer) Jenkins Illus. p. 18
  COMMON NAME: none
  CAP: 2–6¾" (5–17 cm) wide, convex, becoming nearly plane in age, sometimes depressed on the disc; surface faintly striate, especially near the margin; subviscid when moist, glabrous, orange-red to brownish red on the disc, paler toward the margin, fading with age to pale orange to nearly white, with deep yellow to orange yellow or tannish yellow floccose patches or warts; margin entire.
  FLESH: white; odor and taste not distinctive.
  GILLS: free, broad, crowded, white to pale cream; edges even.
  STALK: 2¾–5½" (7–14 cm) long, ⅜–1" (1–2.5 cm) thick, enlarged downward to a whitish or tannish and ovoid basal bulb, dry, solid then densely stuffed, fibrous to fibrous-scaly, white to cream, becoming tannish when handled; partial veil submembranous, creamy white to creamy yellow or creamy tan, leaving a superior ring that soon collapses or tears; basal bulb bearing ascending rings of yellowish to tannish warts or floccose chunks.
  SPORE PRINT COLOR: white.
  MICROSCOPIC FEATURES: spores 10–12 × 7–9 μm, broadly elliptic to elongate, smooth, hyaline, inamyloid.
  FRUITING: solitary, scattered, or in groups on the ground in mixed conifer and broadleaf forests; June–March; widely distributed in the Southeast; fairly common.
  EDIBILITY: poisonous.
  COMMENTS: When this species grows in more open areas that are exposed to sunlight, such as along roadsides, the cap soon fades to pale orange. *Amanita muscaria* var. *formosa*, poisonous, is very similar, but it has a yellow to orangish yellow or orange cap (see *MNE*, p. 273). *Amanita muscaria* var. *persicina* (not illustrated), poisonous, is also very similar, but it has a light orange to melon orange cap and cream gills with a pale pinkish tint.

Amanita mutabilis Beardslee
Illus. p. 18

COMMON NAME: none.

CAP: 2⅜–4⅜" (6–11 cm) wide, convex, becoming broadly convex to nearly plane in age; surface subviscid when moist, glabrous, whitish to creamy tan, with whitish to creamy tan patches, sometimes staining pink when bruised; margin nonstriate, slightly appendiculate.

FLESH: white, typically turning pink when injured; odor variously described as oily or anise-like or not distinctive; taste not distinctive.

GILLS: free or attached with a decurrent line, moderately broad, crowded, white to pale cream, typically staining pink when bruised.

STALK: 2–6⅜" (5–16 cm) long, ⅜–⅞" (1–2.2 cm) thick, enlarged downward to a white, globose to ovoid, marginate or submarginate basal bulb; dry, solid, whitish with occasional pale tan fibrils, sometimes staining pink when bruised; partial veil membranous, white to whitish, leaving a pendant, subapical ring that sometimes detaches.

SPORE PRINT COLOR: white.

MICROSCOPIC FEATURES: spores 10–15 × 5.5–9 μm, elliptic to cylindric, smooth, hyaline, amyloid.

FRUITING: solitary, scattered, or in groups on sandy soil in oak and pine forests; June–November; widely distributed in the Southeast; fairly common.

EDIBILITY: unknown.

COMMENTS: The pink staining reaction is a variable feature. *Amanita anisata, Amanita submutabilis,* and *Amanita abruptiformis* are synonyms.

Amanita parcivolvata (Peck) Gilbert
Illus. p. 19

COMMON NAME: none.

CAP: 1⅛–4¾" (3–12 cm) wide, convex, becoming broadly convex in age, sometimes with a depressed disc; surface viscid when moist, glabrous, crimson to reddish orange, lighter toward the margin, with yellowish to cream warts and patches; margin usually strongly striate to tuberculate-striate.

FLESH: white; odor and taste not distinctive.

GILLS: free, moderately broad, crowded, pale yellow; edges often floccose.

STALK: 1⅛–4¾" (3–12 cm) long, ⅛–⅝" (3–17 mm) thick, enlarged slightly downward to a subglobose, occasionally slightly rooting whitish to pale yellow basal bulb; dry, stuffed or hollow, furfuraceous or mealy near the apex or sometimes overall, more fibrillose toward the base, pale yellow to whitish; partial veil and ring absent.

SPORE PRINT COLOR: white.

MICROSCOPIC FEATURES: spores 9–14 × 6–8 μm, elliptic to cylindric, smooth, hyaline, inamyloid.

FRUITING: solitary, scattered, or in groups on the ground in mixed conifer and broadleaf forests; June–October; widely distributed in the Southeast; occasional to fairly common.

EDIBILITY: toxicity suspected.

COMMENTS: The crimson to reddish orange cap with yellowish warts or patches, pale yellow or whitish stalk, and lack of a partial veil and ring are the distinctive features. Some varieties of *Amanita muscaria,* poisonous, are similar, but they have a partial veil that leaves a persistent ring on the stalk. *Amanita jacksonii = Amanita caesarea,* edible, has a yellowish stalk with darker orange streaks, a superior skirt-like ring, and a conspicuous white volva (see *MNE,* p. 272).

Amanita pelioma Bas Not Illustrated
  COMMON NAME: none.
  CAP: 2–3½" (2–9 cm) wide, convex to broadly convex, nonstriate; surface slightly viscid when moist, becoming dry and shiny, glabrous, grayish olive to brownish gray or whitish, with concolorous warts or sometimes with small bluish green scales; margin appendiculate.
  FLESH: white; odor resembling chlorine; taste not reported.
  GILLS: free, crowded, pale grayish olive to pale brown or brownish gray, sometimes with a pale lavender tint.
  STALK: 3½–6" (9–15.5 cm) long, 5/16–½" (8–12 mm) thick, enlarged slightly downward to a clavate to fusiform, sometimes rooting basal bulb; dry, solid, grayish olive to pale grayish; partial veil membranous, whitish to pale brown, leaving a delicate apical ring that often tears and falls away; basal bulb bearing grayish olive to brownish gray warts on the upper portion and lower stalk, often with a bluish green stain on the top portion.
  SPORE PRINT COLOR: cream to buff or pale gray.
  MICROSCOPIC FEATURES: spores 9–13 × 6–8 μm, elliptic to elongate, smooth, hyaline, amyloid.
  FRUITING: solitary, scattered, or in groups on the ground in mixed conifer and broadleaf forests; July–October; widely distributed in the Southeast; uncommon.
  EDIBILITY: unknown.
  COMMENTS: Although not commonly encountered, the combination of brownish gills and the bluish green stain on the top portion of the basal bulb makes this mushroom easy to identify.

Amanita polypyramis (Berk. and Curt.) Sacc. Illus. p. 19
  COMMON NAME: none.
  CAP: 3–8" (7.5–20 cm) wide, broadly convex to nearly plane in age; surface dry or subviscid when moist, sometimes shiny, white with white conical warts that often disappear; margin nonstriate, appendiculate.
  FLESH: thick, firm, white; odor resembling chlorine; taste not distinctive.
  GILLS: free or very finely attached, crowded, broad, white to creamy white; edges even.
  STALK: 3–8" (7.5–20 cm) long, 3/8–1 3/8" (1–3.5 cm) thick, enlarged slightly downward to a subglobose to broadly clavate or submarginate white basal bulb, dry, solid, with a slight flocculence when young, becoming glabrous or nearly so at maturity, white; partial veil thick, submembranous to floccose, uneven, delicate, white, tearing and typically leaving pieces attached to the cap margin; basal bulb bearing small, whitish pyramidal warts, at least on the upper portion.
  SPORE PRINT COLOR: creamy white.
  MICROSCOPIC FEATURES: spores 9–14 × 5–10 μm, broadly elliptic to elongate, smooth, hyaline, amyloid.
  FRUITING: solitary, scattered, or in groups on the ground under conifers or in mixed conifer and broadleaf forests; August–December; widely distributed in the Southeast; fairly common.
  EDIBILITY: poisonous.
  COMMENTS: *Amanita chlorinosma* (p. 105), poisonous, is similar, and its flesh also has a chlorine odor, but it has smaller spores that measure 8–11 × 5–7 μm.

Amanita ravenelii (Berk. and Curt.) Sacc.   Illus. p. 19
- COMMON NAME: none.
- CAP: 3–7⅜" (7.5–18.5 cm) wide, convex to nearly plane in age; surface dry, smooth, whitish to creamy white, with large, truncate, dull cream to brownish warts; margin nonstriate, slightly appendiculate.
- FLESH: thick, white; odor resembling chlorine.
- GILLS: free, crowded, pale cream to yellowish cream; edges even.
- STALK: 3½–10" (9.5–25.5 cm) long, ⅜–1⅜" (1–3.5 cm) thick, enlarged downward to a large, whitish, ventricose to fusiform, rooting basal bulb, dry, solid, floccose to flocculose-fibrillose or somewhat scaly, white to creamy white; partial veil floccose-membranous, whitish, leaving tiny torn pieces on the cap margin and typically a thick, sometimes collapsed superior ring; basal bulb bearing thick, recurved, whitish to brownish scales that are frequently in irregular rings.
- SPORE PRINT COLOR: white.
- MICROSCOPIC FEATURES: spores 8–11 × 5–8 μm, elliptic to elongate, smooth, hyaline, anyloid.
- FRUITING: solitary, scattered, or in groups on the ground in mixed conifer and broadleaf forests; August–November; widely distributed in the Southeast; occasional to frequent.
- EDIBILITY: unknown.
- COMMENTS: This mushroom was named in honor of American mycologist Henry William Ravenel (1814–1887).

Amanita rhopalopus f. **rhopalopus** Bas   Illus. p. 19
- COMMON NAME: none.
- CAP: 1¾–7¼" (4.5–18 cm) wide, convex to nearly plane in age; surface viscid or dry, often conspicuously cracked, white with sticky, whitish to brownish floccose patches or warts; margin nonstriate, appendiculate.
- FLESH: thick, firm, white; odor strong, resembling chlorine.
- GILLS: free or slightly attached, crowded, white to pale cream; edges even.
- STALK: 2⅜–8" (6–20 cm) long, ⅜–1" (1–2.5 cm) thick, enlarged slightly downward to a whitish or brownish, fusiform, deeply rooting basal bulb; dry, solid, glabrous to fibrillose, white; partial veil membranous, delicate, whitish to brownish, leaving tiny pieces on the cap margin and a delicate superior ring that soon falls away; basal bulb smooth or sometimes with floccose patches or warts.
- SPORE PRINT COLOR: white.
- MICROSCOPIC FEATURES: spores 8–12 × 5–8 μm, elliptic to elongate, smooth, hyaline, amyloid.
- FRUITING: solitary, scattered, or in groups on the ground in mixed conifer and broadleaf forests; July–December; widely distributed in the Southeast; occasional to frequent.
- EDIBILITY: unknown.
- COMMENTS: *Amanita rhopalopus* f. *turbinata* (not illustrated), edibility unknown, is very similar but has a very large turbinate to napiform basal bulb.

Amanita rubescens (Persoon : Fries) S. F. Gray   Illus. p. 19
- COMMON NAME: The Blusher.
- CAP: 1½–8" (4–20 cm) wide, convex to plane in age; surface typically smooth and dry, bronze with brown or red tints, with numerous randomly arranged pinkish or reddish tan to tan floccose warts; margin nonstriate to faintly striate.

FLESH: thick, white, rather slowly bruising reddish.

GILLS: free or very finely attached, crowded, broad, white, often with slight reddish discolorations; edges even.

STALK: 3–8" (7–2.5 cm) long, ¼–1½" (7–40 mm) thick, tapering slightly toward the top, dry, solid to stuffed, smooth to slightly floccose or fibrous, white to pale tan, slowly bruising reddish; partial veil white to tan, leaving a superior, pendant, white to pale tan, membranous, thin, often torn ring; lower portion of stalk infrequently with indistinct rings of floccose universal veil tissue; basal bulb onion-shaped to elongate, normally with prominent reddish stains.

SPORE PRINT COLOR: white.

MICROSCOPIC FEATURES: spores 7–9 × 5–7 m, elliptic, smooth, thin walled, hyaline, amyloid.

FRUITING: scattered to grouped, terrestrial in mixed woods or under trees; June–November; widely distributed in the Southeast; very common.

EDIBILITY: edible when thoroughly cooked.

COMMENTS: An easily identified species by virtue of its reddish stains and nonyellowish warts. Very pale forms, sometimes nearly white, are occasionally encountered. This mushroom is frequently parasitized by *Hypomyces hyalinus*.

Amanita spreta (Peck) Sacc. Illus. p. 19

COMMON NAME: Hated Amanita.

CAP: 2–5⅛" (5–13 cm) wide, convex, becoming broadly convex to nearly plane in age; surface moist or dry, glabrous, sometimes shiny, grayish to grayish brown, usually lacking remnants, or sometimes with white patches; margin faintly to strongly striate.

FLESH: thick, firm, white; odor and taste not distinctive.

GILLS: free, close to somewhat crowded, moderately broad, white.

STALK: 2–6" (5–15.5 cm) long, ⅜–¾" (1–2 cm) thick, slightly enlarged downward, dry, hollow in age, nearly smooth, whitish, lacking a basal bulb; partial veil membranous, white, leaving a superior, persistent, pendant ring; volva present as a thin, white, membranous, cup-like sac.

SPORE PRINT COLOR: white.

MICROSCOPIC FEATURES: spores 10–13 × 5–7 μm, elongate to cylindric, smooth, hyaline, inamyloid.

FRUITING: solitary, scattered, or in groups on sandy soil under oak and pine; June–October; widely distributed in the Southeast; occasional to fairly common.

EDIBILITY: unknown.

COMMENTS: Despite its common name, this mushroom is not known to be poisonous and may have earned a bad reputation owing to confusion with the somewhat similar *Amanita phalloides*, which is deadly. *Amanita phalloides* has greenish tones on its cap, a nonstriate margin, a conspicuously bulbous base sheathed by a white volva, and amyloid spores (see *MNE*, p. 273).

Amanita virosa LaMarck Illus. p. 19

COMMON NAME: Destroying Angel.

CAP: 1⅛–5⅛" (3–13 cm) wide, convex to nearly flat, white but sometimes with a discolored center, smooth, dry to sticky, margin nonstriate; no warts present.

FLESH: white; odor and taste not distinctive.

GILLS: free or very finely attached, crowded, white; edges even.

STALK: 2⅜–8" (6–20 cm) long, ¼–¾" (7–20 mm) thick, tapering slightly toward the top, white, smooth to roughened, solid; partial veil white, thin, membranous, leaving a superior, pendant, readily adhering to stalk, white, membranous, thin, delicate, and often torn ring; bulb small, usually fairly round; volva large, white, saccate, fairly thin, membranous, sometimes multilobed.

SPORE PRINT COLOR: white.

MACROCHEMICAL TESTS: cap surface stains yellow with the application of KOH.

MICROSCOPIC FEATURES: spores 9–11 × 7–9 μm, subglobose to globose, smooth, thin walled, hyaline, amyloid; basidia mostly four-spored.

FRUITING: usually solitary, scattered, or in groups on the ground in mixed woods; June–November; widely distributed in the Southeast; common.

EDIBILITY: deadly poisonous.

COMMENTS: This species, as well as *Amanita bisporigera* (not illustrated) and *Amanita verna* (not illustrated), are equally deadly. *Amanita bisporigera,* has mostly two-spored basidia and also stains yellow in KOH; in the authors' experience, it also is usually less robust, has a thinner, smoother stalk, and is more frequent in early summer. *Amanita verna* does not stain yellow with KOH.

Amanita volvata (Peck) Lloyd   Illus. p. 20

COMMON NAME: none.

CAP: 1⅜–2⅜" (3.5–6 cm) wide, convex, becoming broadly convex to nearly plane in age; surface dry, whitish with an occasional brownish disc, with white to creamy white, thin, floccose patches; margin nonstriate to faintly striate.

FLESH: thick, whitish; odor and taste not distinctive.

GILLS: free, crowded, white.

STALK: 1¾–2¾" (4.5–7 cm) long, ¼–½" (5–12 mm) thick, enlarged slightly downward and covered at the base by a thick, saccate, membranous, white volva, dry, solid, minutely floccose-scaly, whitish; partial veil and ring absent.

SPORE PRINT COLOR: white.

MICROSCOPIC FEATURES: spores 8.5–10 × 5.5–7 μm, elliptic, smooth, hyaline, amyloid.

FRUITING: solitary, scattered, or in groups on the ground in mixed conifer and broadleaf forests; July–November; widely distributed in the Southeast; occasional.

EDIBILITY: toxicity suspected.

COMMENTS: *Amanita peckiana* (not illustrated), edibility unknown, is similar, but it has pinkish to pinkish cream fibrils, squamules, and patches on its cap; a pinkish white volva; longer spores that measure 12–15 × 5–6 μm; and often an appendiculate margin.

Anellaria sepulchralis (Berkeley) Singer   Illus. p. 20

COMMON NAME: none.

CAP: 1½–4" (4–10 cm) wide, convex, becoming broadly convex; surface dry, smooth to wrinkled, often rimose-areolate in dry weather or in age, white to yellowish buff.

FLESH: whitish; odor and taste not distinctive.

GILLS: attached, close, moderately broad, pale grayish at first, becoming mottled with black and finally black at maturity; edges whitish.

STALK: 2¾–7" (7–18 cm) long, ¼–⅝" (5–16 mm) thick, dry, solid, longitudinally twisted-striate, sometimes with moisture beads at the apex when fresh, white to grayish white; partial veil and ring inconspicuous, often absent on mature specimens.

SPORE PRINT COLOR: black.

MICROSCOPIC FEATURES: spores 14–20 × 9–12 μm, elliptic, smooth, truncate, with an apical pore, blackish.

FRUITING: scattered or in groups or clusters on dung, soil, wood shavings, sawdust, or straw mixed with manure; April–December; widely distributed in the Southeast; fairly common.
EDIBILITY: edible.
COMMENTS: Also known as *Panaeolus phalaenarum* and *Panaeolus solidipes*. *Copelandia westii* (see photo, p. 22), edibility unknown, is very similar, but it slowly stains blue when handled, especially on the stalk.

Anthracophyllum lateritium (Berkeley and Curtis) Singer   Illus. p. 20
COMMON NAME: none.
CAP: ¼–1" (5–25 mm) wide, convex, shell to kidney-shaped; surface dry, subtomentose to glabrous, radially rugose, pale reddish brown at first, becoming dark reddish brown to blackish in age; margin incurved at first, remaining so well into maturity, surpassing the gills.
FLESH: pale yellowish brown; odor and taste not distinctive.
GILLS: strongly arched, subdistant to distant, narrow, pale rusty brown to dark reddish brown, becoming blackish in age.
STALK: absent or rudimentary.
SPORE PRINT COLOR: white.
MICROSCOPIC FEATURES: spores 9.5–15 × 5.5–8 μm, oblong-ellipsoid to elongate-ellipsoid with a tapered base, hyaline, inamyloid.
FRUITING: in groups or large clusters on decaying broadleaf trees, especially oak; year-round; fairly common; occurring along the Gulf Coast states from Florida to Texas.
EDIBILITY: unknown.
COMMENTS: The combination of a small, stalkless, dark reddish brown cap; distant brown gills; a white spore print; and the tendency to grow in large clusters is distinctive.

Armillaria mellea (Vahl : Fries) Kummer   Illus. p. 20
COMMON NAME: Honey Mushroom.
CAP: 1½–4" (4–10 cm) wide, convex to nearly flat; surface yellowish brown at first but soon becoming predominantly yellow with a darker center, sometimes dry but typically viscid, usually with a few darkish hairs, especially near the center of the cap, which may or may not be erect.
FLESH: white, moderately thick at the center; odor and taste not distinctive.
GILLS: attached to subdecurrent, typically with fine decurrent lines descending to the ring, white to pale buff; partial veil thick, membranous, whitish on the upper surface, yellowish on the lower surface, leaving a neat disc-like ring.
STALK: 2–6" (5–15 cm) long, about ⅜" (8–10 mm) thick, fibrous, often becoming scaly just below the ring, white at first, becoming yellowish to brown or olive in age, staining brownish where bruised; rhizomorphs flattened.
SPORE PRINT COLOR: pale cream.
MICROSCOPIC FEATURES: spores 8.5–12 × 6–7 μm, broadly elliptic to oval, smooth, hyaline, inamyloid; basidial clamp connections absent.
FRUITING: in clusters on the ground, especially at the base of trees or stumps, sometimes directly on wood; June–December; widely distributed in the Southeast; occasional to common.
EDIBILITY: edible if thoroughly cooked.
COMMENTS: *Armillaria mellea* is a virulent parasite of conifer and broadleaf trees.

Armillaria ostoyae (Romagnesi) Herink Illus. p. 20
  COMMON NAME: none.
  CAP: 2–4" (5–10 cm) wide, variously convex to nearly flat, dry, tan to yellowish brown to more typically dark reddish brown, densely covered with dark reddish brown to blackish scales.
  FLESH: firm, white, rather thick at the center; odor and taste not distinctive.
  GILLS: attached to subdecurrent, close, white to cream at first, becoming grayish orange to cinnamon; partial veil thick, membranous, leaving a whitish ring with a fluffy brown margin.
  STALK: 2–8" (5–21 cm) long, about ⅝" (1.5 cm) thick at the apex, typically quite thickened downward at first, becoming cylindric in age and often quite tapered at the very base, with generally orangish to reddish brown fibers; entire stalk staining mahogany to blackish; often with adhering bits of the partial veil; yellow mycelial growth at the extreme base; rhizomorphs flattened.
  SPORE PRINT COLOR: pale cream.
  MICROSCOPIC FEATURES: spores 8–11 × 5.5–7 μm, broadly elliptic to oval, smooth, hyaline, inamyloid; clamp connections present at the bases of some basidia.
  FRUITING: typically growing in large clusters but sometimes solitary on or about stumps or trees; June–January; widely distributed in the Southeast; occasional to common, more frequent northward.
  EDIBILITY: a fine and often abundant edible if thoroughly cooked.
  COMMENTS: This species can be quite difficult to distinguish from the very similar *Armillaria gemina* (not illustrated), edible, which has cylindric rhizomorphs that are not flattened.

Armillaria tabescens (Scopoli) Emel Illus. p. 20
  COMMON NAME: Ringless Honey Mushroom.
  CAP: 1¼–4" (3–10 cm) wide, convex to broadly convex with a very broad umbo, dry, orangish brown to brown, with darker cottony scales or tufts of fibers.
  FLESH: white to brownish; odor strong, taste astringent.
  GILLS: subdecurrent, whitish at first, becoming pinkish brown.
  STALK: 3–8" (7.5–20 cm) long, ¼–⅝" (5–15 mm) thick, typically tapered toward the base; vertically scurfy-fibrous; whitish near the top, yellowish to brownish below.
  SPORE PRINT COLOR: cream.
  MICROSCOPIC FEATURES: spores 6.5–8 × 4.5–5.5 μm, elliptic, smooth, hyaline, inamyloid.
  FRUITING: in clusters on or about trees or stumps; June–December; widely distributed in the Southeast; common.
  EDIBILITY: edible.
  COMMENTS: Some people get upset stomachs from eating this tasty mushroom; it should be cooked thoroughly. Do not confuse this mushrooms with species of *Galerina*, which have a brown spore print color.

Asterophora lycoperdoides (Bull. : Mérat) Ditmar : Fries Illus. p. 20
  COMMON NAME: Powder Cap.
  CAP: ⅜–¾" (1–2 cm) wide, rather round, dry, cottony, whitish at first, becoming covered with dense brown powder at maturity.
  FLESH: thin, whitish; odor and taste farinaceous.
  GILLS: typically malformed, attached, distant, narrow, thick, often forked; whitish to somewhat pinkish beige.

STALK: ¾–1¼" (2–3 cm) long, ⅛–⅜" (3–10 mm) thick, white, becoming brownish, minutely hairy or silky, stuffed, becoming hollow.
SPORE PRINT COLOR: white.
MICROSCOPIC FEATURES: basidiospores 3–6 × 2–4 μm, elliptic to oval, smooth, hyaline, inamyloid; cap surface also produces chlamydospores, which are strongly warted or with blunt spines, oval to subglobose, 13–20 × 10–20 μm (excluding ornamentation), thick walled, pale brown, inamyloid.
FRUITING: in clusters on decaying mushrooms, especially *Russula* and *Lactarius* species; June–December; widely distributed in the Southeast; occasional and often overlooked.
EDIBILITY: unknown.
COMMENTS: *Asterophora parasitica*, edibility unknown, also grows in clusters on decaying *Russula* and *Lactarius* species, but it has a whitish to grayish or grayish brown cap that does not become powdery at maturity, and it has more well-defined brownish gills (see *MNE*, p. 275).

Baeospora myosura (Fries) Singer Illus. p. 21
COMMON NAME: Conifer-cone Baeospora.
CAP: ¼–¾" (5–20 mm) wide, convex, cinnamon to pale tan, moist, smooth.
FLESH: very thin, whitish; odor and taste not distinctive.
GILLS: attached, quite crowded and narrow, white.
STALK: ⅜–2" (1–5 cm) long, 1/32–1/16" (1–2 mm) thick, whitish to brownish, minutely hairy above, with longer hairs at the base, hollow.
SPORE PRINT COLOR: white.
MICROSCOPIC FEATURES: spores 3–4.5 × 1–2.5 μm, elliptic, smooth, hyaline, amyloid; cheilocystidia clavate to bowling-pin-shaped.
FRUITING: in groups on conifer cones, especially pine and spruce, also on Magnolia cones; August–November; widely distributed in the Southeast; common.
EDIBILITY: unknown.
COMMENTS: Similar gilled mushrooms that grow on cones lack the distinctly crowded gills and very small amyloid spores of this genus. *Baeospora myriadophylla*, edibility unknown, has a larger lavender cap, lavender gills, a lavender to brownish lavender stalk with long hairs on the base, and it grows on decaying conifer and broadleaf wood (see *MNE*, p. 275). Species of *Strobilurus* are similar and also grow on conifer cones, but have more distant gills and differ microscopically.

Bolbitius vitellinus (Person : Fries) Fries Illus. p. 21
COMMON NAME: Yellow Bolbitius.
CAP: ⅜–2¾" (1–7 cm) wide, conic at first, becoming bell-shaped, then nearly plane in age, often with a broad umbo; surface slimy-viscid when fresh, orange-yellow to yellow; margin faintly striate when young, becoming sulcate in age and sometimes splitting.
FLESH: thin, whitish; odor and taste not distinctive.
GILLS: attached at first, becoming free at maturity, pale yellow when young, becoming dark rusty tawny at maturity; gill edges minutely hairy.
STALK: 1⅛–3⅛" (3–8 cm) long, 1/16–¼" (1.5–6 mm) thick, nearly equal or enlarged downward, fragile, somewhat viscid, scurfy, pale yellow.
SPORE PRINT COLOR: rusty ochre.
MICROSCOPIC FEATURES: spores 11–15 × 6–8 μm, elliptic to oval, smooth, truncate, with an apical pore, brownish yellow.
FRUITING: scattered or in groups or clusters on dung, straw, wood shavings, and compost

mixed with manure; May–December; widely distributed in the Southeast; fairly common.

EDIBILITY: reportedly edible, but of little culinary interest.

COMMENTS: *Bolbitius variicolor*, edibility unknown, is similar, but it has a dark brownish olive to pale grayish olive cap that sometimes has orange-yellow tints (see *MNE*, p. 275).

Callistosporium luteo-olivaceum (Berk. and Curt.) Singer   Illus. p. 21

COMMON NAME: none.

CAP: ½–1" (1.2–2.6 cm) wide, convex to somewhat broadly bell-shaped, usually with a low broad umbo, becoming nearly plane in age; surface moist or dry, faintly translucent-striate when moist, smooth when dry, glabrous, pale yellowish brown to yellow-brown or reddish brown, along the margin sometimes retaining a darker band of color; margin with a narrow band of sterile projecting tissue, decurved and remaining so well into maturity.

FLESH: thin, soft, colored like the cap or paler; odor not distinctive; taste disagreeable, more or less bitter.

GILLS: attached and typically notched or sometimes subdecurrent, close to crowded, pale yellow at first, darkening at maturity, sometimes with reddish tints; edges concolorous, even to slightly eroded.

STALK: ⅞–2⅜" (2.2–5.5 cm) long, ⅛–¼" (2–6 mm) thick, nearly equal or slightly enlarged downward, pruinose near the apex, smooth below, hollow in age, colored like the cap or gills, with a white basal mycelium.

SPORE PRINT COLOR: white.

MICROSCOPIC FEATURES: spores 4.5–6.5 × 3–4 μm, ellipsoid, smooth, hyaline, often with reddish pigment in oil drops or near the wall; inamyloid.

FRUITING: scattered or in groups on decaying conifer or broadleaf trees; May–October; widely distributed in the Southeast; occasional.

EDIBILITY: unknown.

COMMENTS: Formerly known as *Collybia luteo-olivaceus*. Compare *Callistosporium purpureomarginatum*, reported from North Carolina and further northward, which has a reddish purple cap with a thin violet band at the margin and reddish purple marginate gills (see *MNE*, p. 275).

Catathelasma ventricosa (Peck) Singer   Illus. p. 21

COMMON NAME: Swollen-stalked Cat.

CAP: 3–6" (7.5–15 cm) wide, convex, dry, smooth at first, becoming patchy-scaly, whitish to gray; margin usually adorned with bits of tissue from the partial veil.

FLESH: thick, very firm, white; odor not distinctive, taste unpleasant.

GILLS: decurrent, close to subdistant, whitish to buff.

STALK: 2–4" (5–10 cm) long, 1–2" (2.5–5 cm) thick, tapering toward the base and usually mostly buried, dry, white above the double ring, brownish yellow below it; having a two-layered ring, the upper one fibrous, the lower one membranous.

SPORE PRINT COLOR: white.

MICROSCOPIC FEATURES: spores 9–11 × 4–6 μm, elliptic, smooth, hyaline, amyloid.

FRUITING: solitary, scattered, or in groups on the ground under conifers; July–November; widely distributed in the Southeast; infrequent.

EDIBILITY: edible, but mediocre at best.

COMMENTS: The two-layered ring is a unique feature of this mushroom.

Chlorophyllum molybdites Illus. p. 21
- COMMON NAME: Green-spored Lepiota, Green-spored Parasol.
- CAP: 2¾–12" (7–30 cm) wide, round at first, becoming convex to nearly flat, often with a broad umbo; surface dry, white, covered with large pinkish brown to cinnamon patches when young that become scale-like in age, usually clustered toward the disc.
- FLESH: white, thick, not staining when cut or bruised, or occasionally slowly staining reddish; odor and taste not distinctive.
- GILLS: free from the stalk, close, broad; white at first, becoming greenish to grayish green in age, staining yellow to brownish when cut or bruised.
- STALK: 4–10" (10–25.5 cm) long, ⅜–1" (1–2.5 cm) thick, nearly equal or enlarged downward, usually with a swollen base, smooth, white, staining brownish when bruised; partial veil white, membranous, forming a superior ring with a fringed or double edge; ring becoming brownish on the underside in age, often moveable.
- SPORE PRINT COLOR: green to grayish green.
- MICROSCOPIC FEATURES: spores 9–13 × 6–9 μm, subovoid to elliptical with an apical germ pore, smooth, hyaline, dextrinoid.
- FRUITING: solitary, scattered, or in groups or fairy rings in lawns, meadows, pastures, and gardens; July–December; widely distributed in the Southeast; occasional to fairly common.
- EDIBILITY: poisonous (see Comments).
- COMMENTS: This species is one of the most frequent causes of serious mushroom poisoning in eastern North America. Its toxins can cause such severe vomiting that hospitalization is required to prevent acute, life-threatening dehydration. The Shaggy Parasol, *Macrolepiota rachodes*, edible, is similar, but it has white stalk flesh that stains yellow-orange or saffron when cut or bruised, especially at the base, white gills that do not become greenish at maturity, and a white spore print (see *MNE*, p. 300).

Chroogomphus rutilus (Schaeffer : Fries) O. K. Miller Jr. Illus. p. 21
- COMMON NAME: Brownish Chroogomphus, Pine Spike.
- CAP: ¾–3⅛" (2–8 cm) wide, obtuse to convex, often with a small, pointed umbo; surface sticky when fresh, dull orange-brown to ochraceous brown or pale reddish brown; margin incurved when young.
- FLESH: pale salmon to pale ochraceous or pinkish tinged; odor and taste not distinctive.
- GILLS: thick, decurrent, close to subdistant, broad, pale ochre when young, becoming pale cinnamon brown in age.
- STALK: 2–6" (5–15.5 cm) long, ¼–1" (6–25 mm) thick, solid, tapered toward the base or nearly equal, ochraceous buff to tawny orange, often with vinaceous tints in age; partial veil pale ochraceous, fibrillose, sometimes leaving a thin layer of fibrils on the upper portion of the stalk.
- SPORE PRINT COLOR: smoky gray to blackish.
- MICROSCOPIC FEATURES: spores 14–22 × 6–7.5 μm, elliptic, smooth, pale gray-brown; cystidia thin walled.
- FRUITING: scattered or in groups on the ground under conifers; August–January; widely distributed; occasional.
- EDIBILITY: edible.
- COMMENTS: *Chroogomphus vinicolor*, edible, is very similar, but it has a darker vinaceous red cap and cystidia with thickened midportion walls (see *MNE*, p. 276).

Claudopus vinaceocontusus Baroni   Not Illustrated

COMMON NAME: none.

CAP: ⅛–⅜" (2–10 mm) wide, convex to broadly convex with a shallowly depressed and low papillate disc, more or less circular to shell-shaped; surface dry, densely appressed, fibrillose, sordid off-white to pale grayish buff, quickly staining light vinaceous when bruised, drying dark reddish brown; margin inrolled.

FLESH: dingy white, quickly staining vinaceous purple on exposure then rapidly fading; garlic odor strong; taste not determined.

GILLS: attached to short-decurrent, whitish at first, soon becoming fleshy pink, edges even.

STALK: ⅛–¼" (3–5 mm) long, about 1/16" 1–2 mm thick, tapered downward or nearly equal, lateral, dry, solid, colored like the cap, densely whitish fibrillose to pubescent overall with a white basal mycelium and fine white rhizomorphs that stain vinaceous purple when bruised.

SPORE PRINT COLOR: pink to salmon.

MICROSCOPIC FEATURES: spores 8–11 × 6–7 μm, ellipsoid with 6–8 sharp or rounded angles, hyaline, inamyloid.

FRUITING: in clusters or scattered on decaying broadleaf trees, especially oak, usually on moss-covered logs, or sometimes on soil; June–September; fairly common; widespread from North Carolina south to Florida and west along the Gulf Coast region to Texas.

EDIBILITY: unknown.

COMMENTS: This mushroom is easily distinguished by its small size, strong garlic odor, and the vinaceous purple staining reaction when bruised.

Clitocybe clavipes (Fries) Kummer   Illus. p. 21

COMMON NAME: Fat-footed Clitocybe, Club-footed Clitocybe.

CAP: ¾–3½" (2–9 cm) wide, convex when young, becoming flat with a shallowly depressed center in age, with or without a low, broad umbo; surface moist, smooth, sometimes with minute matted fibrils over the disc, gray-brown to olive-brown, paler toward the margin, fading to buff in age; margin becoming elevated and wavy in age.

FLESH: whitish; odor fragrant of bitter almond, taste not distinctive.

GILLS: strongly decurrent, close to subdistant, whitish when young, becoming pale yellowish to cream in age, often forked and crossveined.

STALK: 1⅜–2⅜" (1.5–6 cm) long, ⅛–½" (4–12 mm) thick, enlarged downward, stuffed, moist, longitudinally streaked with olive-gray fibrils over a whitish ground color; base bulbous or club-shaped, often covered with white mycelium.

SPORE PRINT COLOR: white.

MICROSCOPIC FEATURES: spores 6–10 × 3.5–5 μm, oval, smooth, hyaline, inamyloid.

FRUITING: solitary, scattered, or in groups on the ground under conifers, broadleaf trees, or in mixed woods; July–November; widely distributed in the Southeast; fairly common to locally abundant.

EDIBILITY: edible when thoroughly cooked; adverse reactions reported when this mushroom is consumed with alcohol.

COMMENTS: *Clitocybe gibba*, edible, has a pinkish cinnamon to ochraceous salmon, flat to funnel-shaped cap; decurrent, close, whitish, forked gills; and a nearly equal, slender, whitish stalk; and it grows on the ground in broadleaf and mixed woods (see *MNE*, p. 277).

Clitocybe nuda (Fries) Bigelow and Smith   Illus. p. 22
  COMMON NAME: Blewit.
  CAP: 1⅝–5⅞" (4–15 cm) wide, convex when young, becoming broadly convex to nearly flat in age; surface smooth, hygrophanous, slightly viscid and shiny when moist, dull when dry, sometimes finely cracked over the disc, with or without a low, broad umbo, violet to lilac-gray to pinkish buff, fading to pinkish tan or buff in age; margin inrolled when young, becoming expanded and occasionally uplifted in age, wavy, faintly striate when moist.
  FLESH: pale violet or lilac; odor fragrant; taste not distinctive.
  GILLS: notched to sinuate, close to crowded, pale lavender to violet or lilac, becoming lilac-buff to brownish in age.
  STALK: 1⅛–3" (3–7.5 cm) long, ⅜–1⅛" (1–3 cm) thick, equal, often bulbous or club-shaped at the base, solid, dry, fibrillose to scurfy, whitish, pale violet or lavender, bruising dark lavender, becoming brownish in age.
  SPORE PRINT COLOR: pinkish buff.
  MICROSCOPIC FEATURES: spores 5.5–8 × 3.5–5 μm, elliptic, roughened or sometimes smooth, hyaline, inamyloid.
  FRUITING: solitary or in groups or clusters on the ground under broadleaf trees and conifers, in meadows and lawns, on decaying vegetable matter, on wood chips, and near compost piles; August–December; widely distributed in the Southeast; common.
  EDIBILITY: edible and good, although some collections are bitter and disagreeable.
  COMMENTS: When collecting this species for the table, you should avoid confusing it with possibly poisonous look-alikes, especially *Cortinarius* species, which have a rusty brown spore print and a web-like partial veil. *Clitocybe irina*, reported as edible but sometimes causing adverse reactions, has a whitish to buff or pinkish buff cap and stalk, whitish to pale pinkish flesh that has a fragrant odor, and whitish gills. It grows on the ground under conifers or in mixed woods (see *MNE*, p. 277).

Clitocybe subconnexa Murrill   Illus. p. 22
  COMMON NAME: none.
  CAP: 1–3½" (2.5–9 cm) wide, convex, becoming nearly flat in age; surface smooth, glabrous, satiny white at first, becoming buff to pale brownish when old, especially on the disc; margin incurved at first, wavy when expanded.
  FLESH: white or tinged pinkish, brittle; odor fragrant or not distinctive; taste mild when young, becoming bitter or astringent when mature.
  GILLS: attached to short-decurrent, more or less separable from the cap flesh, close to crowded, sometimes forked, narrow, whitish to pale pinkish buff.
  STALK: 1¼–4" (3–10 cm) long, ¼–¾" (5–20 mm) thick, nearly equal or slightly tapered upward, dry, hollow in age, dull, silky, tomentose at the base, grayish buff.
  SPORE PRINT COLOR: pale pinkish buff.
  MICROSCOPIC FEATURES: spores 4.5–6 × 3–3.5 μm, elliptical, minutely warted, hyaline.
  FRUITING: densely caespitose on the ground in broadleaf and mixed woods; June–December; widely distributed in the Southeast; common.
  EDIBILITY: edible when young and fresh, unappealing when older.
  COMMENTS: The tightly packed clusters help to distinguish this species from similar ones such as *Clitocybe irina*, reported as edible but sometimes with adverse reactions, which grows in more loosely scattered groups, often in fairy rings, and has larger spores (see *MNE*, p. 277). The fruit bodies of *Clitocybe robusta*, edibility unknown, are typically larger, develop a skunk-like odor in age, and produce a pale yellow spore deposit (see

MNE, p. 277). *Leucopaxillus albissimus,* edibility unknown, is white at first but becomes yellow-brown on the disc as it ages and has anyloid spores (see *MNE*, p. 299).

Clitocybula familia (Peck) Singer   Illus. p. 22
COMMON NAME: Family Collybia.
CAP: ⅜–1⅝" (1–4 cm) wide, bell-shaped when young, becoming convex then nearly plane in age; surface moist, smooth, brownish buff to creamy buff; margin incurved at first, becoming uplifted and torn in age.
FLESH: thin, fragile, whitish; odor and taste not distinctive.
GILLS: attached to nearly free, crowded, narrow, white to grayish.
STALK: 1⅝–3¼" (4–8 cm) long, ¹⁄₁₆–⅛" (1.5–3 mm) thick, moist or dry, smooth, grayish white, with white hairs at the base.
SPORE PRINT COLOR: white.
MICROSCOPIC FEATURES: spores 3.5–4.5 μm, globose, smooth, hyaline, amyloid.
FRUITING: in dense clusters on decaying conifer wood; July–October; widely distributed in the Southeast; occasional.
EDIBILITY: edible.
COMMENTS: Formerly known as *Collybia familia*. Compare *Clitocybula lacerata*, edibility unknown, which has a distinctly streaked pale to dark grayish brown or brownish gray cap that fades to pale dull brown or tan, a conspicuously torn margin at maturity, and larger elliptic spores that measure 6–8 × 4.5–6 μm, and which grows in dense clusters on decaying conifer or broadleaf wood (see *MNE*, p. 278).

Collybia cookei (Bresadola) Arnold   Illus. p. 22
COMMON NAME: none.
CAP: ⅛–⅜" (2–10 mm) wide, convex, becoming broadly convex to nearly plane in age, sometimes depressed on the disc; surface dry or moist, minutely fibrillose, nearly smooth, pale pinkish buff to orangish buff, fading to dingy whitish and darker on the disc in age; margin incurved to inrolled when young, becoming decurved to straight at maturity, sometimes striate.
FLESH: very thin, colored like the cap; odor and taste not distinctive.
GILLS: attached to subdecurrent, close to subdistant, white to pinkish buff.
STALK: ¼–2" (6–50 mm) long, ¹⁄₃₂–¹⁄₁₆" (1–2 mm) thick, nearly equal, flexible, dry, pruinose on the upper portion, covered with a thin coating of hairs near the base, cinnamon buff, attached to a small yellowish to ochraceous orange, rounded to irregular, often wrinkled sclerotium measuring ⅛–⅜" (3–8 mm) in diameter.
SPORE PRINT COLOR: white.
MICROSCOPIC FEATURES: spores 4.5–6 × 3–3.5 μm, short ellipsoid to lacrymoid, smooth, hyaline.
FRUITING: scattered, in groups, or in dense clusters on the blackened remains of decaying mushrooms or sometimes on well-decayed wood; June–October; widely distributed in the Southeast; occasional.
EDIBILITY: unknown.
COMMENTS: *Collybia tuberosa*, edibility unknown, is very similar, but its stalk is attached to a dark reddish brown, shiny, oval, apple-seed-like sclerotium (see *MNE*, p. 279). *Collybia cirrhata* (not illustrated), edibility unknown, is also very similar, but its stalk is not attached to a sclerotium.

Conocybe lactea (Lange) Métrod Illus. p. 22
  COMMON NAME: White Dunce Cap.
  CAP: ⅜–1⅛" (1–3 cm) wide, conic to bluntly conic, becoming bell-shaped; surface dry, smooth, striate from the margin to the disc, whitish to creamy white, tinged ochraceous on the disc.
  FLESH: whitish; odor and taste not distinctive.
  GILLS: nearly free, close, very narrow, whitish when young, becoming tawny to reddish cinnamon with age.
  STALK: 1½–4" (4–10 cm) long, 1/16–⅛" (1–3 mm) thick, nearly equal, usually with a basal bulb, white pruinose or nearly smooth, white to translucent white; partial veil and ring absent.
  SPORE PRINT COLOR: reddish cinnamon.
  MICROSCOPIC FEATURES: spores 10–14 × 6–9 µm, elliptic with an apical pore, smooth, pale brown.
  FRUITING: scattered or in groups on lawns, fields, and other grassy areas; May–September; widely distributed in the Southeast; common.
  EDIBILITY: reportedly edible but of no culinary interest.
  COMMENTS: *Conocybe tenera*, edibility unknown, is similar, but it has a reddish brown to rusty brown or fulvous cap that fades to pale tawny or ochraceous when dry; and it grows on rich humus in woods, in gardens, and on lawns, fields, and fertilized areas (see *MNE*, p. 280).

Coprinus atramentarius (Bulliard : Fries) Fries Illus. p. 23
  COMMON NAME: Alcohol Inky.
  CAP: 1½–3" (4–7.5 cm) wide, oval to egg-shaped when young, becoming convex in age; surface smooth to slightly scurfy, sometimes forming tiny scales on the disc, dry, gray to gray-brown, often with shallow grooves on the margin.
  FLESH: grayish white; odor and taste not distinctive.
  GILLS: free, very crowded, white when very young, soon gray then black and deliquescing to a black inky fluid in age.
  STALK: 1½–6" (4–15.5 cm) long, ⅜–¾" (1–2 cm) thick, nearly equal or tapered downward, silky, hollow in age, white, with a white annular zone toward the base.
  SPORE PRINT COLOR: black.
  MICROSCOPIC FEATURES: spores 8–12 × 4.5–6 µm, elliptic, smooth, with an apical pore, pale grayish black.
  FRUITING: in clusters on grass, decaying wood, wood chips, and tree bases; May–November; widely distributed in the Southeast; common.
  EDIBILITY: edible, with caution; see Comments.
  COMMENTS: Most people who consume this species enjoy it without adverse reactions. However, some individuals experience coprine poisoning if they consume alcoholic beverages while eating this mushroom, or up to 48 hours before or after its consumption. Coprine poisoning signs and symptoms may include nausea, vomiting, marked flushing of the skin, rapid breathing, and severe headache.

Coprinus comatus (Müller : Fries) S. F. Gray Illus. p. 23
  COMMON NAME: Shaggy Mane, Lawyer's Wig.
  CAP: 1⅛–2" (3–5 cm) wide, oval to cylindric at first, becoming broadly conic to nearly plane and 2–3⅛" (5–8 cm) wide in age, fragile; surface dry, white with a brownish disc,

coated with coarse scales that are white to pale reddish brown and usually darkest at the tips.

FLESH: white at first, becoming black as the mushroom deliquesces in age; odor and taste not distinctive.

GILLS: attached at first then free from the stalk, densely crowded, white then pinkish and finally black as the mushroom deliquesces.

STALK: 3–12" (7.5–30 cm) long, ⅜–1" (1–2.5 cm) thick, enlarged downward to a bulbous base, sometimes rooting, hollow, glabrous to silky-fibrillose, white, fragile; partial veil white, submembranous, leaving a thin, inferior ring.

SPORE PRINT COLOR: black.

MICROSCOPIC FEATURES: spores 10–14 × 6–8.5 μm, ellipsoid, truncate, with an apical pore, smooth, purple-brown.

FRUITING: scattered or in groups or clusters in grassy areas, on soil, and in wood chips; May–November; widely distributed in the Southeast; common.

EDIBILITY: edible.

COMMENTS: *Coprinus sterquilinus* (not illustrated), edibility unknown, is a smaller, white to whitish mushroom that grows on dung or manured soil; it has much larger spores that measure 16–22 × 10–13 μm. Also see the brief descriptions for *Coprinus americanus* (see photo, p. 22) and *Coprinus variegatus* (see *MNE*, p. 281).

Coprinus disseminatus (Persoon) Gray   Illus. p. 23

COMMON NAME: Little Helmets, Non-inky Coprinus.

CAP: ¼–¾" (1.3–4 cm) wide, hemispheric to bell-shaped or convex; surface with fine hairs and minute glistening particles when young, nearly glabrous in age, prominently furrowed from the margin nearly to the center; pale yellowish at first, becoming brownish gray with yellow-brown on the disc; margin scalloped.

FLESH: thin, fragile, whitish; odor and taste not distinctive.

GILLS: attached to nearly free, fairly distant, white at first, becoming grayish to blackish in age, not deliquescing, or only slightly so.

STALK: ½–1½" (1.3–4 cm) long, about 1/32" (1 mm) thick, nearly equal, hollow, fragile, smooth to lightly scurfy, white to gray; base downy or with white bristly hairs.

SPORE PRINT COLOR: blackish.

MICROSCOPIC FEATURES: spores 7–10 × 4–5 μm, elliptical with a truncate apex, smooth, blackish.

FRUITING: typically in large troops, densely clustered on decaying wood of broadleaf trees, especially on and around stumps or the bases of dead trees; year-round; widely distributed in the Southeast; occasional to locally common.

EDIBILITY: edible, but of little culinary value.

COMMENTS: Also known as *Pseudocoprinus disseminatus*. Compare with *Coprinus plicatilis* (p. 126), which is larger and has a deeply sulcate cap with an orange-brown disc and larger spores.

Coprinus floridanus Murrill   Illus. p. 23

COMMON NAME: Florida Inky Cap.

CAP: ⅝–1¼" (1.5–3.2 cm) wide, broadly convex; surface distinctly radiately sulcate, gray or grayish yellow, typically with dingy yellow on the disk; margin entire.

FLESH: very thin, white; odor and taste not distinctive.

GILLS: attached and notched, moderately close, entire, gray at first, soon blackening.

STALK: ¾–2" (2–5 cm) long, ⅛–¼" (2–5 mm) thick, nearly equal or enlarged toward the base, smooth, glabrous, whitish to subhyaline.
SPORE PRINT COLOR: black.
MICROSCOPIC FEATURES: spores 9–11 × 4–6 μm, ellipsoid, smooth, black.
FRUITING: gregarious on the ground around stumps or decaying wood; year-round; reported only from Florida; occasional.
EDIBILITY: unknown.
COMMENTS: This mushroom was first collected by W. A. Murrill from Gainesville, Florida, in 1938. Compare with *Coprinus micaceus* (p. 125).

Coprinus laniger Peck   Illus. p. 23
COMMON NAME: none.
CAP: ⅜–1" (1–2.5 cm) wide, oval at first, becoming convex at maturity; surface dry, covered with large reddish brown to dark tawny patches and scales that easily wash off, striate to somewhat sulcate.
FLESH: thin, fragile, yellowish; odor and taste not distinctive.
GILLS: attached, crowded, white when young, becoming gray and finally black at maturity, dissolving with age.
STALK: ⅜–1" (1–2.5 cm) long, about ⅛" (2–4 mm) thick, slightly enlarged downward to a somewhat bulbous base, dry, hollow in age, white to yellowish, with a white fibrous ring; base embedded in a thick mat of orange threads.
SPORE PRINT COLOR: black.
MICROSCOPIC FEATURES: spores 7–10 × 4–4.5 μm, elliptic, smooth, with an apical pore, grayish black.
FRUITING: scattered or in groups or clusters on decaying wood; May–November; widely distributed in the Southeast; occasional.
EDIBILITY: unknown.
COMMENTS: The Orange-mat Coprinus, *Coprinus radians* (not illustrated), edibility unknown, also has a thick mat of orange threads at the stalk base, but it forms a larger cap, and its cap surface soon breaks up into tiny granules.

Coprinus micaceus (Bulliard : Fries) Fries   Illus. p. 23
COMMON NAME: Mica Cap.
CAP: ¾–2" (2–5 cm) wide, oval to egg-shaped when young, becoming bell-shaped to convex in age; surface covered with a sparse coating of tiny whitish granular scales on the button stage and soon becoming smooth, weakly to distinctly sulcate from the margin nearly to the disc; tawny to orange-brown on the disc, yellowish tan toward the margin, becoming grayish to dark grayish brown in age.
FLESH: whitish; odor and taste not distinctive.
GILLS: attached to nearly free, very crowded, white when young, soon gray then black and deliquescing to a black inky fluid in age.
STALK: 1–3⅛" (2.5–8 cm) long, ⅛–¼" (2–5 mm) thick, equal, smooth, hollow in age, white, lacking an annular zone.
SPORE PRINT COLOR: black.
MICROSCOPIC FEATURES: spores 7–10 × 4–5 μm, elliptic with an apical pore, smooth, pale grayish black.
FRUITING: in dense clusters on lawns or decaying wood; April–October; widely distributed in the Southeast; very common.
EDIBILITY: edible.

COMMENTS: *Coprinus variegatus* edible but unpleasant at best, has a larger grayish brown to gray cap covered with conspicuous white to dingy yellow scales that easily wash off. It has an oval to egg-shaped cap when young that becomes bell-shaped in age; pale tan flesh with a strong and unpleasant odor and a typically disagreeable taste; it grows in dense clusters on decaying broadleaf stumps or in grassy areas on buried wood (see *MNE*, p. 281). *Coprinus americanus* (see photo, p. 22), edibility unknown, also has a larger cap with conspicuous scales that easily wash off. It is nearly identical to *Coprinus variegatus* but differs by having slightly broader spores (7.5–9 × 5–6 μm) and paler cap scales. Some mycologists believe *Coprinus americanus* and *Coprinus variegatus* are synonyms.

Coprinus plicatilis (W. Curtis : Fries) Fries
Illus. p. 24

COMMON NAME: Japanese Umbrella Inky.
CAP: ⅜–1" (1–2.5 cm) wide, oval to egg-shaped when young, becoming convex to nearly plane or with a depressed disc; surface deeply sulcate, brown to rusty tawny on the disc, gray to pale grayish brown toward the margin.
FLESH: thin, grayish; odor and taste not distinctive.
GILLS: free and attached to a collar-like ring of tissue at the stalk apex, distant, grayish buff when young, becoming gray then black, not deliquescing.
STALK: 1⅛–2¾" (3–7 cm) long, 1/32–⅛" (1–3 mm) thick, nearly equal or with an enlarged base, smooth, dry, hollow in age, white to somewhat hyaline.
SPORE PRINT COLOR: blackish.
MICROSCOPIC FEATURES: spores 10–13 × 7–10 μm, broadly oval with an apical pore, smooth, pale grayish black.
FRUITING: scattered or in groups on fields and lawns; April–November; widely distributed in the Southeast; fairly common.
EDIBILITY: edible but of little substance.
COMMENTS: Also known as *Pseudocoprinus plicatilis*. Compare with *Coprinus disseminatus* (p. 124), which is smaller and has a prominently furrowed cap and smaller spores.

Cortinarius delibutus Fries
Illus. p. 24

COMMON NAME: none.
CAP: 1–2¾" (2.5–7 cm) wide, hemispherical at first, becoming broadly convex to nearly plane at maturity; surface smooth, viscid, and shiny when fresh, becoming satiny or dull when dry; lemon yellow to golden yellow when young, becoming brownish yellow in age; margin incurved at first and remaining so well into maturity.
FLESH: thin, whitish; odor and taste not distinctive.
GILLS: attached, close, gray-lilac at first, becoming light violet and finally rusty brown at maturity; edges even, typically whitish.
STALK: 1½–2¾" (4–9.5 cm) long, ¼–½" (5–13 mm) thick, enlarged downward or nearly equal, viscid or dry, solid then hollow in age, pale violet over a whitish ground color when young, later pale violet near the apex and whitish below; partial veil filamentous and slimy, leaving a thin annular zone on the upper portion of the stalk.
SPORE PRINT COLOR: rusty brown.
MICROSCOPIC FEATURES: spores 6.5–10 × 5.5–7 μm, broadly elliptic to subglobose, finely warted, yellow.
FRUITING: scattered or in groups on the ground in broadleaf and mixed woodlands; July–November; widely distributed in the Southeast; occasional.
EDIBILITY: unknown.

COMMENTS: Compare with *Cortinarius lewisii* (p. 127), edibility unknown, which has a dry yellow cap, a dry cream-colored stalk with a white basal mycelium, a dense white to yellow partial veil, and flesh that has a musky to radish musky odor.

Cortinarius iodes Berkeley and Curtis　　　　　　　　　　　　　　　　　　　　Illus. p. 24
COMMON NAME: Viscid Violet Cort, Spotted Cort.
CAP: ¾–2⅜" (2–6 cm) wide, bell-shaped when young, becoming broadly convex in age; surface smooth, slimy, dark lilac or purplish, fading in age, developing yellowish spots or becoming yellowish over the disc.
FLESH: lilac to violet, fading in age; odor and taste not distinctive.
GILLS: attached, close, lilac to violet when young, becoming grayish cinnamon in age.
STALK: 1½–2¾" (4–7 cm) long, ¼–⅝" (5–15 mm) thick, nearly equal or enlarged downward, solid, slimy, smooth, violet or purplish, sometimes whitish toward the base; partial veil cortinate, pale violet, leaving a thin, fibrous annular zone on the upper portion of the stalk.
SPORE PRINT COLOR: rusty brown.
MICROSCOPIC FEATURES: spores 8–10 × 5–6.5 μm, elliptic, finely roughened, pale brown.
FRUITING: scattered or in groups or clusters on the ground under broadleaf trees; July–November; widely distributed in the Southeast; fairly common.
EDIBILITY: reported to be edible, but not recommended.
COMMENTS: *Cortinarius violaceus*, edible but not recommended, has a dry, fibrillose to scaly, dark violet to purple cap, dark violet gills that become grayish cinnamon brown at maturity, and a dark violet to purple stalk; and it grows on the ground or among mosses under conifers (see *MNE*, p. 283). *Cortinarius torvus*, edibility unknown, has a moist but not sticky purplish brown to copper brown cap that fades to buff, purplish gray flesh that has a sweet, aromatic odor, dark purplish gills, and a white sheath from the stalk base up to a white membranous ring; and it grows on the ground in broadleaf or mixed woods (see *MNE*, p. 283).

Cortinarius lewisii O. K. Miller　　　　　　　　　　　　　　　　　　　　　　Illus. p. 24
COMMON NAME: none.
CAP: ½–1¾" (1.2–4.5 cm) wide, convex, becoming broadly convex to nearly plane at maturity, with a broad umbo in age; surface dry, with minutely appressed squamules, bright orange-yellow to yellow; margin incurved at first and plane in age, yellow to apricot yellow.
FLESH: firm, deep buff; odor musky to radish musky; taste not distinctive.
GILLS: notched to narrowly attached, close, narrow, whitish at first, becoming yellowish and finally rusty brown at maturity.
STALK: 1½–3⅛" (4–8 cm) long, ⅛–⅜" (3–11 mm) thick, nearly equal or somewhat enlarged downward, dry, fibrillose, pale orange to cream with a white basal mycelium; partial veil dense, white in button stage but soon yellow, leaving a bright yellow ring near the apex that typically becomes an annular zone on mature specimens.
SPORE PRINT COLOR: brown.
MICROSCOPIC FEATURES: spores 6–7.5 × 5–6 μm, subglobose to globose, warted, yellow-brown.
FRUITING: scattered or in groups on the ground in mixed pine and broadleaf woodlands; May–October; widely distributed along the Gulf Coast from Florida to Texas; fairly common.
EDIBILITY: unknown.

COMMENTS: *Cortinarius delibutus* (p. 126), edibility unknown, has a viscid yellow cap and a slimy partial veil, and its flesh lacks a distinctive odor.

Cortinarius marylandensis Ammirati and Smith   Illus. p. 24
COMMON NAME: none.
CAP: 1–2⅜" (2.5–6 cm) wide, bell-shaped when young, becoming broadly convex to flat with a low, broad umbo in age; surface dry, smooth, shiny or dull, somewhat felty, bright reddish orange to reddish brown, changing to rose then purple with KOH; margin somewhat incurved when young, becoming expanded and often split in age.
FLESH: yellowish; odor and taste not distinctive.
GILLS: attached, subdistant, broad, cinnabar red when young, fading to cinnamon brown in age.
STALK: ¾–2¾" (2–7 cm) long, ⅛–⅜" (3–10 mm) thick, equal or enlarged slightly downward toward the base, dry, hollow, streaked with reddish orange to reddish brown fibrils over a reddish yellow ground color, often yellowish to whitish at the base; partial veil cortinate, reddish, leaving a thin, fibrous annular zone on the upper portion of the stalk.
SPORE PRINT COLOR: rusty brown.
MICROSCOPIC FEATURES: spores 7–9 × 4–5 μm, elliptic, slightly roughened, pale brown.
FRUITING: scattered or in groups on the ground under broadleaf trees, especially oak and beech; August–October; widely distributed in the Southeast; occasional.
EDIBILITY: unknown.
COMMENTS: *Cortinarius cinnabarinus* (not illustrated), edibility unknown, is nearly identical, but its cap cuticle does not stain rose then purple when KOH is applied, and it grows in association with conifers.

Cortinarius semisanguineus (Fries) Gillet   Illus. p. 24
COMMON NAME: Red-gilled Cort.
CAP: ¾–2⅜" (2–6 cm) wide, bell-shaped to convex when young, becoming broadly convex to nearly flat, with a low, broad umbo in age; surface dry, covered with tiny matted fibrils, yellow-brown to cinnamon buff, darker over the disc in age; margin incurved when young, becoming expanded.
FLESH: whitish to yellowish; odor not distinctive; taste not distinctive or slightly bitter.
GILLS: attached, crowded, narrow, dark blood red when young, becoming cinnabar to rusty red at maturity.
STALK: ¾–3" (2–7.5 cm) long, ⅛–¼" (3–6 mm) thick, equal, solid, fibrous, dull yellow, often reddish at the base and whitish at the apex, fibrillose; partial veil cortinate, yellowish, leaving a thin, fibrous annular zone on the upper portion of the stalk.
SPORE PRINT COLOR: rusty brown.
MICROSCOPIC FEATURES: spores 5–8 × 3–5 μm, elliptic, roughened, pale brown.
FRUITING: scattered or in groups on the ground or among mosses under hardwoods and conifers; July–February; widely distributed in the Southeast; fairly common.
EDIBILITY: unknown.
COMMENTS: Pigments extracted from this mushroom are used to dye wool. The dark blood red gills of young specimens are a distinctive feature.

Cortinarius sublilacina Murrill   Not Illustrated

    COMMON NAME: none.
    CAP: 2–3½" (5–9 cm) wide, convex to nearly plane at maturity; surface viscid when moist, innately fibrillose, pale violet, with a ferruginous tint in places; margin even, entire.
    FLESH: firm, thick, whitish with yellowish and violet tints; odor somewhat unpleasant; taste unpleasant.
    GILLS: attached, crowded, violet at maturity; edges entire.
    STALK: ¾–2" (2–5 cm) long, ⅝–¾" (1.5–2 cm) thick, typically enlarged downward to a bulbous base, striate, innately fibrillose to subglabrous, violet; bulb up to 1⅛" (3 cm) thick; partial veil cortinate, yellowish, leaving a thin annular zone on the upper portion of the stalk.
    SPORE PRINT COLOR: rusty brown.
    MICROSCOPIC FEATURES: spores 9–10 × 5–6 μm, fusoid, smooth, rusty brown.
    FRUITING: scattered or in groups on the ground in grassy areas or woods, usually with oak nearby; September–January; reported only from Florida; occasional.
    EDIBILITY: unknown.
    COMMENTS: This species was first collected by Murrill in Gainesville, Florida, in 1938.

Crepidotus mollis (Fries) Staude   Illus. p. 24

    COMMON NAME: none.
    CAP: ⅜–3⅛" (1–8 cm) wide, kidney- to fan-shaped or shell-shaped, somewhat convex to nearly plane; surface hygrophanous, color variable, olive brown to ochraceous or ochraceous whitish, coated with short brownish fibers.
    FLESH: thin, whitish; odor not distinctive; taste bitter or not distinctive; stalkless or with a rudimentary stalk.
    GILLS: decurrent, close to crowded, whitish at first, becoming cinnamon to grayish cinnamon; edges entire or fimbriate.
    SPORE PRINT COLOR: brown.
    MICROSCOPIC FEATURES: spores 7–10 × 4.5–6 μm, ellipsoid, smooth, with a double wall, pale brown.
    FRUITING: in groups on decaying broadleaf bark or, rarely, on conifers; June–October; widely distributed in the Southeast; occasional.
    EDIBILITY: unknown.
    COMMENTS: Formerly known as *Crepidotus fulvotomentosus*.

Crinipellis stipitaria (Fries) Patouillard   Not Illustrated

    COMMON NAME: none.
    CAP: ¼–⅝" (7–15 mm) wide, hemispheric to bell-shaped at first, becoming nearly plane with a depressed center and a small central papilla in age; surface dry, dull, with orange-brown to red-brown radially arranged fibers on a cream-colored background; margin often ragged or torn in age.
    FLESH: thin, membranous, cream-colored; odor faintly spicy; taste not distinctive.
    GILLS: attached and notched to nearly free, close, whitish to cream-colored; edges entire or sometimes toothed.
    STALK: ⅜–1⅛" (1–3 cm) long, 1/32–1/16" (0.5–2 mm) thick, tapered downward or nearly equal, dry, solid or hollow, fibrous, densely tomentose, dark reddish brown to rusty brown or grayish brown.
    SPORE PRINT COLOR: white.

MICROSCOPIC FEATURES: spores 5.5–8 × 4–6 μm, amygdaliform to broadly elliptic, smooth, hyaline.

FRUITING: scattered or in groups or clusters on grass roots, plant stems, woody debris, and tree bark; July–November; widely distributed in the Southeast; occasional.

EDIBILITY: unknown.

COMMENTS: Formerly known as *Crinipellis scabella*. Compare with *Crinipellis zonata*, edibility unknown, which has a larger orangish brown cap and grows on decaying broadleaf wood (see *MNE*, p. 284).

Cyptotrama asprata (Berkeley) Redhead and Ginns Illus. p. 25

COMMON NAME: Golden Scruffy Collybia.

CAP: ¼–1" (5–25 mm) wide, convex to broadly convex, golden at first then becoming bright yellow, finely granular coated, often becoming wrinkled or furrowed; margin at first pressed against the upper stalk.

FLESH: thin, pale yellowish; odor and taste not distinctive.

GILLS: attached, usually becoming slightly decurrent, distant, broad, white.

STALK: 1–2" (1–5 cm) long, 1/16–1/8" (1.5–3 mm) thick, solid, fairly tough, scurfy, golden on a pale yellow background.

SPORE PRINT COLOR: white.

MICROSCOPIC FEATURES: spores 8–12 × 6–7.5 μm, lemon-shaped to broadly oval, smooth, hyaline, inamyloid.

FRUITING: solitary or more often in small groups on broadleaf sticks or logs, especially oak; May–October, widely distributed in the Southeast; occasional to fairly common.

EDIBILITY: unknown.

COMMENTS: Previously known as *Cyptotrama chrysopeplum*.

Cystoderma amianthinum var. **rugosoreticulatum** (F. Lorinser) Bon Illus. p. 25

COMMON NAME: Cornsilk Cystoderma, Pungent Cystoderma.

CAP: 1–2" (2.5–5 cm) wide, broadly conical, becoming convex, often with a low umbo; surface dry, granulose, with radial ribs or wrinkles, especially on the disc, tawny to ochraceous or pale brown; margin fringed with tooth-like veil remnants.

FLESH: thin, white; odor of corn silk; taste not distinctive.

GILLS: attached or notched, close to crowded, narrow to moderately broad, white to pale yellowish.

STALK: 1¼–3" (3–7.5 cm) long, 1/8–¼" (3–6 mm) thick, equal or slightly enlarged toward the base, coarsely granulose-scaly from the base up to an evanescent ring, smooth above the ring, hollow, colored like the cap below the ring, whitish above; partial veil leaving a poorly developed, evanescent superior ring.

SPORE PRINT COLOR: white.

MICROSCOPIC FEATURES: spores 5–6 × 3–4 μm, elliptical, smooth, hyaline, amyloid.

FRUITING: sometimes solitary but more often in groups in conifer needle duff and humus or in moss in conifer woods; June–March; widely distributed in the Southeast; common.

EDIBILITY: unknown.

COMMENTS: *Cystoderma amianthinum* var. *amianthinum*, edibility unknown, is nearly identical but is odorless and lacks prominent radial wrinkles on the cap (see *MNE*, p. 284). *Cystoderma fallax* (not illustrated), edibility unknown, has a sheathing stalk base that culminates in a persistent flaring superior ring. *Cystoderma granosum* (not illustrated), edibility unknown, is similar but grows on wood. *Cystoderma terrei* (not illustrated), edibility unknown, is larger and has a brick red to reddish orange cap. Several smaller species of *Lepiota* are similar, but they have gills that are free from the stalk.

Entoloma abortivum (Berkeley and Curtis) Donk   Illus. p. 25
   COMMON NAME: Abortive Entoloma.
   CAP: 2–4" (5–10 cm) wide, typically convex, somewhat fibrous to almost smooth, the edge inrolled and often becoming irregular in age; sometimes with a low broad umbo or becoming slightly depressed at the center, pale gray to grayish brown.
   FLESH: white, thick; odor and taste farinaceous.
   GILLS: typically decurrent, close, grayish at first, soon becoming pinkish.
   STALK: 1–4" (2.5–10 cm) long, ¼–⅝" (5–15 mm) thick, usually enlarged at the base, dry, solid, scurfy, whitish to grayish, the base usually coated with white mycelium.
   SPORE PRINT COLOR: salmon pink.
   MICROSCOPIC FEATURES: spores 8–10 × 4.5–6 μm, elliptic and angular, typically 6-sided, hyaline, inamyloid.
   FRUITING: scattered or in groups, often gregarious, on the ground, especially in humus or near rotting stumps in woods; July–December; widely distributed in the Southeast; fairly common.
   EDIBILITY: edible with caution (see Comments).
   COMMENTS: This mushroom can best be identified, and should be eaten only, if found in immediate proximity to the aborted fruiting bodies. These bodies are 1–2" (2.5–5 cm) high and 1–4" (2.5–10 cm) wide, chalky white masses with pinkish marbling in the spongy white inner flesh. Both the normal gilled form and the aborted fruiting bodies are good edibles. There is concrete evidence to suggest that the aborted form is caused by some interaction with at least one or two species of Honey Mushrooms, *Armillaria mellea*, which are usually found in the vicinity.

Flammulina velutipes (Fries) Karsten   Illus. p. 25
   COMMON NAME: Velvet Foot, Winter Mushroom.
   CAP: ¾–2¾" (2–7 cm) wide, convex, becoming nearly flat, orangish brown to reddish yellow, darker at the center; smooth, slimy, very sticky when dry; margin incurved at first, becoming striate.
   FLESH: thin, yellowish, usually watery; odor and taste not distinctive.
   GILLS: attached, becoming subdecurrent, yellowish, typically close and narrow.
   STALK: usually 1–3" (2.5–7.5 cm) long but sometimes longer, ⅛–¼" (3–7 mm) thick; smooth and yellowish at the top, becoming dark reddish brown and extremely velvety starting at the base; very tough, not easily broken.
   SPORE PRINT COLOR: white.
   MICROSCOPIC FEATURES: spores 7–9 × 3–6 μm, elliptic, smooth, hyaline, inamyloid.
   FRUITING: in clusters on broadleaf trees, logs, and stumps; usually fruiting during early spring or late fall, but also during winter thaws and summer cold spells; widely distributed in the Southeast; fairly common.
   EDIBILITY: edible, though the sticky caps can be difficult to clean, and the tough stalks should be discarded.
   COMMENTS: Previously known as *Collybia velutipes*, this species has been cultivated in Southeast Asia as Enotake or Enoki-take. The taxon described here may represent one form of a complex of closely related species.

Galerina marginata (Fries) Kuhner Illus. p. 25
COMMON NAME: Deadly Galerina.
CAP: 1–2½" (2.5–6.5 cm) wide, convex when young, becoming nearly flat in age, with or without a low and broad umbo; surface viscid, moist, smooth, hygrophanous, dark brown to dark amber, fading to yellowish orange or buff, often remaining darker over the disc; margin translucent-striate when moist.
FLESH: brownish to buff, thin; odor not distinctive to slightly farinaceous; taste not distinctive.
GILLS: attached to very slightly decurrent, close, broad, brown, becoming rusty brown.
STALK: 1⅛–3½" (3–9 cm) long, ⅛–½" (3–8 mm) thick, equal or slightly enlarged at the base, hollow, dry, smooth to slightly pruinose at the apex, covered with grayish white flattened fibrils over a brown to blackish brown ground color; base usually coated with white mycelium; partial veil white, membranous, leaving a more or less persistent, membranous, superior ring.
SPORE PRINT COLOR: rusty brown.
MICROSCOPIC FEATURES: spores 8.5–10.5 × 5–6.5 μm, elliptic, roughened, pale brown.
FRUITING: scattered or in groups or clusters on decaying hardwood and conifer stumps and logs; May–October; widely distributed in the Southeast; fairly common.
EDIBILITY: deadly poisonous.
COMMENTS: *Galerina autumnalis* is a synonym. This mushroom contains the same kinds of toxins that are found in deadly *Amanita* species and causes liver and kidney damage, coma, and death.

Gerronema strombodes (Berk. and Mont.) Singer Illus. p. 26
COMMON NAME: none.
CAP: ¾–2" (2–5 cm) wide, convex with an umbilicate center, becoming more or less funnel-shaped in age; surface dull to somewhat silky, grayish brown with darker brownish radial fibrils, fading to ochre or cream toward the margin, with a darker brown center; margin incurved at first, later expanding and often splitting in age.
FLESH: thin, whitish to cream; odor not distinctive; taste somewhat bitter or not distinctive.
GILLS: decurrent with some forking, close to subdistant, broad, white at first, becoming cream-colored to pale yellow at maturity.
STALK: 1–2½" (2.5–6.5 cm) long, up to ¼" (6 mm) thick, nearly equal, dry, hollow, smooth or with longitudinal fibrils, whitish to grayish brown.
SPORE PRINT COLOR: white.
MICROSCOPIC FEATURES: spores 7–9 × 4–5.5 μm, elliptic, smooth, hyaline, inamyloid.
FRUITING: usually in clusters on decaying wood; July–August; widely distributed in the Southeast; occasional to locally common.
EDIBILITY: unknown.
COMMENTS: Also known as *Chrysomphalina strombodes*. Compare with *Chrysomphalina chrysophylla*, edibility unknown, which is similar but has bright golden yellow to orange-yellow gills and a yellowish spore print, and which grows on conifer wood (see *MNE*, p. 276). *Omphalina ectypoides* (p. 180), formerly known as *Clitocybe ectypoides*, edibility unknown, is also similar, but it has a minutely scaly brownish cap and amyloid spores, and it is restricted to conifer wood.

Gymnopilus fulvosquamulosus Hesler Illus. p. 26
- COMMON NAME: none.
- CAP: 1½–3⅛" (4–8 cm) wide, convex, becoming broadly convex in age; surface conspicuously covered with appressed to slightly recurved fibrillose, tawny scales on a yellow-ochre to brighter yellow ground color; margin incurved at first, becoming nearly plane at maturity.
- FLESH: yellow; odor and taste not distinctive.
- GILLS: attached to decurrent, broad, close, dull yellow, becoming dark rusty brown at maturity; edges very finely scalloped.
- STALK: 1⅛–2⅜" (3–6 cm) long, ¼–½" (6–12 mm) thick, nearly equal or enlarged downward, curved, dry, dingy yellow, becoming rusty brown when handled or in age; partial veil submembranous to fibrillose, leaving a thin superior ring.
- SPORE PRINT COLOR: rusty brown.
- MACROCHEMICAL TESTS: flesh slowly stains olive with the application of $FeSO_4$.
- MICROSCOPIC FEATURES: spores 7–10 × 5–6 μm, ellipsoid, warted, lacking a germ pore, pale rusty brown, dextrinoid; caulocystidia 28–54 × 5–7 μm, colorless, cylindric-clavate, in tufts.
- FRUITING: solitary, scattered, or in groups on decaying broadleaf wood; July–November; Gulf Coast from Florida west to Texas; occasional.
- EDIBILITY: unknown.
- COMMENTS: Compare with *Gymnopilus palmicola* (p. 134), which is similar but has a smaller cap, ¾–2" (2–5 cm) wide, grows on palms and living orchids, and lacks caulocystidia. *Gymnopilus sapineus*, edibility unknown, is similar, but its veil does not form a persistent ring (see *MNE*, p. 285). *Gymnopilus penetrans* (p. 134), edibility unknown, is also similar, but it has a glabrous cap, and its veil does not form a persistent ring.

Gymnopilus liquiritiae (Persoon) Karsten Illus. p. 26
- COMMON NAME: none.
- CAP: ¾–3⅛" (2–8 cm) wide, convex, becoming broadly convex to nearly plane in age, sometimes slightly umbonate; surface moist or dry, not viscid, glabrous or nearly so, fulvous to ochraceous tawny or ochraceous orange to orange-brown.
- FLESH: pale orange to tawny yellow; odor sometimes slightly fragrant or resembling raw potatoes or not distinctive; taste bitter.
- GILLS: attached to notched, seceding, close to crowded, at first ochraceous buff to pale orange-yellow, becoming ochraceous orange to ochraceous tawny or orange, sometimes with reddish brown spots, edges fimbriate.
- STALK: 1⅛–2¾" (3–7 cm) long, ⅛–⅜" (3–10 mm) thick, tapered in either direction or nearly equal, dry, hollow, longitudinally fibrillose or subglabrous, whitish to dingy orange, with a whitish or yellowish apex; partial veil and ring absent.
- SPORE PRINT COLOR: rusty brown.
- MICROSCOPIC FEATURES: spores 7–10 × 4–6 μm, ellipsoid, verruculose, lacking a germ pore, ferruginous, dextrinoid.
- FRUITING: scattered or in groups on decaying conifer wood or sawdust, occasionally on decaying wood of broadleaf trees; June–January; widely distributed in the Southeast; occasional.
- EDIBILITY: unknown.
- COMMENTS: *Gymnopilus rufosquamulosus* (not illustrated), edibility unknown, reported only from Texas, has a buff to tawny cap with small red to orange scales, pale yellowish

flesh with a pleasant odor and a bitter taste, and a dingy yellowish brown stalk with a superior ring; it grows on decaying oak trees.

Gymnopilus palmicola Murrill Illus. p. 26
COMMON NAME: none.
CAP: ¾–2" (2–5 cm) wide, convex, becoming nearly flat at maturity, in age depressed over the disc; surface dry, floccose-squamose when young, becoming subglabrous in age, pale ferruginous to ochraceous; margin even.
FLESH: thin, whitish to yellowish; odor not distinctive; taste bitter.
GILLS: attached, close, broad, yellowish at first, becoming ferruginous at maturity.
STALK: 1⅛–2" (3–5 cm) long, ⅛–¼" (3–5 mm) thick, nearly equal, dry, solid, slightly fibrillose, yellowish or colored like the cap but paler; partial veil strongly developed, pale yellow, sometimes leaving a poorly developed ring.
SPORE PRINT COLOR: rusty brown.
MICROSCOPIC FEATURES: spores 7–10 × 5–7.5 μm, ellipsoid, coarsely warted, lacking a germ pore, ferruginous, dextrinoid; caulocystidia lacking.
FRUITING: solitary, scattered, or in groups on decaying logs, stumps of palms, and living orchids; June–February, sometimes year-round; along the Gulf Coast from Florida to Texas; occasional to fairly common.
EDIBILITY: unknown.
COMMENTS: Compare with *Gymnopilus fulvosquamulosus* (p. 133), which is similar but has a larger cap, 1½–3⅛" (4–8 cm) wide; grows on decaying broadleaf wood; and has tufts of cylindric-clavate caulocystidia.

Gymnopilus penetrans (Fries) Murrill Illus. p. 26
COMMON NAME: none.
CAP: ¾–2" (2–5 cm) wide, campanulate-convex, becoming broadly convex to nearly plane in age; surface moist or dry, glabrous, yellow to golden yellow or pale orange-yellow, fading in age; margin even.
FLESH: white to whitish; odor not distinctive; taste bitter.
GILLS: attached to sinuate, close, pale yellow or yellowish white at first, becoming rust-spotted in age.
STALK: 1–2⅜" (2.5–6 cm) long, ⅛–¼" (3–7 mm) thick, nearly equal or enlarged downward, dry, glabrous or nearly so, pale yellow to yellowish white, with a white tomentum over the base; partial veil fibrillose, white, not leaving a ring or sometimes leaving a faint annular zone.
SPORE PRINT COLOR: orange-brown.
MICROSCOPIC FEATURES: spores 6.5–10 × 4–5.5 μm, ellipsoid, lacking a germ pore, warted, ochraceous, dextrinoid.
FRUITING: solitary, scattered, or in groups on decaying wood, wood chips, and sawdust; July–January; fairly common.
COMMENTS: *Gymnopilus sapineus,* edibility unknown, is very similar, but it has a minutely scaly cap (use a hand lens) that often becomes rimose in age, a scanty yellowish partial veil, and flesh that usually has a pungent odor (see *MNE,* p. 285).

Gymnopilus spectabilis (Fries) Smith Illus. p. 26
COMMON NAME: Big Laughing Gym.
CAP: 3–7" (7.5–18 cm) wide, convex, becoming broadly convex to nearly plane; surface

typically dry, smooth when young, becoming scaly in age, pale orange-yellow to ochre-orange.
FLESH: pale yellow when young, becoming yellow in age; odor spicy or anise- to licorice-like, sometimes not distinctive; taste bitter.
GILLS: attached to decurrent, close to crowded, ochraceous buff at first, becoming yellow to rusty; edges minutely fringed.
STALK: 2–7" (5–18 cm) long, ⅜–1⅛" (1–3 cm) thick, typically enlarged downward to the base then abruptly tapered, streaked with tiny fibrils; colored like the cap and often brownish near the base; partial veil submembranous to fibrous, typically leaving a superior ring.
SPORE PRINT COLOR: rusty orange to rusty brown.
MICROSCOPIC FEATURES: spores 7–10 × 4.5–6 μm, elliptic to oval, roughened, lacking an apical pore, pale brownish.
FRUITING: in clusters or scattered on decaying wood and on the ground attached to buried wood; June–January; widely distributed in the Southeast; occasional to fairly common.
EDIBILITY: hallucinogenic.
COMMENTS: Compare with *Gymnopilus luteofolius*, hallucinogenic, which has a tawny cap covered with conspicuous dark reddish purple to purplish red scales that fade in age, and which grows in dense clusters on decaying wood chips, logs, stumps, and sawdust (see *MNE*, p. 285). Also compare with *Gymnopilus fulvosquamulosus* (p. 133).

Gymnopus confluens (Persoon : Fries) Antonín, Halling, and Nordeloos   Illus. p. 26
COMMON NAME: Tufted Gymnopus, Tufted Collybia.
CAP: ⅜–2" (1–5 cm) wide, convex, becoming nearly plane; surface moist then soon dry, smooth, reddish brown, fading to pale pinkish cinnamon or whitish in age.
FLESH: thin, whitish; odor and taste sometimes slightly like garlic or bitter almonds or not distinctive.
GILLS: free or attached, crowded to close, pinkish buff then cream, fading to whitish in age.
STALK: 1–4" (2.5–10 cm) long, 1/16–¼" (1.5–6 mm) thick, typically flared at the apex and base, otherwise nearly equal, pale pinkish buff near the apex, becoming pale cinnamon toward the base, coated with a dense layer of whitish to pale grayish hairs, covered with white mycelium at the base.
SPORE PRINT COLOR: white.
MICROSCOPIC FEATURES: spores 7–11 × 3–5 μm, narrowly elliptic, smooth, hyaline, inamyloid.
FRUITING: in dense clusters or groups or scattered on leaf litter and humus; July–November; widely distributed in the Southeast; common.
EDIBILITY: edible.
COMMENTS: Also known as *Collybia confluens*. Compare with *Gymnopus spongiosus* = *Collybia spongiosa* (not illustrated), edibility unknown, which has a reddish brown cap and a stalk that is whitish and smooth near the apex, reddish brown below, and coated with dull reddish orange hairs from its spongy base to near the apex. *Gymnopus subnudus* = *Collybia subnuda*, edibility unknown, has a cinnamon brown cap and a dark brown stalk that is buff and smooth near the apex and coated with whitish to grayish hairs below (see *MNE*, p. 279 for these *Collybia* species).

Gymnopus dryophilus (Bulliard : Fries) Murrill   Illus. p. 26
  COMMON NAME: Oak-loving Gymnopus, Oak-loving Collybia.
  CAP: ⅜–2¾" (1–7 cm) wide, convex when young, becoming broadly convex to flat in age; surface smooth, moist, silky, hygrophanous, dark reddish brown when young, soon fading to orange-brown, tan, or yellowish tan, usually remaining darker over the disc; margin often uplifted in age.
  FLESH: whitish to yellowish; odor and taste not distinctive.
  GILLS: attached to nearly free, crowded to close, whitish to pinkish buff or sometimes yellowish in age.
  STALK: 1⅛–3½" (3–9 cm) long, ⅛–¼" (2–8 mm) thick, nearly equal or enlarged downward, base sometimes bulbous and often covered with white mycelium, flexible, hollow, dry, smooth when young, finely striate in age, whitish near the apex, becoming pale yellow to orange-yellow or darker toward the base.
  SPORE PRINT COLOR: white to cream.
  MICROSCOPIC FEATURES: spores 5–7 × 3–3.5 μm, lacrymoid to elliptic, smooth, hyaline, inamyloid.
  FRUITING: scattered or in groups or clusters on humus or decaying wood in oak-pine forests; June–November; widely distributed in the Southeast; common.
  EDIBILITY: edible.
  COMMENTS: Also known as *Collybia dryophila,* this species is often confused with *Rhodocollybia butyracea* (p. 188), edible, which is very similar but has a pale pink spore print. This mushroom is often parasitized by the Collybia Jelly, *Syzygospora mycetophila* (p. 306).

Gymnopus iocephalus (Berkeley and Curtis) Halling   Illus. p. 27
  COMMON NAME: Violet Gymnopus, Violet Collybia.
  CAP: ½–1½" (1.2–4 cm) wide, convex, becoming broadly convex to plane in age, sometimes shallowly depressed; surface moist or dry, wrinkled and striate to sulcate nearly to the disc; color variable, violet, purple, vinaceous, fading to purplish lilac to pinkish lilac; margin inrolled at first, becoming uplifted and wavy in age.
  FLESH: purplish lilac to pinkish lilac; odor pungent and disagreeable; taste unpleasant.
  GILLS: attached to notched, subdistant to distant, reddish purple to violet.
  STALK: ¾–2¾" (2–7 cm) long, 1/16–3/16" (1.5–4 mm) thick, enlarged downward or nearly equal, apex often expanded, dry, hollow, whitish or sometimes tinged pale pinkish lilac, minutely pubescent overall.
  SPORE PRINT COLOR: white.
  MACROCHEMICAL TESTS: cap, gills, and stalk stain bright blue with the application of KOH.
  MICROSCOPIC FEATURES: spores 6.5–9 × 3–4.5 μm, ovoid, smooth, hyaline, inamyloid.
  FRUITING: scattered or in groups on leaf litter and humus, usually in broadleaf or mixed woods; August–December; widely distributed in the Southeast; infrequent to locally common.
  COMMENTS: Also known as *Collybia iocephala.*

Hebeloma crustuliniforme (Bull. : St. Amans) Quélet   Illus. p. 27
  COMMON NAME: Poison Pie.
  CAP: 1⅛–3⅛" (3–8 cm) wide, convex, becoming nearly flat at maturity; margin inrolled when young, becoming uplifted and wavy in age; surface smooth, viscid, creamy white to pale brown with dull yellow-brown to reddish brown tones at the center.

FLESH: white; odor radish-like; taste bitter to disagreeable.
GILLS: notched, close to crowded, whitish at first, becoming grayish then finally grayish brown at maturity; edges white, finely fringed, with watery to brownish droplets when fresh, brown dotted when dry.
STALK: 1⅛–3⅛" (3–8 cm) long, ¼–¾" (7–20 mm) thick, nearly equal or bulbous at the base, white-pruinose to scurfy at least on the upper portion, white to buff and becoming brownish from the base upward in age or when handled, hollow, often with white mycelium at the base; partial veil and ring absent.
SPORE PRINT COLOR: yellow-brown to brown.
MICROSCOPIC FEATURES: spores 9–13 × 5–7.5 μm, elliptic, slightly roughened, pale brown; dextrinoid.
FRUITING: scattered, in groups, or sometimes in arcs or fairy rings on the ground under hardwoods or conifers; often found under shrubs and hedges or on lawns in residential areas; July–November; widely distributed in the Southeast; fairly common.
EDIBILITY: poisonous, causing gastric distress.
COMMENTS: *Hebeloma sinapizans* (p. 137), poisonous, is very similar but with a larger cap, up to 6" (15.5 cm) wide; a radish odor and taste; a stouter stalk, up to 1" (2.5 cm) thick, that becomes scaly on the upper portion; and spores that measure 10–14 × 6–8 μm. *Hebeloma mesophaeum*, poisonous, has a smaller reddish brown to orange-brown cap that is brownish yellow to yellowish tan toward the margin, and it lacks the watery to brownish droplets and dots on its gills (see *MNE*, p. 285).

Hebeloma sinapizans (Fr.) Sacc. — Illus. p. 27

COMMON NAME: Rough-stalked Hebeloma.
CAP: 1½–6" (4–15.5 cm) wide, convex to broadly convex, becoming flattened in age, with or without a low umbo; surface moist to viscid, glabrous, smooth, whitish tan to cinnamon tan, also pinkish brown or reddish brown, sometimes with grayish brown areas, especially on the disc; margin incurved at first, later expanded to uplifted, minutely cottony when very young.
FLESH: thick, whitish; odor and taste radish-like.
GILLS: attached and notched, close, broad, whitish when young, becoming yellowish cinnamon to pale brown at maturity; edges minutely fringed or serrated.
STALK: 2–5" (5–12.5 cm) long, ½–1" (1.3–2.5 cm) thick, nearly equal, typically with a swollen base, scurfy to flaky-scaly, granulose at the apex, whitish, stuffed at first, becoming hollow in age but leaving a hanging tooth of tissue in the cavity at the apex; partial veil absent.
SPORE PRINT COLOR: pale pinkish brown.
MICROSCOPIC FEATURES: spores 10–14 × 6–8 μm, broadly elliptic, minutely roughened, pale brown, dextrinoid.
FRUITING: usually gregarious, sometimes in fairy rings on the ground in broadleaf and conifer woods; August–December; widely distributed in the Southeast; fairly common.
EDIBILITY: poisonous.
COMMENTS: The "hanging tooth" in the stalk cavity is a good field character. *Hebeloma crustuliniforme* (p. 136), poisonous, is similar but smaller and paler overall. Some *Cortinarius* species are superficially similar, but they have a web-like partial veil and rusty-colored spores.

Hebeloma syriense Karsten Illus. p. 27
COMMON NAME: Corpse Finder.
CAP: 1–2" (2.5–5 cm) wide, convex, becoming nearly flat at maturity; surface slimy-viscid, smooth, orange-brown, fading to ochre-brown; margin incurved at first.
FLESH: compact, thick, whitish; odor not distinctive; taste not determined.
GILLS: attached, close, fairly broad, whitish when young, becoming reddish brown to cinnamon at maturity; edges fringed.
STALK: 1½–2½" (4–6.5 cm) long, ⅛–¼" (3–6 mm) thick, nearly equal, dry, hollow at maturity, scurfy near the apex and longitudinally striate below, whitish, sometimes with yellowish brown stains.
SPORE PRINT COLOR: pale cinnamon.
MICROSCOPIC FEATURES: spores 8–10.5 × 5–6 μm, elliptic, faintly roughened, pale brown.
FRUITING: solitary or in groups or clusters on the ground in broadleaf woods; August–November; widely distributed in the Southeast; occasional.
EDIBILITY: unknown.
COMMENTS: This mushroom is reported to occur in the vicinity of decaying animal carcasses and human remains, hence the common name.

Hohenbuehelia petaloides (Bulliard : Fries) Schulzer Illus. p. 27
COMMON NAME: Leaf-like Oyster.
CAP: 1–3" (2.5–7.5 cm) wide, roughly fan-shaped or somewhat funnel-shaped, minutely hairy (use a hand lens) to smooth, tan to pale brown or grayish brown, sometimes with a whitish bloom.
FLESH: thin, whitish, gelatinous; odor and taste not distinctive.
GILLS: decurrent to base of stalk, crowded, narrow, whitish to grayish, with minutely fringed edges (use hand lens).
STALK: short, stubby, lateral, hairy, whitish.
SPORE PRINT COLOR: white.
MICROSCOPIC FEATURES: spores 7–9 × 4–5 μm, elliptic, smooth, hyaline, inamyloid; cystidia very large and thick-walled, with rounded, incrusted ends.
FRUITING: solitary or in groups or clusters on wood and woody debris; June–January; widely distributed in the Southeast; fairly common.
EDIBILITY: edible.
COMMENTS: *Hohenbuehelia geogenia* (not illustrated), edible, has a pale yellow-brown to hazel brown cap, a farinaceous odor and taste, and smaller spores that measure 5–6 × 4–5 μm.

Hygrocybe acutoconica (Clements) Singer Illus. p. 27
COMMON NAME: none.
CAP: ¾–4" (2–10 cm) wide, obtusely to sharply conic when young, soon bell-shaped; surface viscid to glutinous, glabrous, orange-red, orange, ochraceous orange, or yellow; margin often upturned, wavy, and splitting radially at maturity.
FLESH: soft, yellow; odor and taste not distinctive.
GILLS: attached and notched at first, becoming nearly free at maturity, moderately broad, close to subdistant, yellow; edges entire.
STALK: 2⅜–4¾" (6–12 cm) long, ⅛–½" (3–12 mm) thick, nearly equal or enlarged toward the base, rounded or compressed, slippery, solid or hollow, fibrillose or glabrous, striate, sometimes twisted-striate, often splitting lengthwise, colored like the cap or paler,

typically with a white base, not blackening when bruised, but often blackening at the base in age.

SPORE PRINT COLOR: white.

MICROSCOPIC FEATURES: spores 9–15 × 5–9 μm, ellipsoid, smooth, hyaline.

FRUITING: solitary, scattered, or in groups on soil, among grasses in fields and roadsides, and in broadleaf and mixed woodlands; April–January; widely distributed in the Southeast; occasional.

EDIBILITY: edible.

COMMENTS: Also known as *Hygrophorus acutoconicus* and *Hygrocybe persistens*. Compare with *Hygrocybe conica* (p. 140), which is very similar, but all parts of which stain black when bruised.

Hygrocybe andersonii Cibula and N. S. Weber Illus. p. 27

COMMON NAME: Clustered Dune Hygrocybe.

CAP: ½–1¼" (1.3–3.3 cm) wide, convex with a flattened to depressed center; surface smooth or often with a scurfy disc, color variable from orange-yellow, orange, or deep reddish orange to scarlet, becoming reddish brown to nearly black in age; disc often blackish; margin incurved.

FLESH: thin, orange-yellow to reddish orange; odor and taste not distinctive.

GILLS: attached to sinuate or subdecurrent, subdistant, fairly broad, yellow-orange to brownish orange or deep orange, becoming blackish in age.

STALK: 1–1¾" (2.5–4.2 cm) long, ¼–⅜" (6–9 mm) thick, cylindrical to flattened, glabrous smooth, often with longitudinal furrows, solid or hollow, colored like the cap or paler on the upper portion, becoming yellowish orange to pale yellow at the base, lacking a veil.

SPORE PRINT COLOR: white.

MICROSCOPIC FEATURES: spores 16–19 × 3.8–5.6 μm, rod-shaped with a distinct projection, smooth, hyaline, inamyloid.

FRUITING: occasionally single but usually gregarious, caespitose in sand near the shrub Seaside Rosemary *(Ceraticola ericoides)*; November–March; coastal areas and barrier islands along the Gulf Coast from Florida to Mississippi; locally common.

EDIBILITY: unknown.

COMMENTS: This brightly colored waxy cap is typically found in old, tree-colonized sand dunes. It is distinctive for its clustered manner of growth and because it invariably occurs in the vicinity of the evergreen shrub Seaside Rosemary. The association with Seaside Rosemary may be mycorrhizal, but the relationship is not fully understood.

Hygrocybe chlorophana (Fries : Fries) Wünsche Illus. p. 27

COMMON NAME: none.

CAP: ¾–2¾" (2–7 cm) wide, hemispherical, becoming broadly convex to nearly plane in age, sometimes with a low and broad umbo or depressed on the disc; surface viscid, glabrous, orange-yellow to lemon yellow; margin weakly translucent-striate.

FLESH: thin, pale yellow to yellow; odor and taste not distinctive.

GILLS: attached and notched, close, broad, whitish at first, becoming yellowish to yellow-orange at maturity; edges even.

STALK: 1⅛–3⅛" (3–8 cm) long, 3⁄16–½" (4–12 mm) thick, nearly equal, often compressed with one or two longitudinal grooves, dry, moist or viscid, hollow at maturity, colored like the cap or slightly paler.

SPORE PRINT COLOR: white.

MICROSCOPIC FEATURES: spores 6–10 × 4–6.5 μm, ellipsoid to ovoid or oblong, smooth, hyaline.

FRUITING: solitary, scattered, or in groups on the ground in woodlands; June–November; widely distributed in the Southeast; occasional.

EDIBILITY: edible.

COMMENTS: Also known as *Hygrophorus chlorophanus*. Although *Hygrocybe flavescens* is sometimes listed as a distinct species, it is now considered a synonym by most authors.

Hygrocybe conica (Schaeffer : Fries) Kummer Illus. p. 28

COMMON NAME: Witch's Hat.

CAP: ¾–3½" (2–9 cm) wide, sharply conic to bell-shaped, usually with an umbo; surface smooth, slightly sticky when moist, otherwise dry, orange to dark orange-red to red, lighter orange near the margin, sometimes yellow overall, often with olive green tints, quickly staining black when bruised or in age.

FLESH: thin, fragile, colored like the cap, bruising black; odor and taste not distinctive.

GILLS: free from the stalk, close, broad, waxy, light yellow to greenish orange, becoming black in age or where bruised.

STALK: ¾–4" (2–10 cm) long, ⅛–⅜" (3–10 mm) thick, equal, hollow at maturity, fragile, smooth, not sticky, often longitudinally striate or twisted-striate, yellow to yellow-orange, pale yellow near the base, staining black when bruised or in age; partial veil and ring absent.

SPORE PRINT COLOR: white.

MICROSCOPIC FEATURES: spores 8–10 × 5–5.6 μm, elliptic, smooth, hyaline, inamyloid.

FRUITING: solitary to scattered on the ground in broadleaf woods, in open areas, and under conifers; June–January; occasional to fairly common.

EDIBILITY: poisonous.

COMMENTS: Also known as *Hygrophorus conicus*. A very similar species is *Hygrocybe conicoides* (p. 140), but it differs in soon developing salmon orange to reddish gills; larger, more elongated spores; and a sand dune habitat. *Hygrocybe acutoconica* = *Hygrocybe persistens*) (p. 138), edible, has a sharply conic, orange-red or orange or ochraceous orange or yellow cap and a similarly colored stalk that is longitudinally twisted or striate, often blackening at the base in age, but not blackening when bruised. *Hygrocybe coccinea* = *Hygrophorus coccineus*, edible, has a dry, bright red cap that fades to orange-red in age, red-orange to creamy peach or yellowish gills, and a yellow-orange to orange-red stalk that is paler yellow downward and often coated with white mycelium on the base (see *MNE*, p. 287).

Hygrocybe conicoides (Orton) P. D. Orton and Watling Illus p. 28

COMMON NAME: Dune Witch's Hat.

CAP: 1–2½" (2.5–6.5 cm) wide, bell-shaped to convex or conical, becoming more or less flattened with an acute umbo in age; surface smooth, satiny when dry, scarlet to cherry red, slowly turning black in age.

FLESH: thin, fragile, colored like the cap, slowly turning black on exposure; odor and taste not distinctive.

GILLS: attached to notched, close, broad, waxy, yellow at first, soon infused with salmon orange to reddish tones, eventually becoming black.

STALK: 1–3" (2.5–7.5 cm) long, ⅛–⅜" (3–10 mm) thick, equal, longitudinally striate or twisted striate, yellow to orange, often paler at the base, slowly turning black when handled or in age; partial veil and ring absent.

SPORE PRINT COLOR: white.
MICROSCOPIC FEATURES: spores 9–14 × 4.5–7 μm, elongated-ellipsoid, smooth, hyaline, inamyloid.
FRUITING: solitary or in small groups in sand dunes; September–January; occasional to fairly common locally.
EDIBILITY: unknown, possibly poisonous.
COMMENTS: This attractive, waxy cap is often buried "up to its neck" in sand. Some authors refer to this species as *Hygrocybe conica* var. *conicoides*. *Hygrocybe conica* (p. 140) is very similar, but its gills do not develop salmon orange to reddish tones; it also has smaller spores and typically grows on soil in hardwoods, conifer woods, and open areas.

Hygrocybe nitida (Berk. and Curtis) Murrill   Illus. p. 28
COMMON NAME: none.
CAP: ⅜–1½" (1–4 cm) wide, broadly convex or flattened when young, disc very soon becoming depressed, in age deeply funnel-shaped, sometimes perforating; surface smooth, glabrous, sticky, bright yellow to apricot yellow, fading to pale yellow or whitish, striatulate when moist; margin incurved and remaining so well into maturity.
FLESH: soft, fragile, very thin, yellowish to whitish; odor and taste not distinctive.
GILLS: long-decurrent, subdistant to distant, yellow to pale yellow, often with darker yellow edges, crossveined; edges even.
STALK: 1⅛–3⅛" (3–8 cm) long, ⅛–¼" (2–5 mm) thick, nearly equal or slightly enlarged above, smooth, sticky, colored and fading like the cap, hollow.
SPORE PRINT COLOR: white.
MICROSCOPIC FEATURES: spores 6.5–9 × 4–6 μm, ellipsoid to subovoid, smooth, hyaline.
FRUITING: scattered or in groups on humus, in wet soil, and among mosses in conifer or broadleaf forests and in bogs; June–December; widely distributed in the Southeast; occasional to fairly common.
EDIBILITY: unknown.
COMMENTS: Also known as *Hygrophorus nitidus*. *Hygrophorus cantharellus*, edible, is similar, but it has a dry, not viscid, orange to orange-red or scarlet cap and stalk (see *MNE*, p. 287).

Hygrocybe ovina (Bulliard : Fries) Kühner   Illus. p. 28
COMMON NAME: none.
CAP: ¾–3⅛" (2–8 cm) wide, hemispheric to convex, becoming broadly convex to somewhat expanded in age; surface dry, smooth at first, becoming pitted and scurfy in age, dark grayish brown when young, becoming dull gray to dark brownish gray in age; margin inrolled at first, becoming decurved and wavy at maturity.
FLESH: moderately thick, whitish to pale grayish brown, staining reddish when cut or bruised; odor variously described as nitrous or fruity, often unpleasant in older specimens; taste slightly alkaline to somewhat unpleasant or sometimes not distinctive.
GILLS: attached and deeply notched, subdistant, broad, whitish at first, becoming pale gray, staining reddish when bruised.
STALK: 1⅛–3½" (3–9 cm) long, ¼–¾" (6–20 mm) thick, often enlarged downward or nearly equal, frequently compressed with one or two longitudinal grooves, dry, smooth, hollow at maturity, colored like the cap or paler, sometimes staining pinkish brown then blackish when handled or bruised.
SPORE PRINT COLOR: white.

MICROSCOPIC FEATURES: spores 6–10 × 4.5–7 μm, broadly ellipsoid to subovoid, smooth, hyaline.

FRUITING: scattered or in groups on the ground in broadleaf or mixed woods; July–October; widely distributed in the Southeast; fairly common.

EDIBILITY: unknown.

COMMENTS: Also known as *Hygrophorus ovinus*. The reddish staining of the flesh and gills is distinctive. *Hygrophorus subovinus* (not illustrated), edibility unknown, is similar but is darker umber brown and has a sweet, burned sugar odor.

Hygrophorus hypothejus (Fries) Fries   Illus. p. 28

COMMON NAME: none.

CAP: ¾–2⅜" (2–6 cm) wide, convex then depressed at maturity, sometimes with a small umbo; surface glutinous when young and fresh, becoming viscid and finally dry in age, olive brown on young specimens, fading to pale olive brown to yellow-brown or dull yellow on the margin on mature specimens and remaining olive brown on the disc; margin sterile and incurved at first, becoming decurved and remaining so well into maturity, often wavy in age.

FLESH: white, unchanging; odor and taste not distinctive.

GILLS: subdistant, attached to decurrent, with three to four tiers of attenuate lamellulae, white at first, soon pale yellow, often developing orange stains in age, typically veined on the faces and sometimes crossveined.

STALK: 1⅛–2⅛" (3–5.5 cm) long, ¼–⅜" (7–11 mm) thick, tapered slightly downward, weakly fibrillose, glutinous when fresh, with a thin fibrillose band near the apex, whitish at first, soon yellowish to pale ochraceous orange on the upper portion and remaining white toward the base.

SPORE PRINT COLOR: white.

MICROSCOPIC FEATURES: spores 7–10 × 4–6 μm, ellipsoid, smooth hyaline, inamyloid.

FRUITING: scattered or in groups on sandy soil among mosses under conifers or mixed pines and oaks; August–November; widely distributed in the Southeast; fairly common.

EDIBILITY: edible.

COMMENTS: In the Southeast, this mushroom is one of the most common *Hygrophorus* species found under two-needle pines during the fall.

Hygrophorus marginatus var. marginatus Peck   Illus. p. 28

COMMON NAME: Orange-gilled Waxy Cap.

CAP: ⅜–2" (1–5 cm) wide, obtusely conic at first, becoming convex to bell-shaped and sometimes nearly plane at maturity; surface moist, lubricous, hygrophanous, orange to yellow-orange, fading to pale yellowish or nearly white in age; margin incurved and remaining so well into maturity, sometimes faintly striate.

FLESH: thin, fragile, colored like the cap surface; odor and taste not distinctive.

GILLS: attached and notched, becoming nearly free in age, subdistant, with crossveins, brilliant orange; edges even.

STALK: 1½–4" (4–10 cm) long, ⅛–¼" (3–6 mm) thick, nearly equal, often curved, round or compressed, moist or dry but not sticky, hollow, pale orange-yellow.

SPORE PRINT COLOR: white.

MICROSCOPIC FEATURES: spores 7–10 × 4–6 μm, ellipsoid to suboblong, smooth, hyaline.

FRUITING: solitary, scattered, or in groups on humus and soil in mixed woods; June–December; widely distributed in the Southeast; occasional.

EDIBILITY: edible, but of little culinary interest.

COMMENTS: Also known as *Hygrocybe marginata* and *Humidicutis marginata*. *Hygrophorus marginatus* var. *concolor,* edible, is similar, but it has an orange-yellow to bright golden yellow cap, and its gills are also orange-yellow (see photo, p. xvi and *MNE,* p. 288). *Hygrophorus marginatus* var. *olivaceus* (not illustrated), edibility unknown, has ochraceous orange gills and an olive brown to dark olive cap when young that develops pale dull orange coloration near the margin at maturity.

Hygrophorus pratensis Fries Illus. p. 28

COMMON NAME: Salmon Waxy Cap.

CAP: ¾–2¾" (2–7 cm) wide, obtuse to convex when young, becoming broadly convex, typically with a low and broad umbo at maturity; surface smooth, moist then dry and often cracked on the disc in age, reddish orange to salmon orange or dull orange when young, fading to pale orange to orange-yellow or pale tawny in age; margin incurved at first, sometimes wavy at maturity.

FLESH: thick, brittle, whitish to pale reddish cinnamon; odor and taste not distinctive.

GILLS: decurrent, subdistant to distant, often crossveined, salmon buff to pale orange.

STALK: 1⅛–3⅛" (3–8 cm) long, ¼–¾" (5–20 mm) thick, usually tapered downward or nearly equal, often curved, dry, smooth, whitish to pale salmon buff.

SPORE PRINT COLOR: white.

MICROSCOPIC FEATURES: spores 5–8 × 3–5 μm, ellipsoid, subovoid or subglobose, smooth, hyaline.

FRUITING: solitary, scattered, or in groups in grassy areas and on humus in woods; May–December; widely distributed in the Southeast; occasional.

EDIBILITY: edible.

COMMENTS: Also known as *Camarophyllus pratensis* and *Hygrocybe pratensis*.

Hygrophorus roseibrunneus Murrill Illus. p. 29

COMMON NAME: Rosy-brown Waxy Cap.

CAP: ¾–3½" (2–9 cm) wide, convex, becoming nearly plane at maturity, typically with a broad, low central umbo; surface agglutinated-fibrillose, especially near the margin, sticky when wet, soon dry, dark pinkish brown to pinkish cinnamon, paler and often with rosy pink tints toward the margin at maturity, paler overall in age; margin incurved at first, becoming uplifted and wavy in age.

FLESH: thick, soft, white; odor and taste not distinctive.

GILLS: attached or slightly decurrent, close to crowded, white; edges even.

STALK: 1⅛–4¾" (3–12 cm) long, ¼–¾" (6–20 mm) thick, nearly equal or tapered downward near the base, usually curved, dry, solid, densely pruinose to scurfy, at least on the upper half, white.

SPORE PRINT COLOR: white.

MICROSCOPIC FEATURES: spores 6–9 × 3.5–5 μm, ellipsoid, smooth, hyaline.

FRUITING: scattered or in groups on the ground under conifers and broadleaf trees, especially oak; September–March; widely distributed in the Southeast; occasional.

EDIBILITY: edible.

COMMENTS: A thin, whitish partial veil may be evident on very young specimens but soon disappears as the mushroom expands.

Hypholoma fasciculare (Hudson : Fries) Kummer Illus. p. 29
  COMMON NAME: Sulphur Tuft.
  CAP: ¾–3⅛" (2–8 cm) wide, convex, becoming broadly convex to nearly flat, often with an umbo; surface smooth, moist or dry, orange-yellow to sulfur yellow or greenish yellow, with a darker orange to brownish orange disc; margin occasionally rimmed with veil remnants.
  FLESH: pale yellow, bruising brownish; odor not distinctive; taste bitter.
  GILLS: attached, close, yellow to greenish yellow when young, becoming grayish then tinged pale purple-brown at maturity.
  STALK: 2–4¾" (5–12 cm) long, ⅛–⅜" (3–10 mm) thick, nearly equal or slightly tapered in either direction, fibrillose, pale yellow to yellow, becoming fulvous from the base upward; partial veil fibrous to cortinate, whitish, usually leaving a thin, superior annular zone.
  SPORE PRINT COLOR: purple-brown.
  MICROSCOPIC FEATURES: spores 6.5–8 × 3.5–4 μm, elliptic with an apical pore, smooth, pale purple-brown.
  FRUITING: in clusters on conifer and broadleaf logs and stumps or on the surrounding soil; May–November; widely distributed in the Southeast; fairly common.
  EDIBILITY: poisonous.
  COMMENTS: Formerly known as *Naematoloma fasciculare*, this species causes gastric distress. *Hypholoma subviride* is a synonym.

Hypholoma sublateritium (Fries) Quélet Illus. p. 29
  COMMON NAME: Brick Cap, Brick Tops.
  CAP: 1–4" (2.5–10 cm) wide, convex, becoming broadly convex to nearly flat; surface smooth, moist or dry, with scattered yellowish fibrils, brick red with paler yellow-orange to nearly white at or near the margin.
  FLESH: dull yellow to pale yellowish brown, thick, firm; odor not distinctive; taste mild or bitter.
  GILLS: attached, close, narrow, whitish to pale greenish yellow, becoming purplish gray to purplish brown at maturity; not staining when cut or bruised.
  STALK: 2–4" (5–10 cm) long, ¼–⅝" (6–15 mm) thick, equal, hollow in age, pale yellow to whitish above the ring, dull brown or grayish below, covered with reddish brown fibrils; partial veil fibrous to cortinate, leaving a sparse superior ring or annular zone.
  SPORE PRINT COLOR: purple-brown.
  MICROSCOPIC FEATURES: spores 6–7 × 3.5–4.5 μm, elliptical with an apical pore, smooth, pale brown.
  FRUITING: in dense clusters or scattered on broadleaf stumps or logs; July–November; widely distributed in the Southeast; fairly common.
  EDIBILITY: edible.
  COMMENTS: Formerly known as *Naematoloma sublateritium*. Compare with *Hypholoma capnoides* (not illustrated), edible, which has a yellowish orange to reddish orange or cinnamon orange cap, mild-tasting flesh, and which grows on decaying conifer wood.

Hypsizygus tessulatus (Bulliard : Fries) Singer Illus. p. 29
  COMMON NAME: Elm Oyster.
  CAP: 2–6" (5–15 cm) wide, convex at first, becoming more or less flat, often with a sunken

center in age, smooth to very finely hairy, dull white to yellowish buff, with distinct "water spots," usually becoming cracked in age.

FLESH: thick, white, firm to hard; odor and taste not distinctive.

GILLS: attached, often appearing notched, sometimes with fine and barely subdecurrent lines, close to subdistant, whitish.

STALK: 2–4" (5–10 cm) long, ⅜–1" (1–2.5 cm) thick, off-center to almost lateral, dry, solid, very tough, smooth to finely hairy, whitish.

SPORE PRINT COLOR: white to pale buff.

MICROSCOPIC FEATURES: spores 5–7 μm, globose, smooth, hyaline, inamyloid.

FRUITING: solitary or in groups or clusters of two or three, usually on living broadleaf trees, especially elm and box elder, often high up on the trunks; August–December; North Carolina west to Tennessee and northward, southern distribution limits yet to be determined; occasional to fairly common.

EDIBILITY: edible but tough unless collected when young.

COMMENTS: The nomenclature for this species has been terribly confused. According to mycologist Scott Redhead, *Hypsizygus marmoreus, Hypsizygus ulmarius,* and *Hypsizygus elongatipes* are synonyms. The water spots on the cap are more diagnostic than the cracks. The species name is sometimes spelled *tesselatus*.

Inocybe geophylla var. lilacina (Peck) Gillet Illus. p. 29

COMMON NAME: Lilac Fiber Head.

CAP: ⅝–1½" (1.5–4 cm) wide, bluntly conical to bell-shaped or somewhat flattened with a broad umbo; surface dry, silky-glossy to smooth, pale to dark lilac when young, becoming pale pink to lilac-gray or tan in age, often retaining lilac tones on the disc; margin incurved and entire when young, becoming cracked and lacerated in age.

FLESH: whitish, thick in the disc, thinner toward the margin; odor mildly to distinctly unpleasant; taste not distinctive or somewhat disagreeable.

GILLS: attached and sometimes notched, close, moderately broad, whitish to pale lilac when young, becoming pale grayish brown at maturity.

STALK: 1½–2½" (4–6.5 cm) long, ¼–⅜" (6–10 mm) thick, equal, solid, lilac when young, pale pink to pinkish brown in age, whitish and finely powdery at the apex, silky below; partial veil cortinate, soon disappearing.

SPORE PRINT COLOR: brown.

MICROSCOPIC FEATURES: spores 7–9 × 4.5–5.5 μm, elliptic to oval, smooth, thin-walled, pale brown.

FRUITING: scattered or in groups on the ground in conifer and broadleaf forests; August–November; widely distributed in the Southeast; fairly common.

EDIBILITY: poisonous.

COMMENTS: The bluntly conical lilac cap, pinkish stalk, and brown spore print are important features for the identification of this mushroom. *Inocybe geophylla* var. *geophylla* (not illustrated), is nearly identical but has a whitish cap that lacks lilac tones. Similarly colored species of *Cortinarius* are larger and have rusty brown spore prints.

Inocybe rimosa (Bulliard : Fries) Kummer Illus. p. 29

COMMON NAME: Straw-colored Fiber Head.

CAP: ¾–3½" (2–9 cm) wide, broadly conic when young, becoming broadly convex to nearly flat with a distinct conic umbo in age; surface slippery when wet, silky and shiny when dry, covered overall with radial fibers, golden or straw yellow to honey brown, often darker over the disc; margin incurved when young, whitish, becoming expanded, cracked, and yellowish in age.

FLESH: white; odor and taste not distinctive.

GILLS: attached, close, narrow, whitish when young, becoming grayish then coffee brown at maturity.

STALK: 1½–3½" (4–9 cm) long, ⅛–½" (3–12 mm) thick, nearly equal or tapered slightly downward, smooth, silky-fibrillose, longitudinally striate, whitish, developing yellowish tinges in age; partial veil and ring absent.

SPORE PRINT COLOR: brown.

MICROSCOPIC FEATURES: spores 9–15 × 5–8 μm, elliptic, smooth, pale brown.

FRUITING: solitary, scattered, or in groups on the ground and among mosses under conifers and broadleaf trees; June–December; widely distributed in the Southeast; fairly common.

EDIBILITY: poisonous, causing gastric distress.

COMMENTS: Formerly known as *Inocybe fastigiata*. Compare with *Inocybe rimosa* var. *microsperma* (not illustrated), poisonous, which is nearly identical but has a somewhat unpleasant odor.

Laccaria laccata (Scopoli : Fries) Berk. and Broome Illus. p. 30

COMMON NAME: Common Laccaria.

CAP: ⅜–2¼" (1–6 cm) wide, finely fibrous-scaly, orangish brown at first, fading eventually to buff; margin striate or not, sometimes translucent-striate or plicate-striate.

FLESH: thin, colored more or less like the cap; odor and taste not distinctive.

GILLS: attached, occasionally notched or subdecurrent to decurrent, close to distant, broad, thin to thick, slightly waxy, pinkish color, sometimes becoming slightly purplish in age.

STALK: ⅝–4" (1.5–10 cm) long, ⅛–⅜" (3–10 mm) thick, often twisted or flattened, sometimes slightly enlarged downward or with a slight bulb, fibrous, longitudinally striate or not, orangish brown fading eventually to buff.

SPORE PRINT COLOR: white.

MICROSCOPIC FEATURES: spores 6.5–13 × 6–11.5 μm, not including the well-spaced to crowded spines that are mostly 1–2.5 μm long, usually globose to subglobose; hyaline, inamyloid; basidia mostly 4-spored; cheilocystidia absent to abundant.

FRUITING: solitary to gregarious, occasionally in small clusters, on the ground in a wide variety of habitats, sometimes among mosses; May–December; widely distributed in the Southeast; common.

EDIBILITY: edible.

COMMENTS: This species is the central one in a rather large complex. *Laccaria bicolor* (not illustrated), edible but of poor quality, is very similar but has lilac-colored mycelium at the base of the stalk.

Laccaria ochropurpurea (Berkeley) Peck Illus. p. 30

COMMON NAME: Purple-gilled Laccaria.

CAP: 2–5" (5–12.5 cm) wide, nearly smooth to very finely fibrous or occasionally slightly scaly or cracking in age, distinctly purplish (light violet-brown or pinkish buff) at first, fading to buff or nearly whitish in age, margin not striate.

FLESH: thick, violet-buff; odor and taste not distinctive.

GILLS: attached to somewhat decurrent, thick and waxy, close to subdistant, dark purple or violet to violet-gray.

STALK: 2–6" (5–15 cm) long, usually ⅝–1¼" (1.5–3.5 cm) thick, sometimes enlarged at the base, coarsely fibrous, usually with brownish to reddish brown longitudinal striations or scales or both, ground color light violet-brown or pinkish buff at first, fading to buff in age.

SPORE PRINT COLOR: white to pale violet.

MICROSCOPIC FEATURES: spores 6.5–11 × 6.5–9.5 μm, not including crowded spines that are mostly 1–1.5 μm long, mostly globose to subglobose; hyaline, inamyloid.

FRUITING: solitary to scattered, rarely in clusters of two to four, sometimes gregarious, on the ground under conifers and in mixed woods, especially with oaks and pines; June–November; widely distributed in the Southeast; fairly common.

EDIBILITY: edible, but of poor quality.

COMMENTS: An unusually robust and distinctive mushroom for the genus. Compare the spores of *Laccaria ochropurpurea* with those of *Laccaria trullisata*.

Laccaria trullisata (Ellis) Peck

Illus. p. 30

COMMON NAME: Sandy Laccaria.

CAP: 1¼–3" (3.5–7.5 cm) wide, convex to plane, sometimes depressed; surface dry, fibrillose to finely scaly, grayish purple when very young, becoming red-brown to brown or buff to pinkish buff; margin incurved to decurved, not striate.

FLESH: thick, pale purple to pale purple-gray; odor and taste not distinctive.

GILLS: attached to somewhat decurrent, thick and waxy, close to subdistant, purple to dark violet-gray, becoming dull reddish violet in age.

STALK: 1½–3½" (4–9 cm) long, typically ⅜–1" (1–2.5 cm) thick, enlarged at the base, dry, fibrillose, longitudinally striate, brown to buff or pinkish buff, covered with sand on the lower portion or overall.

SPORE PRINT COLOR: white.

MICROSCOPIC FEATURES: spores 14–21 × 5.5–8 μm, subfusiform to fusiform-ellipsoid, smooth to very finely roughened, hyaline.

FRUITING: occasionally solitary, more often scattered or in groups, sometimes abundant, in sand and in very sandy soil, especially in dunes; July–February; widely distributed in the Southeast; fairly common.

EDIBILITY: edible but of poor quality.

COMMENTS: The elliptical smooth spores separate this mushroom from all other *Laccaria* species. When growing in dunes, it is often buried up to the cap in sand. Compare the spores of *Laccaria trullisata* with those of *Laccaria ochropurpurea*.

Lactarius alachuanus var. *alachuanus* Murrill

Illus. p. 30

COMMON NAME: none.

CAP: 2–3" (5–7.5 cm) wide, convex to nearly plane, often shallowly depressed and sometimes with a slight umbo; margin entire; surface azonate, glabrous, moist to viscid when fresh, pale pinkish cinnamon to pale yellowish cinnamon when young, becoming pinkish buff in age.

FLESH: moderately thick, firm, whitish; odor not distinctive or sometimes aromatic; taste slightly bitter then moderately acrid.

LATEX: white on exposure, unchanging, not staining the gills; taste slowly and moderately acrid.

GILLS: narrowly attached to slightly decurrent, moderately broad, close, sometimes forking, pinkish buff, slowly darkening when bruised.

STALK: 1–2¾" (2.5–7 cm) long, ⅜–⅝" (10–16 mm) thick, nearly equal or tapered downward, often curved at the base, dry, solid, pinkish buff, with a white canescence, pinkish cinnamon when rubbed.

SPORE PRINT COLOR: white to creamy white.

MICROSCOPIC FEATURES: spores 7.5–9 × 6–7.5 μm, broadly ellipsoid, ornamented with warts and ridges that sometimes form a partial reticulum, prominences up to 1.5 μm high, hyaline, amyloid.

FRUITING: scattered or in groups on sandy soil and rotting wood in mixed oak and pine woods; October–February; Tennessee and North Carolina south to Alabama and Florida; infrequent to common.

EDIBILITY: unknown.

COMMENTS: Large fruitings are very common in central Florida following rainy periods. Considerable variation in cap color may be observed depending on moisture content. *Lactarius alachuanus* var. *amarissimus* (not illustrated), edibility unknown, a questionable variety, also has a pale pinkish cinnamon to pale yellowish cinnamon cap, very bitter flesh that is not acrid, and sparse watery latex that is assumed not to taste acrid; it grows on the ground under laurel oaks in Florida.

Lactarius allardii Coker   Illus. p. 30

COMMON NAME: none.

CAP: 2⅜–6" (6–15.5 cm) wide, convex, becoming broadly convex with a depressed center; surface dry, somewhat velvety to nearly glabrous, azonate, whitish when very young or when covered by fallen leaves, soon pale pinkish cinnamon to pinkish buff, often with grayish red tints, then becoming pale cinnamon to orange-brown and finally dull brick red in age, typically streaked and spotted white, especially on the margin; margin incurved and remaining so well into maturity.

FLESH: firm, compact, white, slowly staining pinkish then olivaceous when exposed; odor pungent in age or not distinctive; taste acrid.

LATEX: white on exposure, slowly becoming greenish olive then brownish, staining gills dull green to olive then slowly dull brown; taste acrid.

GILLS: attached to decurrent, close to subdistant at first, becoming nearly distant at maturity, white to ivory yellow.

STALK: ¾–2" (2–5 cm) long, ⅜–1⅛" (1–3 cm) thick, tapered downward to nearly equal, hollow, dry, glabrous, colored like the cap or paler.

SPORE PRINT COLOR: white to creamy white.

MACROCHEMICAL TESTS: flesh rapidly stains vinaceous red with the application of $FeSO_4$.

MICROSCOPIC FEATURES: spores 8–11 × 5–8 μm, ellipsoid to subglobose, ornamented with warts and ridges that do not form a reticulum, prominences up to 0.2 μm high, hyaline, amyloid.

FRUITING: scattered or in groups on the ground in broadleaf and mixed woods; widely distributed in the Southeast; occasional.

EDIBILITY: unknown.

COMMENTS: *Lactarius similis* (not illustrated), edibility unknown, reported only from Mississippi, is very similar, has a smaller, pale vinaceous cinnamon cap that often fades to pale yellow, and has mild-tasting latex.

Lactarius atroviridis Peck   Illus. p. 30

COMMON NAME: none.

CAP: 2⅜–5⅝" (6–15 cm) wide, broadly convex, becoming plane with a depressed center; surface dry or somewhat viscid in wet weather, scurfy to minutely scaly, pitted, various shades of green ranging from olive to dark grayish green, usually zoned with dark green to nearly black spots arranged concentrically; margin incurved when young, remaining so into maturity.

FLESH: whitish to pinkish brown, not staining; odor not distinctive; taste acrid.

LATEX: white on exposure, very slowly turning greenish, staining gills dark grayish green to greenish brown; taste acrid.

GILLS: attached to slightly decurrent, close, cream to pinkish white, occasionally forked near the stalk, staining greenish gray to brownish; edges typically olive brown to greenish.

STALK: ¾–3⅛" (2–8 cm) long, ⅜–1⅛" (1–3 cm) thick, nearly equal, hollow, dry, colored and streaked like the cap, scrobiculate.

SPORE PRINT COLOR: cream to buff.

MICROSCOPIC FEATURES: spores 7–10 × 6–9 μm, subglobose to elliptic, ornamented with warts and ridges that form a partial reticulum, prominences up to 0.5 μm high, hyaline, amyloid.

FRUITING: solitary to scattered on the ground under conifers or broadleaf trees; July–November; widely distributed in the Southeast; occasional.

EDIBILITY: unknown.

COMMENTS: *Lactarius plumbeus* (not illustrated), edibility unknown, is similar, but it has an olive brown or darker brown cap, sometimes with a dull honey yellow margin, and it lacks the dark green coloration.

Lactarius chelidonium var. chelidonium Peck   Illus. p. 30

COMMON NAME: Celandine Lactarius.

CAP: 2–3" (5–8 cm) wide, convex, becoming broadly convex to nearly plane, disc depressed, slightly viscid, glabrous, grayish green to grayish yellow or yellow-brown, with bluish green tints in age, sometimes with a few narrow zones on the margin.

FLESH: yellow at first, becoming blue overall, staining green where bruised; taste mild.

LATEX: scant, yellow on exposure, changing to dingy brownish yellow to dingy yellowish brown, staining gills green; taste mild.

GILLS: attached to slightly decurrent, narrow, close, forked and wavy near the stalk, grayish yellow.

STALK: 1–1½" (2.5–4 cm) long, ¼–¾" (7–20 mm) thick, nearly equal, dry, glabrous, hollow in age, colored like the cap.

SPORE PRINT COLOR: yellowish.

MICROSCOPIC FEATURES: spores 8–10 × 6–7 μm, ellipsoid, ornamented with warts and ridges that form a partial reticulum, prominences up to 0.5 μm high, hyaline, amyloid.

FRUITING: on sandy soil under pines; July–January; widely distributed in the Southeast; fairly common.

EDIBILITY: edible.

COMMENTS: *Lactarius chelidonium* var. *chelidonioides* (not illustrated), edibility unknown, reported from northeastern North America, is similar, but it has slowly and slightly acrid-tasting flesh that is azure blue in the upper half of the cap and paler to dingy yellowish near the gills and that typically has an odor similar to the Yellow Morel, *Morchella esculenta*; it also has scant dingy yellow to yellow-brown latex. Compare to *Lactarius paradoxus* (p. 157), edible, which often grows in the same habitat.

Lactarius chrysorheus Fries   Illus. p. 31
  COMMON NAME: Gold Drop Milk Cap.
  CAP: 1⅛–3½" (3–9 cm) wide, convex then soon slightly depressed over the disc, becoming broadly funnel-shaped in age; surface moist to lubricous or subviscid, soon dry and whitish canescent, glabrous, azonate or with watery spots more or less arranged in zones, subhygrophanous, whitish to pale yellowish cinnamon with darker spots when young, fading to nearly whitish overall in age; margin incurved at first, becoming decurved.
  FLESH: thin, whitish, soon yellow when cut; odor not distinctive; taste distinctly acrid (slowly at times).
  LATEX: copious, white on exposure, quickly changing to yellow, not staining mushroom tissues; taste acrid (slowly at times).
  GILLS: attached to short-decurrent, narrow, close, sometimes forked near the stalk, whitish to pale orange-buff or slightly darker, not discoloring or spotting vinaceous or brown.
  STALK: 1⅛–3⅛" (3–8 cm) long, ⅜–¾" (1–2 cm) thick, nearly equal, moist or dry, glabrous, hollow in age, whitish, sometimes flushed orange-buff in age, at times somewhat strigose at the base.
  SPORE PRINT COLOR: pale yellow.
  MICROSCOPIC FEATURES: spores 6–9 × 5.5–6.5 μm, broadly ellipsoid, ornamented with warts and ridges that sometimes form a partial reticulum, prominences up to 0.5 μm high, hyaline, amyloid.
  FRUITING: solitary, scattered, or in groups on the ground in broadleaf and mixed woods, especially with oak; July–October; widely distributed in the Southeast; fairly common.
  EDIBILITY: reported to be poisonous.
  COMMENTS: *Lactarius vinaceorufescens,* poisonous, is similar, but it has a darker cap and stalk and grows on the ground under pine, and its gills stain vinaceous to brown (see *MNE*, p. 296). *Lactarius moschatus* (not illustrated), edibility unknown, is also similar, but it has watery to watery white latex that does not change color and white to whitish flesh that tastes musky.

Lactarius corrugis Peck   Illus. p. 31
  COMMON NAME: Corrugated Milk Cap, Corrugated-cap Milky.
  CAP: 1½–7⅞" (4–20 cm) wide, convex, becoming broadly convex to nearly plane with a depressed center in age; surface velvety, dry, azonate, distinctly wrinkled to finely corrugated in a concentric pattern, reddish brown to vinaceous brown, paler and sometimes orange-brown on the margin, often with a whitish bloom, especially when young.
  FLESH: whitish, staining brown; odor slightly to strongly fishy in mature mushrooms, often not distinctive in young specimens; taste mild.
  LATEX: copious, white on exposure, unchanging, staining gills and flesh tawny brown; taste mild.
  GILLS: attached, close, occasionally forked, pale cinnamon to pale golden brown or darker orange-brown, staining darker brown when bruised.
  STALK: 2–4⅜" (5–11 cm) long, ⅝–1⅛" (1.5–3 cm) thick, equal, solid, dry, velvety, colored like the cap or paler, sometimes with a whitish bloom.
  SPORE PRINT COLOR: white.
  MICROSCOPIC FEATURES: spores 9–12 × 8.5–12 μm, subglobose, ornamented with warts

and ridges that form a partial reticulum, prominences up to 0.8 μm high, hyaline, amyloid.

FRUITING: solitary to scattered on the ground in broadleaf and mixed woods, usually with oak present; July–November; widely distributed in the Southeast; fairly common.

EDIBILITY: edible.

COMMENTS: *Lactarius volemus* var. *volemus* (p. 162), edible, is similar but has somewhat smaller spores and a less wrinkled, dark orange-brown to cinnamon brown cap that is paler orange-brown toward the margin.

Lactarius croceus Burlingham Illus. p. 31

COMMON NAME: none.

CAP: 2–4" (5–10 cm) wide, convex, becoming broadly convex to nearly plane with a depressed center; surface glabrous, slimy when wet, shiny when dry, bright orange to saffron yellow initially, fading to paler yellow-orange or yellowish tan in age, azonate or sometimes distinctly zoned with darker orange bands; margin incurved and somewhat downy when young.

FLESH: thick, whitish, staining sulfur yellow before the latex changes color; odor not distinctive; taste bitter to acrid.

LATEX: scant, white immediately on exposure but very quickly becoming sulfur yellow to yellow-orange, staining gills and flesh sulfur yellow; taste bitter to acrid.

GILLS: attached to slightly decurrent, broad, close to subdistant, creamy buff to pale yellowish tan, staining sulfur yellow where cut or bruised.

STALK: 1⅛–2⅜" (3–6 cm) long, ⅜–¾" (1–2 cm) thick, nearly equal, stuffed but becoming hollow, glabrous, at times velvety at the base, colored like the cap or paler, sometimes spotted brownish.

SPORE PRINT COLOR: yellowish.

MICROSCOPIC FEATURES: spores 7.5–10 × 5.5–7.5 μm, elliptic, ornamented with warts and ridges that form a partial reticulum, prominences up to 0.6 μm high, hyaline, weakly amyloid.

FRUITING: solitary or scattered on the ground under broadleaf trees, especially oak; July–November; widely distributed in the Southeast; occasional.

EDIBILITY: unknown.

COMMENTS: The orange cap and scant sulfur yellow latex are the distinctive features. *Lactarius agglutinatus* (see photo, p. 30), edibility unknown, is similar, but it has a paler stalk; its paler orange to orange-tan cap has tufts of hyphae that resemble small scales embedded in a slime layer; and its acrid latex is white and dries creamy yellow. *Lactarius yazooensis* (see photo, p. 34), edibility unknown, is also similar, but its sticky to dry cap has conspicuous orange and orange-yellow concentric rings and lacks scale-like tufts, and its latex is white, unchanging, and extremely acrid.

Lactarius deceptivus Peck Illus. p. 31

COMMON NAME: Deceptive Milk Cap, Deceptive Milky.

CAP: 3–10" (7.5–25.5 cm) wide, convex, becoming broadly funnel-shaped; surface dry, glabrous when young, whitish, often with yellowish or brownish stains, becoming coarsely scaly and darkening to dull brownish ochre in age; margin distinctly inrolled and cottony when young.

FLESH: thick, white; odor pungent or not distinctive; taste strongly acrid.

LATEX: white on exposure, unchanging, staining tissues brownish; taste strongly acrid.

GILLS: attached, close to subdistant, white at first then cream to pale ochre.

STALK: 1½–4" (4–10 cm) long, up to 1⅛" (3 cm) thick, nearly equal or tapered downward, dry, scurfy to nearly glabrous, white, staining brownish in age.

SPORE PRINT COLOR: white to whitish.

MACROCHEMICAL TESTS: dull brownish ochre coarse cap scales stain reddish cinnamon with the application of KOH; flesh stains reddish cinnamon with the application of $FeSO_4$ and is negative with KOH.

MICROSCOPIC FEATURES: spores 9–13 × 7–9 μm, broadly ellipsoid, ornamented with warts and spines that do not form a reticulum, prominences up to 1.5 μm high, hyaline, amyloid.

FRUITING: solitary, scattered, or in groups on the ground in conifer and broadleaf woods, often under oak or hemlock; June–November; widely distributed in the Southeast; common.

EDIBILITY: reported to be edible when thoroughly cooked.

COMMENTS: *Lactarius caeruleitinctus*, (not illustrated), edibility unknown, is similar, but it has a milk white stalk with blue tints that become more intense after picking, and it lacks a cottony inrolled margin.

Lactarius floridanus Beardslee and Burlingham Illus. p. 31

COMMON NAME: none.

CAP: up to 4¾" (12 cm) wide, umbilicate with an arched and incurved margin at first, becoming deeply depressed in age; surface viscid, azonate to faintly zoned, tomentose with long agglutinated fibers, but not extending as a beard along the margin; color variable, brownish pink to brownish orange or buff with an ochraceous to dingy light yellow center.

FLESH: firm, thick, white to pale yellowish pink; odor aromatic and persisting when dried; taste acrid.

LATEX: scant, white on exposure, unchanging; taste acrid.

GILLS: attached to subdecurrent, moderately broad, close, sometimes forked, whitish at first, becoming honey yellow at maturity.

STALK: ¾–1½" (2–4 cm) long, ½–⅞" (1.4–2.3 cm) thick, nearly equal, dry, firm, solid, pruinose at the apex, sometimes scrobiculate, pinkish buff to pale cinnamon buff, developing yellowish tints in age.

SPORE PRINT COLOR: not recorded in the original description, presumed to be white.

MICROSCOPIC FEATURES: spores 7.5–9 × 5–6 μm, ellipsoid to nearly oblong, ornamented with warts and ridges that form a partial reticulum, prominences up to 0.2 μm high, hyaline, amyloid.

FRUITING: in sandy soil under mixed live oak and pine; December–January; reported only from Florida, distribution limits and frequency yet to be determined.

EDIBILITY: unknown.

COMMENTS: The variable cap colors, agglutinated fibers, aromatic persisting odor, and narrow spores characterize this species.

Lactarius glaucescens Crossland Illus. p. 31

COMMON NAME: none.

CAP: 1½–4¾" (4–12 cm) wide, convex, becoming nearly plane with a depressed center; surface dry, dull to somewhat shiny, glabrous, sometimes rimose-areolate in dry weather, white to pale cream when young, becoming dingy yellow-brown in age, especially over the center; margin inrolled to incurved at first, becoming decurved.

FLESH: thick, hard, pale cream, unchanging or very slowly staining pale bluish green when cut; odor faintly pungent or not distinctive; taste acrid.
LATEX: white to whitish on exposure, not staining tissues, drying pale bluish green; taste very acrid.
GILLS: subdecurrent to decurrent, broad, crowded, often forked, whitish to dull cream, very slowly staining brown when bruised.
STALK: 1⅛–3½" (3–9 cm) long, ⅜–1⅛" (1–3 cm) thick, tapered downward or sometimes nearly equal, dry, solid, glabrous, white to pale cream.
SPORE PRINT COLOR: white to yellowish.
MACROCHEMICAL TESTS: latex stains yellow to orange with the application of KOH.
MICROSCOPIC FEATURES: spores 6.5–9 × 5–7 μm, broadly ellipsoid, ornamented with warts and fine lines that do not form a reticulum, prominences up to 0.2 μm high, hyaline, amyloid.
FRUITING: solitary, scattered, or in groups on the ground under broadleaf trees, especially oak; July–December; widely distributed in the Southeast; occasional to fairly common.
EDIBILITY: poisonous as reported by Florida mycologist Robert S. Williams.
COMMENTS: *Lactarius piperatus* var. *piperatus* (not illustrated), inedible, is very similar, but it has white latex that does not change to bluish green when dry.

Lactarius hygrophoroides var. **rugatus** (Kühner and Rom.) Hesler and A. H. Smith
Illus. p. 31

COMMON NAME: none.
CAP: 1¾–3½" (4.5–9 cm) wide, convex, becoming broadly convex and deeply depressed over the disc in age; surface dull, somewhat velvety, conspicuously wrinkled and concentrically wrinkled near the margin, rusty orange-brown, paler toward the margin and darker on the disc; margin irregular, often curved and wavy.
FLESH: whitish, moderately thick; odor faintly fishy or not distinctive; taste not distinctive.
LATEX: abundant, white to watery white on exposure, unchanging, not staining tissues; taste mild.
GILLS: attached, subdecurrent, distant, fairly narrow, creamy white at first, becoming pale ochraceous to pale golden yellow at maturity, not staining when bruised.
STALK: 1⅜–2⅜" (3.5–6 cm) long, ⅜–⅝" (1–1.3 cm) thick, nearly equal to somewhat spindle-shaped, dry, solid, finely tomentose, orange-cream to pale vinaceous cinnamon.
SPORE PRINT COLOR: white.
MACROCHEMICAL TESTS: cap surface stains rose to dull brownish red with the application of $FeSO_4$.
MICROSCOPIC FEATURES: spores 7.5–9 × 6–7.5 μm, ellipsoid, ornamented with a partial reticulum and isolated warts and short ridges, prominences up to 0.5 μm, hyaline, amyloid.
FRUITING: solitary, scattered, or in groups on the ground under broadleaf trees, especially oak; July–November; widely distributed in the Southeast; occasional.
EDIBILITY: edible.
COMMENTS: *Lactarius hygrophoroides* var. *hygrophoroides*, edible, is very similar, but its cap surface is smooth or nearly so (see *MNE*, p. 294). *Lactarius hygrophoroides* var. *lavendulaceus* (not illustrated), edible, is also very similar, but its gills stain pink to lavender when cut or bruised. Compare with *Lactarius volemus* var. *volemus* (p. 162), which has close gills, latex that stains tissue brown, and a strong fishy odor.

Lactarius imperceptus Beardslee and Burlingham Illus. p. 31
COMMON NAME: none.
CAP: 1⅛–3½" (3–9 cm) wide, broadly convex to nearly plane, usually slightly depressed on the disc with a small umbo at maturity, glabrous, azonate; surface dry to slightly sticky when wet, pale cinnamon to pinkish cinnamon or pale reddish brown, often pitted and stained with darker pinkish cinnamon; margin sometimes striate in age.
FLESH: whitish, sometimes staining yellow when exposed; odor not distinctive; taste acrid, sometimes slowly.
LATEX: white to creamy white on exposure, sometimes slowly darkening to yellow on exposure; taste either bitter then acrid or acrid only, sometimes slowly.
GILLS: attached to slightly decurrent, close, whitish to pinkish white, usually staining brownish, occasionally forked at the stalk.
STALK: 1⅛–3½" (3–9 cm) long, ¼–¾" (6–20 mm) thick, tapered downward or nearly equal overall, solid, dry to moist, pinkish white, darkening in age, glabrous, with a whitish tomentose base.
SPORE PRINT COLOR: white to pale cream.
MACROCHEMICAL TESTS: flesh stains pale olive with the application of $FeSO_4$.
MICROSCOPIC FEATURES: 8–11 × 7–8.5 μm, broadly ellipsoid, ornamented with heavy bands that do not form a complete reticulum, prominences up to 1.5 μm high, hyaline, amyloid.
FRUITING: solitary, scattered, or in groups on the ground or in needle litter in conifer or broadleaf woods, especially with oak or pine; July–November; widely distributed in the Southeast; occasional to fairly common.
EDIBILITY: unknown.
COMMENTS: The staining of the flesh and the latex color change are variable features, especially if the fruit bodies are soaked following heavy rainfall. *Lactarius quietus* var. *incanus* (see photo, p. 33), edibility unknown, is similar, but it has a darker red-brown cap when young that becomes paler pinkish brown toward the margin at maturity; pale pinkish buff flesh with a sweet, fragrant odor that resembles burned sugar and slowly develops a mildly acrid taste; white latex in young specimens that soon becomes watery as specimens mature; and a stalk that darkens from the base upward as it ages.

Lactarius indigo var. **diminutivus** Hesler and A. H. Smith Illus. p. 32
COMMON NAME: Smaller Indigo Milk Cap.
CAP: 1⅛–2¾" (3–7 cm) wide, convex, becoming broadly convex to nearly plane at maturity, depressed to deeply depressed and funnel-shaped in age; surface glabrous, viscid to slimy-viscid, azonate to slightly zonate, dark blue at first, becoming gray-blue to gray-green with olive green tints in age; margin inrolled at first, becoming decurved to uplifted at maturity, often weakly striate in age.
FLESH: white but quickly changing to dark blue on exposure and slowly becoming green; odor and taste not distinctive.
LATEX: copious, dark blue on exposure, unchanging, staining tissues dark blue; taste mild.
GILLS: decurrent to long-decurrent, close, colored like the cap, becoming green in age, quickly staining dark indigo where cut or bruised, then slowly changing to gray-green or olive.
STALK: ⅝–1½" (1.5–4 cm) long, ⅛–⅜" (3–10 mm) thick, tapered downward, viscid or dry, solid at first, becoming hollow at maturity, colored like the cap, sometimes scrobiculate.
SPORE PRINT COLOR: white.

MICROSCOPIC FEATURES: spores 7–9 × 6–7.5 μm, subglobose to broadly ellipsoid, ornamented with warts and ridges that sometimes form a reticulate pattern, prominences up to 0.5 μm high, hyaline, amyloid.

FRUITING: scattered or in groups on the ground, typically among grasses in wet areas under pine trees; August–January; along the Gulf Coast from Florida to Texas; occasional.

EDIBILITY: edible.

COMMENTS: *Lactarius indigo* var. *indigo* (p. 155), edible, has a much larger cap, a much larger and typically scrobiculate stalk, grows on the ground under broadleaf and conifer trees, and is much more common and widely distributed in the Southeast.

Lactarius indigo var. **indigo** (Schweinitz) Fries — Illus. p. 32

COMMON NAME: Indigo Milk Cap, Indigo Milky.

CAP: 2–6" (5–15.5 cm) wide, convex to convex-depressed, becoming broadly funnel-shaped; surface viscid, glabrous, zonate to nearly azonate, dark blue to silvery blue when fresh, fading to grayish blue to grayish with a silvery luster, often with dark green areas where bruised; margin inrolled at first.

FLESH: thick, firm, whitish, rapidly staining dark blue on exposure, then slowly greenish; odor not distinctive; taste mild or rarely slightly bitter then slightly acrid.

LATEX: scant, dark blue on exposure, slowly becoming dark green; taste mild.

GILLS: attached to short-decurrent, broad, close, sometimes forked near the stalk, dark blue at first, becoming paler blue at maturity, developing yellowish tints in age, staining green when bruised.

STALK: ¾–3⅛" (2–8 cm) long, ⅜–1" (1–2.5 cm) thick, nearly equal or tapered downward, hollow in age, viscid but soon dry, dark blue when young, paler blue in age, sometimes with a sheen, often scrobiculate.

SPORE PRINT COLOR: creamy white.

MICROSCOPIC FEATURES: spores 7–9 × 5.5–7.5 μm, broadly ellipsoid to subglobose, ornamented with ridges that form a partial to complete reticulum, prominences up to 0.5 μm high, hyaline, amyloid.

FRUITING: solitary, scattered, or in groups on the ground in conifer or broadleaf woods; July–January; eastern Canada south to Florida, west to Michigan and Texas; fairly common in the southern part of its range, occasional northward.

EDIBILITY: edible.

COMMENTS: The name *indigo* refers to the dark blue color of this mushroom. *Lactarius indigo* var. *diminutivus* (p. 154), edible, reported along the Gulf Coast from Florida to Texas, is similar, but it has a much smaller (1⅛–2" [3–5 cm] wide), viscid to slimy cap that is dark blue at first and soon becomes gray-blue then gray-green to olive green in age; has copious dark blue latex on exposure that is unchanging; and grows on the ground, often among grasses and weeds, usually in pine woods. Also compare with *Lactarius paradoxus* (p. 157).

Lactarius luteolus Peck — Illus. p. 32

COMMON NAME: Buff Fishy Milk Cap, Buff Fishy Milky.

CAP: 1–3⅛" (2.5–8 cm) wide, convex to nearly plane, often shallowly depressed in age; surface dry, slightly velvety, azonate, whitish to buff with a white bloom, becoming brownish in age.

FLESH: white, staining brown; odor not distinctive in young specimens, soon becoming fishy or unpleasant and resembling spoiled crab; taste mild.

LATEX: copious, sticky, watery white to white on exposure, unchanging, staining tissues brown; taste mild.

GILLS: attached to slightly decurrent, close, white becoming cream, staining yellow-brown to brown.

STALK: 1–2⅜" (2.5–6 cm) long, ¼–¾" (6–20 mm) thick, tapered downward or nearly equal, solid to stuffed, dry, slightly velvety, whitish to buff, staining brown.

SPORE PRINT COLOR: white to cream.

MACROCHEMICAL TESTS: flesh stains bluish olive with the application of $FeSO_4$ and slightly orange-buff with KOH.

MICROSCOPIC FEATURES: spores 7–9 × 5.5–7 μm, elliptic, ornamented with warts and ridges that sometimes form a partial reticulum, prominences up to 0.8 μm high, hyaline, amyloid.

FRUITING: solitary, scattered, or in groups on the ground in broadleaf and mixed woods, usually with oak present; June–November; widely distributed in the Southeast; fairly common.

EDIBILITY: edible.

COMMENTS: The slightly velvety whitish cap, copious sticky latex that stains tissues brown, the strong fishy to spoiled-crab odor of the mature specimens' flesh, and the small spores are the distinctive features.

Lactarius maculatipes Burlingham   Illus p. 32

COMMON NAME: none.

CAP: 2–3½" (5–9 cm) wide, broadly convex, becoming nearly plane with a depressed center, often broadly funnel-shaped in age; margin incurved at first, becoming expanded and uplifted; surface glabrous, slimy and sticky when fresh, whitish to cream overall, becoming tawny on the disc in age, with pale yellow zones and spots that darken in age.

FLESH: white, slowly staining yellowish on exposure; odor not distinctive; taste slowly acrid.

LATEX: white on exposure, unchanging, staining gills and flesh yellowish; taste slowly acrid.

GILLS: decurrent, narrow, crowded, sometimes forked, whitish, becoming pinkish buff, staining yellowish to pale ochre when cut or bruised.

STALK: 1¼–3⅛" (3–8 cm) long, ⅜–1" (1–2.5 cm) thick, tapering downward, slimy, colored like the cap, spotted and streaked with darker tan, staining yellow when bruised or in age, scrobiculate.

SPORE PRINT COLOR: pinkish buff to yellowish.

MICROSCOPIC FEATURES: spores 6.5–8 × 6–7.5 μm, subglobose to broadly ellipsoid, ornamented with warts and ridges that sometimes form a partial reticulum, prominences up to 1 μm high, hyaline, amyloid.

FRUITING: scattered or in groups on the ground under oak; July–January; West Virginia, North Carolina, and Tennessee south to Florida, west to Texas; fairly common in the southern part of its range.

EDIBILITY: unknown.

COMMENTS: *Lactarius dunfordii* (not illustrated), edibility unknown, reported only from Tennessee, is very similar, but it has white latex that does not stain the gills and flesh, lacks spots on the stalk, and grows on the ground under pine. *Lactarius carolinensis* (not illustrated), edibility unknown, is similar but has a tan cap with darker yellowish brown zones, white flesh with a faintly alkaline odor, and a white spore deposit.

Lactarius paradoxus Beardslee and Burlingham　　　　　　　　　　Illus. p. 32
　COMMON NAME: Silver-blue Milk Cap, Silver-blue Milky.
　CAP: 2–3⅛" (5–8 cm) wide, broadly convex, becoming plane with a depressed center, often funnel-shaped in age; surface glabrous, slimy to sticky when wet, with a silvery sheen when young, zoned with bands of grayish blue, grayish purple, green and blue, staining green when bruised; margin incurved at first, becoming expanded to uplifted.
　FLESH: thick, whitish with greenish to bluish tints, slowly staining greenish when cut or bruised; odor not distinctive; taste mild or slightly acrid.
　LATEX: scant, dark vinaceous brown on exposure, staining tissues green; taste mild or slightly acrid.
　GILLS: attached to slightly decurrent, narrow to broad, close, occasionally forked near the stalk, vinaceous to pinkish orange, staining blue-green when bruised.
　STALK: 1–1¼" (2–3 cm) long, ⅜–⅝" (1–1.5 cm) thick, nearly equal or tapered downward, hollow, dry, colored like the cap, staining green when bruised or in age.
　SPORE PRINT COLOR: cream to yellow.
　MICROSCOPIC FEATURES: spores 7–9 × 5.5–6.5 μm, broadly ellipsoid, ornamented with warts and ridges that form a partial reticulum, prominences up to 1 μm high, hyaline, amyloid.
　FRUITING: solitary, scattered, or in groups on the ground and in grass under pine, oak, and palmetto palm; August–February; widely distributed in the Southeast; occasional in its northern range, common along the Gulf Coastal Plain.
　EDIBILITY: edible.
　COMMENTS: *Lactarius chelidonium* var. *chelidonium* (p. 149), edibility unknown, is similar, but it has a grayish green cap with blue and yellow tints and yellow latex. *Lactarius chelidonium* var. *chelidonioides* (not illustrated), edibility unknown, is also similar but has a sordid azure blue cap that has dingy orange-brown areas or zones, loses its blue color in age, and becomes dull orange-brown to reddish brown; it also has scant dingy yellow to yellow-brown latex. Also compare with *Lactarius indigo* var. *indigo* (p. 155).

Lactarius peckii var. **peckii**　　　　　　　　　　Illus. p. 32
　COMMON NAME: Peck's Milk Cap, Peck's Milky.
　CAP: 2–6" (5–15.5 cm) wide, broadly convex with a depressed center; surface dry, velvety when young then scurfy to nearly glabrous; color variable, typically dull brick red or brownish red but also orange-brown to orange-red or dull reddish orange, often paler toward the margin, usually distinctly zoned with darker bay red to orange-red bands, often fading in age; margin inrolled at first, becoming decurved and remaining so well into maturity, sometimes striate on young specimens.
　FLESH: firm, pale vinaceous brown, not staining when cut; odor not distinctive; taste extremely acrid.
　LATEX: copious, white on exposure, unchanging, not staining tissues; taste extremely acrid.
　GILLS: decurrent, close, narrow, pale cinnamon buff at first, soon tinged pinkish cinnamon and darkening in age, staining rusty brown to reddish brown.
　STALK: ¾–2⅜" (2–6 cm) long, ⅜–1" (1–2.5 cm) thick, nearly equal, hollow in age, covered with a whitish bloom when young, colored like the cap but usually paler, often spotted reddish brown or dull orange in age, but not scrobiculate.
　SPORE PRINT COLOR: white.
　MICROSCOPIC FEATURES: spores 6–7.5 μm, globose to subglobose, ornamented with

warts and ridges that form a partial to complete reticulum, prominences up to 0.8 μm high, hyaline, amyloid.

FRUITING: solitary, scattered, or in groups on the ground in oak woods; July–November; widely distributed in the Southeast; occasional to fairly common.

EDIBILITY: unknown.

COMMENTS: This species has gills that are among the darkest in the genus. *Lactarius peckii* var. *glaucescens* (not illustrated), edibility unknown, reported from Tennessee south to Florida, west to Mississippi, is nearly identical except that the latex dries pale bluish green. *Lactarius peckii* var. *lactolutescens* (not illustrated), edibility unknown, reported from Tennessee and Florida, is also nearly identical, but it has very acrid white latex that quickly changes to ivory yellow on exposure and dries pale yellowish green.

Lactarius proximellus Beardslee and Burlingham   Illus. p. 32

COMMON NAME: none.

CAP: 1–2⅜" (2.5–6 cm) wide, broadly convex-umbilicate, then expanding to shallowly funnel-shaped; surface viscid when wet, whitish pruinose in the center, distinctly zonate, zones tawny to brownish orange, with darker brownish orange to cinnamon zones; margin arched, at length uplifted and more or less fluted, sometimes striate.

FLESH: pale cream; odor not distinctive.

LATEX: scant, white on exposure, unchanging; taste very acrid.

GILLS: attached to slightly decurrent, narrow to moderately broad, close, sometimes forked near the stalk, whitish at first, becoming pale brownish yellow.

STALK: ½–¾" (1.3–2 cm) long, 5/16–⅜" (8–10 mm) thick, nearly equal or enlarged slightly at the apex, pruinose when young, pale yellow, becoming pale bluish green where bruised.

SPORE PRINT COLOR: very pale yellow.

MICROSCOPIC FEATURES: spores 7.5–9 × 6–7.5 μm, broadly ellipsoid, ornamented with warts and ridges that form a partial reticulum, prominences up to 0.5 μm high, hyaline, amyloid.

FRUITING: on sandy soil under oaks; August–January; widely distributed in the Southeast; occasional to locally common.

EDIBILITY: unknown.

COMMENTS: The stocky stature, tawny to brownish orange zonate cap, acrid latex, and growth under oaks are the distinctive features. During the winter months, this mushroom can be fairly abundant in the oak-pine woods of the coastal dunes.

Lactarius psammicola f. psammicola A. H. Smith   Illus. p. 32

COMMON NAME: none.

CAP: 1½–5½" (4–14 cm) wide, convex to broadly convex and deeply depressed, expanding to broadly vase-shaped; surface glutinous, coarsely fibrillose, the fibrils agglutinated, ochraceous buff to ochraceous orange, becoming dingy yellow-brown in age, with conspicuous light buff zones especially when young; margin inrolled at first, becoming decurved, coarsely strigose at first, becoming matted-fibrillose in age.

FLESH: thick, dingy buff, sometimes staining pinkish lilac; odor not distinctive; taste acrid.

LATEX: copious, white on exposure, slowly changing to pale dingy pinkish lilac or not changing or staining this color; taste very acrid.

GILLS: decurrent, narrow, close, sometimes forked near the stalk, whitish to light buff at first, becoming darker and sordid ochraceous in age, sometimes staining pinkish lilac.

STALK: ⅜–1⅛" (1–3 cm) long, ⅜–¾" (1–2 cm) thick, tapered downward, dry, stuffed to hollow, whitish to grayish, often scrobiculate.

SPORE PRINT COLOR: yellowish.

MACROCHEMICAL TESTS: flesh stains pale dull bluish gray with the application of $FeSO_4$.

MICROSCOPIC FEATURES: spores 7.5–9 × 6–7.5 μm, ellipsoid, ornamented with warts and ridges that sometimes form a partial reticulum, prominences up to 0.5 μm high, hyaline, amyloid.

FRUITING: on the ground in broadleaf and mixed woods, usually with oak present; July–January; widely distributed in the Southeast; occasional.

EDIBILITY: unknown.

COMMENTS: *Lactarius psammicola* f. *glaber* (not illustrated), edibility unknown, is very similar, but it has a glabrous, viscid cap with light buff to pale tan and dull orange zones, and whitish flesh with a slightly fragrant odor.

Lactarius pseudodeliciosus var. **pseudodeliciosus** Beardslee and Burlingham
Illus. p. 33

COMMON NAME: none.

CAP: 2⅜–3⅜" (6–8.5 cm) wide, broadly convex-umbilicate, expanding to funnel-shaped; surface viscid when wet, azonate to faintly zonate, nearly white at first, becoming yellowish with age, central portion pale pinkish buff to ochraceous buff; margin incurved at first, becoming decurved, with agglutinated fibrils when young (use a hand lens).

FLESH: turning grayish green when cut; odor not distinctive; taste not reported.

LATEX: scant, ochraceous orange to apricot orange; taste slowly acrid.

GILLS: attached to subdecurrent, honey yellow with orange tones, intervenose, gill spacing and width not stated in the original description.

STALK: ½–1" (1.2–2.5 cm) long, ⅜–¾" (1–2 cm) thick, solid, somewhat scrobiculate, white tomentose on lower half.

SPORE PRINT COLOR: ochraceous.

MICROSCOPIC FEATURES: spores 7–9 × 6–7.5 μm, ellipsoid, ornamented with broad and narrow bands that form a reticulum, prominences up to 0.5 μm high, hyaline, amyloid.

FRUITING: on soil under oak and pine; November–February; reported only from Florida, distribution limits and frequency yet to be determined.

EDIBILITY: edible.

COMMENTS: *Lactarius pseudodeliciosus* var. *paradoxiformis* (not illustrated), edibility unknown, reported only from Florida, has a cream-colored cap that becomes partly or entirely bluish green, slightly fragrant flesh that has a distinctly acrid taste, and ochraceous gills and stalk that become spotted bluish green in age or when bruised; it grows on the ground under broadleaf trees, especially red oak.

Lactarius salmoneus var. **salmoneus** Peck
Illus. p. 33

COMMON NAME: none.

CAP: 1–3⅛" (2.5–8 cm) wide, convex, becoming nearly plane and depressed at the center; surface dry, subvelvety, azonate, at first white then becoming orange nearly overall as the white layer wears away in age, staining orange when bruised; margin inrolled to incurved at first, becoming decurved and remaining so well into maturity.

FLESH: moderately thick, firm, light orange, staining darker orange especially near the gills when exposed; odor slightly fragrant or not distinctive; taste slightly peppery or mild.

LATEX: scant, dark orange on exposure, staining gills dark orange; taste slightly peppery or mild.

GILLS: attached to decurrent, narrow, close to subdistant, typically not forked, light orange at first, becoming paler orange to ochraceous buff in age, staining dark orange when bruised.

STALK: ½–1" (1.2–2.5 cm) long, ¼–½" (6–13 mm) thick, tapered downward, dry, glabrous, hollow in age, white at first, becoming light orange overall as the white layer wears away in age or sometimes retaining white areas, usually not scrobiculate.

SPORE PRINT COLOR: creamy white.

MICROSCOPIC FEATURES: spores 7.5–9 × 5–6 μm, ellipsoid, ornamented with warts and ridges that form a partial reticulum, prominences up to 0.4 μm high, hyaline, amyloid.

FRUITING: solitary, scattered, or in groups on the ground or in wet areas under pine or in mixed oak and pine woods; July–December; South Carolina south to Florida, west to Texas; occasional.

EDIBILITY: unknown.

COMMENTS: *Lactarius salmoneus* var. *curtisii* (not illustrated), edibility unknown, is very similar if not identical, grows on moist low-lying areas under pine and mixed pine-hardwood forests from Tennessee and North Carolina south to Florida, west to Texas, and differs from var. *salmoneus* in that the cap surface of old specimens of var. *curtisii* stains bluish green, the gills readily stain green when injured, and the stalk stains green when bruised.

Lactarius subplinthogalus Coker

Illus. p. 33

COMMON NAME: none.

CAP: 1¼–2" (3–5 cm) wide, nearly plane, becoming broadly funnel-shaped in age; surface nearly glabrous when young, becoming finely wrinkled in age, whitish when young, becoming yellowish buff to pale tan and finally brownish yellow in age; margin uplifted and conspicuously plicate.

FLESH: whitish, staining rosy salmon when cut or bruised; odor not distinctive; taste acrid.

LATEX: white on exposure, unchanging, but staining gills and flesh rosy salmon; taste acrid.

GILLS: attached to slightly decurrent, appearing more decurrent as the margin becomes uplifted, broad, distant, colored like the cap, staining rosy salmon when cut and bruised.

STALK: 1¼–3⅛" (3–8 cm) long, ¼–⅝" (7–15 mm) thick, equal or tapered slightly downward, solid, becoming hollow, glabrous, dry, cream with orangish or orange-brown stains in age.

SPORE PRINT COLOR: pinkish buff to cinnamon buff.

MICROSCOPIC FEATURES: spores 7.5–9.5 × 7–8 μm, subglobose to broadly ellipsoid, ornamented with warts and ridges that do not form a reticulum, prominences up to 2.5 μm high, hyaline, amyloid.

FRUITING: solitary to scattered on the ground under oak or in oak-pine woods; July–September; Maryland south to Florida, west to Texas; occasional.

EDIBILITY: unknown.

COMMENTS: *Lactarius sumstinei* (not illustrated), edibility unknown, is similar, but it has white latex that does not stain tissues. *Lactarius subvernalis* var. *cokeri* (not illustrated), edibility unknown, is also similar and has white unchanging latex that stains gills and flesh pinkish to pinkish orange, but the gills are crowded.

Lactarius tomentoso-marginatus Hesler and A.H. Smith Illus. p. 33
> COMMON NAME: none.
> CAP: 1½–3½" (4–9 cm) wide, convex-depressed, becoming vase-shaped; surface dry, appressed-fibrillose to silky, white at first, becoming flushed dull grayish pink and finally pale dingy orange-brown; margin incurved and long remaining arched.
> FLESH: thick, hard, white, slowly darkening to more or less pinkish buff where cut; odor not distinctive; taste acrid.
> LATEX: scant, white on exposure, unchanging, staining tissues cinnamon buff; taste acrid.
> GILLS: decurrent, narrow, crowded, often forked especially near the stalk, white when young, becoming pinkish buff then cinnamon buff and darkening to dull brown in age.
> STALK: 1⅛–2" (3–5 cm) long, ½–1⅛" (1.3–3 cm) thick, nearly equal or narrowed downward, dry, solid, firm, somewhat velvety, white.
> SPORE PRINT COLOR: unknown.
> MACROCHEMICAL TESTS: cap surface is negative with the application of KOH; flesh stains vinaceous cinnamon with the application of $FeSO_4$ and is negative with KOH.
> MICROSCOPIC FEATURES: spores 9–11 × 7–8.5 μm, ellipsoid, ornamented with isolated warts and indistinct lines that do not form a reticulum, prominences up to 0.7 μm high, hyaline, amyloid.
> FRUITING: solitary, scattered, or in groups on the ground in broadleaf and mixed woods, usually with oak; July–November; widely distributed in the Southeast; fairly common, especially in the southern part of its range.
> EDIBILITY: unknown.
> COMMENTS: *Lactarius deceptivus* (p. 151), reportedly edible when thoroughly cooked, is similar, but it has a larger cap and stalk, a conspicuous cottony inrolled margin on young specimens, and larger spore prominences that measure up to 1.5 μm high.

Lactarius volemus var. *flavus* Hesler and A. H. Smith Illus. p. 33
> COMMON NAME: none.
> CAP: 1½–3½" (4–9 cm) wide, convex, becoming broadly convex to nearly plane and somewhat depressed in age; surface dry, velvety, azonate, buff-yellow to yellow or orange-yellow, staining brownish where bruised; margin incurved at first, becoming uplifted and wavy in age.
> FLESH: whitish to ivory; odor not distinctive at first, becoming somewhat fishy or unpleasant at maturity; taste not distinctive.
> LATEX: white on exposure, unchanging, staining tissues brown, sticky; taste mild.
> GILLS: attached, close, narrow to moderately broad, whitish at first, becoming creamy white at maturity, frequently forked near the stalk.
> STALK: 2–4" (5–10 cm) long, ¼–⅝" (5–16 mm) thick, nearly equal, dry, solid, somewhat velvety, cream to pale yellow.
> SPORE PRINT COLOR: white.
> MICROSCOPIC FEATURES: spores 6.5–9 × 6.5–8 μm, globose to subglobose or broadly ellipsoid, ornamented with broad and narrow bands that form a partial reticulum, prominences up to 0.5 μm high, hyaline, amyloid.
> FRUITING: solitary, scattered, or in groups on the ground in broadleaf and mixed woods; June–December; widely distributed in the Southeast; occasional.
> EDIBILITY: edible.
> COMMENTS: *Lactarius volemus* var. *volemus* (p. 162), edible, has an orange to dark orangish brown or cinnamon brown cap with a paler orange-brown margin.

Lactarius volemus var. **volemus** (Fries) Fries　　　　　　　　　　　　　　　Illus. p. 33
　COMMON NAME: Apricot Milk Cap, Bradley, Voluminous-latex Milky, Weeping Milk Cap.
　CAP: 2–4" (5–10 cm) wide, broadly convex, becoming nearly plane with a depressed center to broadly funnel-shaped in age; surface dry, pruinose to velvety when young, becoming glabrous to finely wrinkled at maturity, orange to dark orange-brown or cinnamon brown at the center, often paler toward the margin, fading in age to pale orange-brown then honey yellow; margin incurved at first, expanding and becoming uplifted in age.
　FLESH: thick, brittle, white, staining brownish when cut or bruised; odor not distinctive in very young specimens, soon becoming distinctly fishy as the mushrooms mature; taste mild.
　LATEX: copious, white on exposure, becoming creamy white then brownish or grayish and staining gills and flesh tawny brown; taste mild.
　GILLS: attached to slightly decurrent, broad, close, often forked, whitish to cream, slowly bruising tawny brown.
　STALK: 2–4½" (5–11.5 cm) long, ¼–¾" (5–20 mm) thick, nearly equal or tapered at the base, solid, sometimes hollow in age, nearly glabrous, colored like the cap.
　SPORE PRINT COLOR: white.
　MACROCHEMICAL TESTS: flesh instantly stains dark blue-green with the application of $FeSO_4$.
　MICROSCOPIC FEATURES: spores 7.5–10 × 7.5–9 μm, globose to subglobose, ornamented with warts and ridges that form a reticulum, prominences up to 1 μm high, hyaline, amyloid.
　FRUITING: solitary, scattered, or in groups on the ground in broadleaf and mixed woods; June–September; widely distributed throughout eastern North America, also reported from California and western Canada; fairly common.
　EDIBILITY: edible.
　COMMENTS: *Lactarius volemus* var. *flavus* (p. 161), edible, is nearly identical but has a yellow to orange-yellow cap, cream to pale yellow stalk, and mild-tasting white and unchanging latex that stains all tissues brown. *Lactarius subvelutinus* (not illustrated), edibility unknown, is similar, but it has an ochraceous orange to bright golden orange-brown cap, white latex that is unchanging and does not stain tissues, and narrow gills, and it does not develop a distinctly fishy odor at maturity.

Lactarius xanthydrorheus Singer　　　　　　　　　　　　　　　　　　　　Illus. p. 33
　COMMON NAME: none.
　CAP: ⅜–1" (9–25 mm) wide, broadly convex to nearly plane and sometimes shallowly depressed at maturity; surface dry, subglabrous, papillate, conspicuously rugose-venose over the disc, dingy yellowish olive or more brownish, especially on the disc; margin arched, sometimes short-sulcate.
　FLESH: fragile, white, not changing when cut; odor not distinctive; taste mild.
　LATEX: watery on exposure, instantly changing to yellow, not staining tissues; taste mild.
　GILLS: arcuate-decurrent, narrow, subdistant, cream-colored, unchanging, and not staining when bruised.
　STALK: ½–⅞" (11–22 mm) long, ⅛" (3 mm) thick, nearly equal, dry, glabrous, subglabrous, solid at first, becoming hollow in age, colored like the gills, with a whitish basal mycelium.
　SPORE PRINT COLOR: not stated in the original description.
　MICROSCOPIC FEATURES: spores 7.5–9 × 6–7.5 μm, ellipsoid to broadly ellipsoid, orna-

mented with spines and ridges that form a partial reticulum, prominences up to 1.5 μm high, hyaline, amyloid.

FRUITING: scattered or in groups on the ground under broadleaf trees, especially oak; July–October; Florida west along the Gulf Coastal Plain to Texas; occasional.

EDIBILITY: unknown.

COMMENTS: *Lactarius louisii* (not illustrated), reported only from Pennsylvania, is similar, but it has cream-colored flesh that tastes bitter and scant watery latex that is unchanging, and it grows on buried wood and among sphagnum mosses in wet areas of broadleaf woods that are predominantly oak and maple.

Lentinellus ursinus (Fries) Kühner   Illus. p. 34

COMMON NAME: Bear Lentinus.

CAP: 1–4" (2.5–10 cm) wide, semicircular, convex to nearly flat, distinctly velvety to fuzzy at least at the center, dark brown overall, paler toward the margin.

FLESH: firm, fairly thick, whitish to pale brownish; odor fruity or not distinctive; taste exceedingly acrid.

GILLS: radiating from the point of attachment to the substrate, broad, close to subdistant, color rather variable but typically pinkish to brownish.

STALK: absent.

SPORE PRINT COLOR: white.

MICROSCOPIC FEATURES: spores 3–4.5 × 2–3.5 μm, subglobose to oval, with minute amyloid spines, hyaline.

FRUITING: typically in clusters on decaying broadleaf wood, especially oak, maple, and beech; June–December; widely distributed in the Southeast; occasional.

EDIBILITY: inedible.

COMMENTS: *Lentinellus vulpinus,* inedible, is similar, but each cap in a cluster has a short lateral stalk and is covered with short white hairs over a pale pinkish to pale cinnamon ground color (see *MNE,* p. 297). *Lentinellus cochleatus,* inedible, forms a cluster of shoehorn- to vase-shaped pinkish to cinnamon smooth caps that arise from a single basal stalk (see *MNE,* p. 297). *Lentinellus omphalodes,* inedible, has smooth pinkish to brownish caps with a sunken center and a distinct stalk (see *MNE,* p. 297).

Lentinula raphanica (Murrill) Mata and R. H. Petersen   Illus. p. 34

COMMON NAME: none.

CAP: ¾–3" (2–7.5 cm) wide, convex, becoming broadly convex to nearly plane in age; surface moist or dry, smooth, matted-fibrillose when young, often becoming wrinkled at maturity, especially on the disc; dark reddish brown at first, becoming pale reddish brown to dull ochre or dingy yellow in age; margin entire, with a sterile band of tissue that extends beyond the gills, often adorned with torn patches of partial veil, especially when young.

FLESH: whitish, rubbery; odor of garlic or radish or sometimes not distinctive; taste resembling garlic.

GILLS: attached, very crowded, moderately narrow, white, staining reddish then brown when bruised; edges finely scalloped to serrate.

STALK: ¾–3⅛" (2–8 cm) long, ⅛–⅜" (3–10 mm) thick, enlarged slightly downward or nearly equal, often curved near the base, dry, conspicuously scaly, white to yellowish or ochre; partial veil membranous, white, tearing and remaining attached to the cap margin, typically not forming a ring.

SPORE PRINT COLOR: white.

MICROSCOPIC FEATURES: spores 4.5–6 × 2–3 μm, elliptic, smooth, hyaline; cheilocystidia mostly clavate, rarely sphaeropedunculate, lobed or knobbed at the apex.

FRUITING: scattered or in dense groups on decaying wood, especially of broadleaf trees; July–November; widely distributed in the Southeast, especially along the Gulf Coast; common.

EDIBILITY: unknown.

COMMENTS: Previously known as *Gymnopus alliaceus*. In some field guides, this mushroom has been incorrectly identified as *Lentinus detonsus* = *Lentinula detonsa*, edibility unknown, a species that lacks the garlic odor and has larger spores that measure 6–8 × 3–4 μm. *Lentinula boryana* is a synonym.

Lentinus crinitis (L.: Fries) Fries   Illus. p. 34

COMMON NAME: none.

CAP: 1–3" (2.5–7.5 cm) wide, convex with a depressed center to funnel-shaped; surface dry, with dense radiating hairs, silky shining in the center; hairs pale yellowish brown to dark reddish brown, sometimes with purplish hues over a pale tan ground color; margin incurved at first, becoming somewhat elevated and wavy at maturity, usually adorned with loose, projecting hairs.

FLESH: thin, white to buff, fibrous-tough, leathery; odor and taste not distinctive.

GILLS: deeply decurrent, moderately to densely crowded, narrow, whitish to cream; edges finely serrate.

STALK: ¾–1½" (2–4 cm) long, ⅛–¼" (2–6 mm) thick, nearly equal or sometimes tapered in either direction, dry, solid, scurfy, colored like the cap but usually paler, especially toward the apex.

SPORE PRINT COLOR: white.

MICROSCOPIC FEATURES: spores 5.5–8 × 1.8–3 μm, elliptic, smooth, hyaline, inamyloid.

FRUITING: solitary, scattered, or in groups or clusters on decaying broadleaf wood; June–December, often persisting year-round; widely distributed along the Gulf Coast from Florida to Texas; common.

EDIBILITY: inedible.

COMMENTS: Formerly known as *Panus crinitis*. *Lentinus tigrinus* (not illustrated), edibility unknown, is similar, but its cap has grayish brown to blackish scales on a pale yellowish white ground color that become more conspicuous as the cap expands, and it has slightly larger spores that measure 6–9.5 × 2.5–3.5 μm. *Lentinus suavissimus*, edibility unknown, has a smooth, pale yellow to tawny ochraceous cap, and its flesh has a strongly anise-like odor (see *MNE*, p. 298).

Lentinus levis (Berkeley and Curtis) Murrill   Illus. p. 34

COMMON NAME: none.

CAP: 1–8" (2.5–21 cm) wide, convex-umbilicate to nearly plane or slightly depressed; surface dry, coated with short erect or matted hairs, at least over the disc, becoming nearly smooth in age; white at first, discoloring yellowish to salmon buff in age; margin thin, straight, more or less sulcate, sometimes wavy.

FLESH: up to ¾" (2 cm) thick, firm to tough, white; odor fragrant, citrus-like, resembling grapefruit, or not distinctive; taste somewhat bitter to unpleasant or not distinctive.

GILLS: deeply decurrent, moderately crowded to subdistant, broad, white or colored like the cap; edges entire, not serrate, becoming lacerated in age.

STALK: 1½–6" (4–15.5 cm) long, ⅜–1½" (1–4 cm) thick, typically enlarged downward,

sometimes with a rooting base, eccentric or central, sometimes nearly lateral, dry, solid, white, tomentose or nearly glabrous toward the apex, villose to strigose below.
SPORE PRINT COLOR: pale cream.
MICROSCOPIC FEATURES: spores 9–16 × 4–6 μm, cylindric, smooth, hyaline, inamyloid.
FRUITING: solitary or in groups on decaying tree trunks, logs, and stumps of a variety of broadleaf trees and sometimes conifers; June–December; widely distributed in the Southeast; occasional.
EDIBILITY: inedible.
COMMENTS: Formerly known as *Panus levis.*

Lentinus strigosus (Schweinitz) Fries   Illus. p. 34
COMMON NAME: Hairy Panus, Ruddy Panus.
CAP: 1–3" (2.5–7.5 cm) wide, kidney- to fan-shaped or broadly funnel-shaped, densely hairy and velvety to fuzzy, pinkish tan to reddish brown with purplish tints when young, becoming pale reddish brown to pale ochraceous or tan in age.
FLESH: tough, thin, white; odor not distinctive; taste slightly bitter.
GILLS: decurrent, close, narrow, violaceous to whitish or tan.
STALK: short, stubby, lateral, pinkish tan to light reddish brown.
SPORE PRINT COLOR: white to cream.
MICROSCOPIC FEATURES: spores 4.5–7 × 2.5–3 μm, elliptic, smooth, hyaline, inamyloid.
FRUITING: solitary to clustered on broadleaf wood; May–January; widely distributed in the Southeast; fairly common.
EDIBILITY: inedible.
COMMENTS: Formerly known as *Panus rudis. Lentinus torulosus,* edible but tough, is similar but larger, and its cap is violet to purplish when young but soon fades to pale yellowish brown to tan and sometimes cracks and forms tiny scales in age (see *MNE,* p. 298). *Lentinus lepideus,* has a large whitish to tan cap with brown scales, a partial veil that leaves a ring on the stalk, flesh with an odor of anise, and gill edges that are distinctly serrate (see *MNE,* p. 297).

Lentinus tephroleucus Mont. in Tijds.   Illus. p. 34
COMMON NAME: none.
CAP: ¼–1½" (6–40 mm) wide, convex, deeply umbilicate to infundibuliform; surface dry, appearing velvety, covered with short, erect bundles of hairs that become somewhat scale-like toward the margin; dark purple at first, becoming paler purple to lilac then fading to reddish brown to grayish brown and sometimes yellowish; disc typically glabrous and often finely striate; margin inrolled on young specimens, expanding and becoming nearly plane at maturity.
FLESH: very thin, up to ¹⁄₁₆" (1.5 mm) thick, creamy white, leathery; odor and taste not distinctive.
GILLS: deeply decurrent, moderately distant, narrow, creamy white, edges entire.
STALK: ¾–4" (2–10 cm) long, ¹⁄₁₆–½" (1.5–12 mm) thick, nearly equal or enlarged at either end, central or eccentric, dry, solid, uniformly velvety to distinctly hairy, colored like the cap.
SPORE PRINT COLOR: white.
MICROSCOPIC FEATURES: spores 6–8 × 2.5–4 μm, oblong-cylindric to lacrymoid, smooth, hyaline, inamyloid.
FRUITING: solitary, scattered, or in groups on the ground arising from a sclerotium and

from decaying wood; March–August; along the Gulf Coast from Florida to Texas; occasional.

EDIBILITY: inedible.

COMMENTS: Formerly known as *Lentinus siparius* and *Panus siparius*.

Lepiota clypeolaria (Fries) Quélet   Illus. p. 35

COMMON NAME: Shaggy-stalked Lepiota.

CAP: 1–2¾" 2.5–7 cm) wide, broadly bell-shaped, becoming convex to nearly plane, typically with a conspicuous darkly colored umbo; surface dry, coated with ochraceous to brown or reddish brown scales concentrated on the disc over a whitish ground color; margin paler than the disc, ragged with remnants of a partial veil.

FLESH: thin, white; odor and taste slightly unpleasant or not distinctive.

GILLS: free from the stalk, close, narrow, whitish.

STALK: 1½–4½" (4–11.5 cm) long, ⅛–⅜" (3–10 mm) thick, nearly equal or slightly enlarged downward, dry, hollow at least on the upper portion, covered with shaggy, cottony fibers or scales from the base up to a poorly defined ring, smooth above the ring zone, white on the upper portion, yellowish brown toward the base; partial veil fibrous, white, leaving an evanescent fragile ring that is often integrated with the fibers and scales on the stalk.

SPORE PRINT COLOR: white.

MICROSCOPIC FEATURES: spores 14–20 × 4–6 μm, spindle-shaped, smooth, hyaline, dextrinoid.

FRUITING: solitary, scattered, or in small groups on the ground in conifer and broadleaf woods; June–December; widely distributed in the Southeast; occasional to locally common.

EDIBILITY: poisonous.

COMMENTS: The Shaggy-stalked Lepiota is variably colored and may represent a complex of closely related species or varieties. Because some small *Lepiota* species contain deadly toxins, none should be eaten. *Lepiota acutesquamosa*, reported to be edible, is similar, but it has pyramidal erect scales on its cap, a web-like partial veil, and a nearly smooth stalk (see *MNE*, p. 298). Compare *Lepiota clypeolaria* with *Lepiota cristata* (p. 166), which has a smooth stalk with a persistent membranous ring and smaller wedge-shaped spores with a small corner spur. Also compare with *Macrolepiota subrachodes* (p. 173), which has a relatively smooth stalk with a bulbous base that bruises reddish orange and a thick, movable ring.

Lepiota cristata (Alb. and Schwein. : Fr.) Kummer   Illus. p. 35

COMMON NAME: Malodorous Lepiota.

CAP: ⅜–2¾" (1–7 cm) wide, convex, becoming broadly convex to nearly plane, sometimes with an umbo; surface dry, smooth at first, soon breaking up into concentric rings of tiny reddish brown scales, reddish brown to yellowish brown over the disc, paler toward the margin, with white flesh showing between the scales.

FLESH: white; odor unpleasant to spicy or absent; taste not distinctive.

GILLS: nearly free, close, narrow, white; edges even to somewhat toothed.

STALK: ¾–3⅛" (2–8 cm) long, ⅛–¼" (2–6 mm) thick, nearly equal, sometimes bulbous at the base, hollow in age, smooth, white to pale pinkish buff, often brownish at the base; partial veil membranous, white, leaving a persistent, small, white superior ring.

SPORE PRINT COLOR: white or green.

MICROSCOPIC FEATURES: spores 5.5–8 × 3–4.5 μm, wedge-shaped with a small corner spur, smooth, hyaline, dextrinoid.

FRUITING: scattered or in groups on the ground in woods and grassy areas; June–December; widely distributed in the Southeast; fairly common.

EDIBILITY: unknown.

COMMENTS: *Lepiota acutesquamosa*, reported to be edible, is a much larger mushroom that has a whitish cap with small, erect, pointed, reddish brown or cinnamon brown scales (see *MNE*, p. 298).

Lepiota meleagris (Sow.) Quelét   Illus. p. 35

COMMON NAME: none.

CAP: ¾–2⅛" (2–5.5 cm) wide, ovate to hemispherical at first, becoming broadly convex at maturity, often with a low and broad umbo; surface moist when fresh, dark brown to cinnamon brown over the disc, with small dark brown scales scattered toward the margin and sometimes in concentric rings over a whitish ground color, sometimes with pinkish brown tones near the disc; margin striate, often split in age.

FLESH: white, staining orange-red to dull red then slowly dull brown, somewhat fragile; odor and taste not distinctive.

GILLS: free, fairly crowded, narrow, white to yellowish, often stained pinkish brown to reddish brown in older specimens; edges smooth, entire, sometimes with reddish drops adhering on young fresh specimens.

STALK: 1¾–3¼" (4.5–8.3 cm) long, ⅛–½" (0.4–1.2 cm) thick, enlarged downward and spindle-shaped with a tapered base, anchored by a mass of branching white to grayish white mycelial strands, hollow, becoming stuffed at maturity, whitish on the upper portion, coated with fine dark brown to cinnamon scales, near the base colored like the cap disc; partial veil membranous, whitish, delicate, leaving a whitish to brown somewhat persistent nonmovable ring near the center of the stalk.

SPORE PRINT COLOR: white.

MICROSCOPIC FEATURES: spores 8–12 × 5.5–8.5 μm, broadly elliptic to ovate, smooth, hyaline, dextrinoid.

FRUITING: typically caespitose, sometimes in scattered groups or solitary on decaying leaf litter and compost piles; June–December; reported from North Carolina to Florida, west to Texas, also reported from Illinois; occasional to locally fairly common.

EDIBILITY: unknown.

COMMENTS: Also known as *Leucocoprinus meleagris* and *Leucoagaricus meleagris*. The typically caespitose growth habit, relatively small size of the fruit bodies, white mycelial strands at the stalk base, and the small blackish brown granular scales on the cap are a distinctive combination of features.

Lepiota phaeostictiformis Murrill   Illus. p. 35

COMMON NAME: none.

CAP: ⅜–1⅛" (1–3 cm) wide, convex, becoming nearly plane, with a pointed umbo on the disc; surface dry, covered with blackish scales over a white ground color, disk blackish; margin entire, even.

FLESH: thin, white, unchanging when exposed; odor not distinctive; taste not reported.

GILLS: free from the stalk, crowded, rather broad, finely fringed, white, not staining when bruised.

STALK: 1⅛–2⅜" (3–6 cm) long, 1/16–⅛" (1.5–3 mm) thick, tapered upward from a somewhat clavate base, finely pruinose, white; partial veil membranous, white, leaving a persistent superior double ring with a fringed margin.

SPORE PRINT COLOR: white.

MICROSCOPIC FEATURES: spores 5–8 × 3–4 μm, narrowly ellipsoid to slightly ovoid, smooth, hyaline.

FRUITING: scattered or in groups or clusters on decaying wood and leaf litter; July–November, reported from Florida, distribution limits yet to be determined; occasional.

EDIBILITY: unknown.

COMMENTS: This species name may be a synonym for *Lepiota phaeosticta*.

Leucoagaricus americanus (Peck) Singer   Illus. p. 35

COMMON NAME: Reddening Lepiota.

CAP: 1⅛–6" (3–15 cm) wide, narrowly convex at first, becoming convex and finally nearly plane in age, typically with a low and broad central umbo; surface dry, dull reddish brown overall when young, breaking up into concentric rings of dull reddish brown scales on a white ground color at maturity; white ground color bruising yellow and then slowly reddish brown; margin incurved at first, uplifted at maturity, finely striate.

FLESH: thin, soft, fragile, white, bruising yellow and then slowly reddish brown; odor and taste not distinctive.

GILLS: free, closely spaced, fragile, white, bruising like the cap and flesh.

STALK: 2–5½" (5–14 cm) long, ¼–⅞" (6–22 mm) thick, tapered both above and at the base or enlarged downward, dry, hollow in age, white, bruising and aging dull reddish; cut flesh often yellow to orange, especially at the base; partial veil white, membranous, typically leaving a double-edged, white superior ring.

SPORE PRINT COLOR: white.

MICROSCOPIC FEATURES: spores 8–14 × 5–10 μm, elliptic with a thick-walled pore, smooth, hyaline, dextrinoid.

FRUITING: scattered or in clusters near stumps, among wood chips, on compost piles, in sawdust, in gardens, and on lawns; June–November; widely distributed in the Southeast; occasional.

EDIBILITY: edible with caution; be certain of the identification because other similar species may be toxic.

COMMENTS: Formerly known as *Lepiota americana*. The Shaggy Parasol, *Macrolepiota rachodes*, edible, has very coarse cap scales; white stalk flesh that stains yellow-orange or saffron, then reddish brown when cut; and a stalk that is enlarged downward, often has a bulbous base, and is white near the apex and brown below (see *MNE*, p. 300). *Lepiota besseyi* (see photo, p. 35), edibility unknown, is similar and also stains reddish when bruised, but its cap has smaller scales; its stalk has small reddish brown scales at least on the lower half and is not typically enlarged in the middle; and it grows on lawns and mulches composed of wood and bark chips. *Lepiota cortinarius* (see photo, p. 35), edibility unknown, is somewhat similar but has a web-like partial veil that leaves an evanescent ring on the stalk.

Leucoagaricus hortensis (Murrill) Pegler   Illus. p. 35

COMMON NAME: none.

CAP: 1½–4" (4–10 cm) wide, ovoid when young, becoming convex with a persistent umbo; surface dry, fibrillose to squamulose on a white ground color, tawny olive to light brown on the disk, becoming dingy cream-colored to light buff outward toward the margin; squamules typically concentrically arranged; margin thick and rounded, striate.

FLESH: thin, soft, whitish, discoloring vinaceous brown on exposure; odor and taste not distinctive.

GILLS: attached when young, becoming free at maturity, rather crowded; edges entire.

STALK: 1½–2¾" (4–7 cm) long, ⅛–¼" (3–6 mm) thick, dry, solid then hollow, nearly equal or enlarged downward, white, smooth and glabrous above the ring, brownish and slightly fibrillose below; partial veil submembranous, brownish, fragile, leaving a median ring.

SPORE PRINT COLOR: white.

MICROSCOPIC FEATURES: spores 8–10 × 6–7 μm, ellipsoid, smooth, hyaline, dextrinoid.

FRUITING: scattered or in groups or clusters on the ground in grassy areas and in gardens; September–January; reported from Alabama and Florida; occasional.

EDIBILITY: unknown.

COMMENTS: Formerly known as *Lepiota humei*. Although not mentioned in Murrill's original description, Pegler, in his monograph on the genus *Lentinus*, states that the stalk immediately stains bright pink or dark vinaceous red when bruised.

Leucoagaricus viridiflavoides Akers and Angels — Not Illustrated

COMMON NAME: none.

CAP: ⅝–1⅜" (1.6–3.5 cm) wide, convex, becoming nearly plane in age, obtusely umbonate; surface glabrous or nearly so, subviscid when moist, color highly variable, pinkish brown to dull sooty brown, dull violaceous brown, greenish brown, gray, or blackish, darkest on the disk, paler toward the margin, staining dark bluish green when bruised; margin nonstriate to occasionally minutely striate, often splitting radially, sometimes appendiculate.

FLESH: sulfur yellow, staining dark bluish green when cut, drying greenish gray, sometimes with a yellowish tint; odor pleasant; taste not determined.

GILLS: free, broad, close to subdistant, pale sulfur yellow, staining dark bluish green when bruised, drying yellow with greenish gray edges.

STALK: ¾–1¾" (1.8–4.5 cm) long, up to ⅛" (3 mm) thick, nearly equal, glabrous, hollow, sulfur yellow, staining dark bluish green where handled; partial veil membranous, sulfur yellow, leaving marginal remnants or a poorly formed superior ring or both, occasionally evanescent.

SPORE PRINT COLOR: pale yellow.

MICROSCOPIC FEATURES: spores 5.3–11(–13) × 3.4–5.1 (to 6) μm, ellipsoidal to ovoid or amygdaliform to arachiform, smooth, hyaline, hilar appendix prominent, germ pore indistinct to conspicuous, sometimes truncate.

FRUITING: scattered or in groups on the ground in a hammock of an old sinkhole containing water elms; July–August; reported only from the San Felasco Hammock State Preserve, Florida; rare.

EDIBILITY: unknown.

COMMENTS: The sulfur yellow flesh and stalk and the dark blue staining reactions are a most distinctive combination. This mushroom is related to *Lepiota viridiflava* (not illustrated), edibility unknown, described from Sri Lanka.

Leucocoprinus birnbaumii (Corda) Singer — Illus. p. 36

COMMON NAME: Lemon-yellow Lepiota.

CAP: ¾–2⅜" (2–6 cm) wide, campanulate to broadly conic, typically with an umbo;

surface dry, powdery or with tiny scales, bright to pale yellow, becoming pale yellow to whitish in age; margin distinctly striate to the disc.

FLESH: very thin, whitish to pale yellow; odor and taste not distinctive.

GILLS: free, crowded, whitish to pale yellow; edges fibrillose.

STALK: 1½–4½" (4–11.5 cm) long, 1/16–¼" (1.5–6 mm) thick, enlarged downward or sometimes nearly equal, dry, smooth or powdery, colored like the cap; partial veil fibrous-cottony, bright yellow, leaving a movable persistent or evanescent ring.

SPORE PRINT COLOR: white.

MICROSCOPIC FEATURES: spores 8–13 × 5–8 µm, elliptic, smooth, thick-walled, hyaline, dextrinoid.

FRUITING: scattered or in groups or clusters on rich soils, among wood chips used for landscaping, in greenhouses, and on soil in potted plants; June–December, year-round indoors; widely distributed in the Southeast; common.

EDIBILITY: poisonous.

COMMENTS: This mushroom was formerly known as *Lepiota lutea*. Compare with *Leucocoprinus fragilissimus* (p. 170).

Leucocoprinus cepaestipes (Sowerby : Fries) Pat. Illus. p. 36

COMMON NAME: Onion-stalk Lepiota.

CAP: ¾–2½" (2–7 cm) wide, ovoid at first, becoming bell-shaped to broadly convex with an umbo; surface dry, granular to scaly, usually smooth on the disc, white to whitish overall or sometimes with a pinkish tan to grayish brown disc, sometimes yellowish in age; margin with deep radial grooves, sometimes splitting.

FLESH: thin, fragile, white; odor not distinctive; taste bitter or not distinctive.

GILLS: free from the stalk, crowded, broad, white, becoming dingy white in age; edges floccose.

STALK: 1½–4½" (4–12 cm) long, ⅛–¼" (3–6 mm) thick, slightly enlarged downward or sometimes nearly equal, often with a swollen base, dry, hollow, white or sometimes with pinkish tints, often bruising yellowish, coated with powdery white scales at least on the lower half, often smooth above the ring; partial veil membranous, white, leaving a white membranous ring on the upper stalk.

SPORE PRINT COLOR: white.

MICROSCOPIC FEATURES: spores 7–10 × 6–8 µm, broadly elliptic with a germ pore, smooth, hyaline, weakly dextrinoid.

FRUITING: usually in groups or clusters but sometimes scattered on wood mulch, straw, organic compost, and rich soil; sometimes grows in greenhouses and flowerpots.

EDIBILITY: possibly poisonous.

COMMENTS: Formerly known as *Lepiota cepaestipes*. *Leucocoprinus lilacinogranulosus* (see photo, p. 36), edibility unknown, is similar, and its cap margin has deep radial grooves, but its cap disc is dull purple to purplish brown, and the cap surface is covered with small dull purple to purplish brown scales at maturity. Compare with *Leucocoprinus birnbaumii* = *Lepiota lutea* (p. 169), which is similar but has a yellow cap and stalk.

Leucocoprinus fragilissimus (Ravenel in Berk. and Curtis) Patouillard Illus. p. 36

COMMON NAME: Fragile Leucocoprinus.

CAP: ⅜–1¾" (1–4.5 cm) wide, ovate to bell-shaped when young, becoming broadly convex with a small umbo; surface translucent-striate to sulcate from the margin to the disc, tawny to ochre-brown on the umbo, pale bright yellow over the remainder or pale yellowish white on the margin, fading in age; margin incurved at first, becoming uplifted in age.

FLESH: very thin, whitish; odor and taste not distinctive.

GILLS: attached, becoming nearly free, narrow, white.

STALK: 1½–6" (4–15.5 cm) long, ¹⁄₁₆–⅛" (1–3 mm) thick, enlarged downward, extremely fragile, pale grayish yellow, darkening in age, coated with tiny yellowish scales; partial veil cottony, whitish to pale yellow, leaving a superior ring.

SPORE PRINT COLOR: white.

MICROSCOPIC FEATURES: spores 9–13 × 7–8 μm, elliptic with an apical pore, smooth, hyaline, dextrinoid; cheilocystidia 12–25 × 9–15 μm; pleurocystidia absent.

FRUITING: solitary or scattered on the ground in mixed woods; July–November; widely distributed in the Southeast; fairly common.

EDIBILITY: unknown.

COMMENTS: *Leucocoprinus magnicystidiosus* (not illustrated), edibility unknown, is very similar but has a darker brown disc and very large cheilocystidia and pleurocystidia. It has been reported only from Tennessee. Compare with *Leucocoprinus birnbaumii* (p. 169).

Limacella illinita var. **illinita** (Fries) Earle   Illus. p. 36

COMMON NAME: none.

CAP: ¾–2¾" (2–7 cm) wide, convex, becoming nearly plane in age; surface coated with hyaline slime when fresh, smooth, white to creamy; margin sometimes dripping slime during rainy periods.

FLESH: white; odor and taste not distinctive.

GILLS: free or nearly so, close, broad, white.

STALK: 2–4" (5–10 cm) long, ⅛–⅜" (3–10 mm) thick, slimy, whitish, staining brownish where bruised, typically lacking a ring or sometimes with a sparse fibrillose annular zone.

SPORE PRINT COLOR: white.

MICROSCOPIC FEATURES: spores 4–6.5 μm, globose to broadly elliptic, smooth, hyaline, inamyloid.

FRUITING: scattered or in groups in various habitats, including woods, fields, and sand dunes; July–January; widely distributed in the Southeast; rare to locally common.

EDIBILITY: unknown.

COMMENTS: *Limacella illinita* var. *argillacea* (not illustrated), edibility unknown, is nearly identical but has a grayish brown cap when young. *Limacella glioderma*, edibility unknown, has a dark reddish brown to brown slimy cap, a light brown stalk, and a strongly farinaceous odor, especially when its flesh is crushed (see *MNE*, p. 299).

Macrocybe titans (Bigelow and Kimbrough) Pegler, Lodge, and Nakasone   Illus. p. 37

COMMON NAME: none.

CAP: 3⅛–32" (8–80 cm) wide, convex, becoming nearly plane in age; surface dry or moist, often with watery spots or small scales on the disc, glabrous elsewhere, warm buff to creamy buff near the margin, darker on the disc, darkening somewhat in age; margin incurved and remaining so well into maturity, often conspicuously undulating and sometimes upturned, disc sometimes depressed in age.

FLESH: white; odor variously described as resembling cyanide or green pecans; taste either musty and somewhat disagreeable or not distinctive.

GILLS: attached and strongly sinuate, crowded, grayish buff to pale grayish yellow, sometimes with brownish stains.

STALK: 3–10" (7.5–25.5 cm) long, 1–3⅛" (2.5–8 cm) thick, clavate to bulbous, usually tapered to a point at the base when specimens growing in clusters, dry, solid, fibrous-tough, whitish at first then becoming dingy buff, darkening when bruised or handled, surface squamulose or with concentric scabrous rings in age.

SPORE PRINT COLOR: white to creamy white.

MICROSCOPIC FEATURES: spores 5.5–8 × 4–5.5 μm, broadly ellipsoid to ovoid, smooth, inamyloid.

FRUITING: typically in dense clusters, sometimes solitary or in groups on the ground in woodlands, grassy areas, and disturbed areas, including near the foundations of houses; July–September; reported only from Florida; uncommon.

EDIBILITY: reported to be edible when thoroughly cooked (see Comments).

COMMENTS: Formerly known as *Tricholoma titans*. The genus *Macrocybe* is tropical and subtropical, with only one species known to occur in North America. In his recent book *Common Florida Mushrooms* (2000), Kimbrough states that this species is tough and "should be cooked longer than normal in a well ventilated room, and that the water should be poured off."

Macrolepiota procera (Scopoli) Singer Illus. p. 37

COMMON NAME: Parasol.

CAP: 2¾–10" (7–25 cm) wide, egg-shaped when young, becoming bell-shaped then broadly convex to nearly flat in age, with an umbo; surface dry, reddish brown and nearly smooth when young, soon breaking up into coarse reddish brown scales and patches with white flesh showing between, fading to dull pale brown in age; margin incurved at first, often wrinkled on mature specimens.

FLESH: thick, white, sometimes slightly tinged reddish, not staining yellow-orange when cut or bruised; odor and taste not distinctive.

GILLS: free, close, broad, white at first, darkening in age; edges somewhat woolly.

STALK: 6–12" (15.5–30 cm) long, ⅜–½" (1–1.3 cm) thick, enlarged downward to a bulbous base, dry, coated with tiny brownish scales on a white ground color; partial veil white, membranous, leaving a persistent, thick-edged, movable superior ring.

SPORE PRINT COLOR: white.

MICROSCOPIC FEATURES: spores 15–20 × 10–13 μm, broadly elliptic, with a large apical pore, smooth, hyaline, dextrinoid.

FRUITING: solitary, scattered, or in groups on the ground in woodlands and grassy areas; July–December; widely distributed in the Southeast; fairly common.

EDIBILITY: edible.

COMMENTS: *Macrolepiota gracilenta* (not illustrated), edible, is similar but smaller, and it has a thin funnel-shaped ring and spores that measure 10–13 × 7–8 μm. *Macrolepiota prominens* (not illustrated), edible, is also similar, but it has a white cap with scattered tiny brown scales, a thick double-edged ring, and spores that measure 9–10 × 6–7 μm. The Shaggy Parasol, *Macrolepiota rachodes,* edible, has coarse cinnamon brown to pinkish brown cap scales, a darker brown stalk, white cap flesh that stains orange then reddish brown when cut or bruised, white stalk flesh that stains yellow-orange or saffron when cut or bruised, and spores that measure 6–9.5 × 6–7 μm (see *MNE*, p. 300). Compare also with *Chlorophyllum molybdites* (p. 119), which is poisonous.

Gilled Mushrooms / 173

Macrolepiota subrachodes (Murrill) Akers Illus. p. 37
  COMMON NAME: none.
  CAP: 2½–4" (7–10 cm) wide, more or less globose at first, soon bell-shaped, then becoming nearly flat with an umbo at maturity; surface dry, prominently scaly, with scales concentrated on the disc, a white to cream or pale tan ground color beneath the brown scales.
  FLESH: thick, whitish, staining orange to reddish when bruised or exposed, especially in the stalk base; odor and taste not distinctive.
  GILLS: free from the stalk, close to crowded, broad, white, becoming brownish in age.
  STALK: 2½–5" (6.5–13 cm) long, up to ⅝" (1.5 cm) thick, slender, tapering upward from a bulbous base, dry, hollow in age, pale grayish with a pinkish tint, staining brownish when bruised; partial veil thick, white, membranous, leaving a prominent, movable, double-edged superior ring.
  SPORE PRINT COLOR: white.
  MICROSCOPIC FEATURES: spores 7.2–10.6 × 5–6.3 μm, ovoid to almond-shaped with a distinct germ pore, smooth, dextrinoid.
  FRUITING: solitary to scattered in woods, usually associated with oaks; widely distributed in the Southeast; occasional.
  EDIBILITY: edible with caution; correct identification is essential so as not to confuse it with smaller species of *Lepiota*, some of which contain dangerous toxins.
  COMMENTS: The Shaggy Parasol, *Macrolepiota rachodes,* edible, has coarse cinnamon brown to pinkish brown cap scales, a darker brown stalk, white cap flesh that stains orange then reddish brown when cut or bruised, white stalk flesh that stains yellow-orange or saffron when cut or bruised, and spores that measure 6–9.5 × 6–7 μm (see *MNE,* p. 300). Compare *Macrolepiota subrachodes* with *Leucoagaricus americanus* (p. 168), which is similar but has larger spores and reddish brown cap scales, and which often grows in wood chips and sawdust piles. Also compare with *Macrolepiota procera* (p. 172), which is larger overall and whose flesh does not stain orangish when bruised. *Chlorophyllum molybdites* (p. 119), poisonous, has a more robust stature, grows in grassy places, has gills that turn dull green when mature, and produces a grayish green spore print.

Marasmiellus albuscorticis (Secr.) Singer Illus. p. 37
  COMMON NAME: White Marasmius.
  CAP: ⅜–1¼" (1–3 cm) wide, convex, becoming nearly plane in age; surface dry, minutely hairy, conspicuously pleated, shiny white, often forming reddish spots in age; margin thin, wavy, grooved.
  FLESH: thin, whitish, pliant; odor and taste not distinctive.
  GILLS: attached to decurrent, distant, narrow to moderately broad, white to pinkish, with conspicuous crossveins.
  STALK: ⅜–¾" (1–2 cm) long, 1/32–1/16" (1–1.5 mm) thick, nearly equal down to a somewhat enlarged base, central or eccentric, typically curved, dry, solid, shiny white, darkening to blackish from the base upward in age.
  SPORE PRINT COLOR: white.
  MICROSCOPIC FEATURES: spores 10–15 × 5–6 μm, narrowly drop-shaped, smooth, hyaline.
  FRUITING: in dense groups, in rows, or scattered on decaying branches, twigs, and stems of broadleaf trees, shrubs, and other woody plants; June–November; widely distributed in the Southeast; occasional.
  EDIBILITY: unknown.

COMMENTS: Formerly known as *Marasmius candidus* and *Marasmius magnisporus*. The Black-footed Marasmius, *Marasmiellus nigripes*, edibility unknown, is similar, but it has a longer black stalk with a coating of minute white hairs (use a hand lens) and triangular spores that measure 7–9 μm (see *MNE*, p. 300). The Pointed-stalked Marasmiellus, *Marasmiellus praeacutus*, edibility unknown, has a reddish brown cap, a reddish brown stalk with a white pointed base, and lacrymoid to narrowly elliptic spores that measure 5.5–8.5 × 2.5–3.5 μm (see *MNE*, p. 300).

Marasmius cohaerens (Pers.: Fr.) Cooke and Quélet   Illus. p. 37

COMMON NAME: Fused Marasmius.

CAP: ⅜–1½" (1–4 cm) wide, broadly conical to bell-shaped, becoming flat or depressed in the center, often with an umbo; surface dry, velvety, smooth to slightly wrinkled, yellow-brown to pale cinnamon or reddish brown, darker in the center; margin somewhat striate when moist.

FLESH: thin, white; odor not distinctive or slightly pungent; taste mild or slightly bitter.

GILLS: attached to nearly free, subdistant, fairly broad, dull white to yellowish white or pale brownish with slightly darker edges.

STALK: 1–3" (2.5–7.5 cm) long, 1/32–⅛" (1–3 mm) thick, nearly equal, dry, smooth, shiny, hollow, fibrous-tough, white to pale yellowish above, becoming darker reddish to nearly blackish brown toward the base.

SPORE PRINT COLOR: white.

MICROSCOPIC FEATURES: spores 7–10 × 3–5.5 μm, broadly elliptic, smooth, hyaline.

FRUITING: typically in clusters but occasionally solitary on decaying leaves, twigs, and other woody debris of broadleaf trees; March–December; widely distributed in the Southeast; fairly common.

EDIBILITY: inedible.

COMMENTS: *Marasmius scorodonius* (not illustrated), inedible, is similar but has a pronounced garlic odor. Compare with *Collybia acervata* = *Gymnopus acervatus*, edibility unknown, which is also similar but has a thicker, more uniformly reddish brown stalk and grows in dense clusters on decaying conifer wood, among sphagnum mosses, and on rich humus (see *MNE*, p. 278).

Marasmius fulvoferrugineus Gilliam   Illus. p. 38

COMMON NAME: none.

CAP: ¾–1¾" (2–4.5 cm) wide, cushion- to bell-shaped at first, becoming convex, often with a small umbo on the disc; surface dry, dull, slightly velvety, radially grooved, tawny brown to rusty brown; margin deeply pleated.

FLESH: thin, tough; odor and taste mildly farinaceous or not distinctive.

GILLS: attached and notched or sinuate to nearly free from the stalk, distant, yellowish white.

STALK: 1–2½" (2.5–6.5 cm) long, up to 1/16" (1.5 mm) thick, slender, round, smooth, shiny, cartilaginous, hollow at maturity, pinkish near the apex, brown to blackish brown on the lower portion, with a tuft of white mycelium at the base.

SPORE PRINT COLOR: white.

MICROSCOPIC FEATURES: spores 15–18 × 3–4.5 μm, oblanceolate, curved-clavate or fusoid-clavate, smooth, hyaline.

FRUITING: in small groups or gregarious on decaying leaves and humus in woods; June–December; widely distributed in the Southeast; locally common.

EDIBILITY: unknown.

COMMENTS: Easily confused with *Marasmius siccus*, edibility unknown, which is typically smaller and more orange (see *MNE*, p. 301). *Marasmius pulcherripes* (not illustrated), edibility unknown, is smaller and more delicate, and it has a reddish to pinkish brown cap.

Marasmius rotula (Scopoli : Fries) Fries Illus. p. 38
COMMON NAME: Pinwheel Marasmius.
CAP: ⅛–¾" (3–20 mm) wide, convex with a small depression at the center, prominently sulcate; white to yellowish or rarely pale orange, typically with a dark center.
FLESH: very thin at the edge, visibly white to yellowish at the center; odor not distinctive, taste mild or with a slight bitter aftertaste.
GILLS: attached to a collarium, which may be either free from or collapsed onto the upper stalk; distant or nearly so, narrow to rather broad, nearly white to pale yellow.
STALK: ⅝–3½" (1.5–8.5 cm) long, less than 1/32" (0.3–1.0 mm) thick, yellowish white to brownish at the apex, reddish brown to dark brown to blackish brown overall at maturity, sometimes with a tiny bulb at the base; rhizomorphs usually present but minimal, brown to blackish brown, typically wiry, curled or not.
SPORE PRINT COLOR: white.
MICROSCOPIC FEATURES: spores 6–9.5 × 3–4.5 μm, more or less narrowly teardrop-shaped, smooth, hyaline, inamyloid.
FRUITING: usually growing in clusters or dense colonies on decaying wood; year-round; very common.
EDIBILITY: inedible.
COMMENTS: *Marasmius capillaris* (not illustrated), edibility unknown, is very similar, but it grows on oak leaves and is never clustered.

Melanoleuca melaleuca (Pers.: Fr.) Murrill Illus. p. 38
COMMON NAME: Changeable Melanoleuca.
CAP: 1–3" (2.5–7.5 cm) wide, broadly convex to flat, usually with an umbo; surface smooth, moist, smoky brown to dark brown, becoming paler on drying; margin upturned and wavy in age.
FLESH: thin, whitish to grayish or ochraceous near the stalk base; odor not distinctive; taste slightly disagreeable or not distinctive.
GILLS: attached, crowded, broad, white to cream.
STALK: 1–3" (2.5–7.5 cm) long, ⅛–⅜" (3–10 mm) thick, slender, nearly equal or enlarged toward the base, dry, solid, smooth, sometimes slightly scurfy at the apex, whitish, coated with brownish longitudinal fibrils.
SPORE PRINT COLOR: white to pale cream.
MICROSCOPIC FEATURES: spores 6–8.5 × 4–4.5 μm, elliptic, minutely roughened, hyaline, amyloid.
FRUITING: solitary, scattered, or in groups on the ground in grassy areas, open woods, and disturbed ground; June–December; widely distributed in the Southeast; fairly common.
EDIBILITY: reported to be edible.
COMMENTS: Species of *Melanoleuca* typically have stalks that are slender compared to the diameter of their caps. *Melanoleuca alboflavida*, edible, is taller and more robust, with a whitish to yellowish brown cap (see *MNE*, p. 301). *Tricholoma* species typically have a more stocky stature and inamyloid spores. Some *Gymnopus* and *Rhodocollybia* species are similar in stature, but they have inamyloid spores.

Mycena epipterygia var. **epipterygioides** (Pearson) Kühner   Illus. p. 38
- COMMON NAME: none.
- CAP: ⅜–1" (1–2.5 cm) wide, ovoid to obtusely conic, becoming broadly conic to convex or nearly plane in age, often with a low and broad umbo or flattened disc; surface pruinose or covered with a hoary sheen at first, becoming bald at maturity, viscid, sulcate-striate in age, olive gray to olive buff, not fading to whitish in age.
- FLESH: dark olive brownish; odor and taste farinaceous.
- GILLS: bluntly attached and often forming a collar, subdistant to close, pale grayish to whitish, sometimes developing reddish brown spots in age, with two tiers of attenuate lamellulae.
- STALK: 1½–3⅛" (4–8 cm) long, 1/16–⅛" (1.5–3 mm) thick, nearly equal, viscid, pruinose at least near the apex and smooth below, pale greenish yellow, base often coated with scattered fibrils and becoming reddish in age.
- SPORE PRINT COLOR: white.
- MICROSCOPIC FEATURES: spores 8–12 × 4–6 μm, ellipsoid, smooth, hyaline, amyloid.
- FRUITING: scattered or in groups among mosses and on humus in pine and oak woods; August–December; widely distributed in the Southeast; occasional to fairly common.
- EDIBILITY: unknown.
- COMMENTS: Compare with *Mycena epipterygia* var. *viscosa* (p. 176).

Mycena epipterygia var. **viscosa** (Maire) Ricken   Illus. p. 38
- COMMON NAME: none.
- CAP: 5/16–⅜" (8–10 mm) wide, cuticle beneath powdery layer yellowish or yellowish gray or greenish gray, becoming brownish in age; distinctly coated with a fine white powdery layer at first.
- FLESH: thin, yellowish to grayish; odor strongly farinaceous to pumpkin-like; taste strongly rancid-farinaceous and very unpleasant.
- GILLS: yellowish to greenish yellow or sometimes whitish, often developing reddish brown spots; edges pale.
- STALK: 1¼–3" (3–8 cm) long, 1/32–⅛" (1–3 mm) thick, lemon yellow to greenish yellow, becoming reddish near the base; distinctly white pruinose at first.
- SPORE PRINT COLOR: white.
- MICROSCOPIC FEATURES: spores 8–11 × 5–8 μm, teardrop-shaped to elliptic, smooth, hyaline, amyloid.
- FRUITING: scattered or in groups or clusters on conifer wood, often at the base of standing trees; July–December; widely distributed in the Southeast; occasional to fairly common.
- EDIBILITY: unknown.
- COMMENTS: Compare with *Mycena epipterygia* var. *epipterygioides* (p. 176).

Mycena haematopus (Pers.) Quélet   Illus p. 38
- COMMON NAME: Bleeding Mycena.
- CAP: ⅜–1½" (1–4 cm) wide, oval at first, becoming bell-shaped to convex, typically with an umbo; surface pruinose when young and dry, smooth and shiny when moist, reddish brown or grayish brown to pinkish beige, paler toward the margin; margin toothed and extending beyond the gills, striate.
- FLESH: thin, fragile, watery, reddish brown, exuding blood-like reddish brown juice when broken; odor not distinctive; taste mild or bitter.

GILLS: attached to sinuate, close to subdistant, narrow to moderately broad, pinkish white to pale vinaceous.

STALK: 1½–4" (4–10 cm) long, 1/16–3/16" (1.5–4.5 mm) thick, nearly equal, pruinose above at first, soon glabrous, hollow, fragile; when fresh exuding a reddish brown juice if broken, colored more or less like the cap or darker.

SPORE PRINT COLOR: white.

MICROSCOPIC FEATURES: spores 8–11 × 5–7 μm, oval to elliptical with a small apiculus, smooth, hyaline, anyloid.

FRUITING: usually in clusters on wood of broadleaf trees; March–November; widely distributed in the Southeast; common.

EDIBILITY: unknown.

COMMENTS: This handsome mushroom is easily recognized by the colored juice that seeps out when the flesh is broken, though this feature may not be apparent on older, dryer specimens. *Mycena sanguinolenta* (not illustrated), edibility unknown, is similar but smaller overall, has vinaceous brown gill edges, and typically grows on soil or humus. The Bleeding Mycena is often infected with *Spinellus fusiger*, a parasitic mold that forms long, projecting hairs on the cap (see *MNE,* p. 302).

Mycena luteopallens (Peck) Saccardo   Illus. p. 39

COMMON NAME: Walnut Mycena.

CAP: ¼–⅝" (8–15 mm) wide, oval, becoming convex to nearly flat, sometimes with a slight umbo in age; surface smooth and glabrous except when very young, brilliant orange-yellow to rich yellow, fading as it dries to yellowish white; margin translucent-striate in age.

FLESH: thin, yellowish to whitish; odor and taste not distinctive.

GILLS: variously attached, subdistant, fairly broad, yellowish to slightly pinkish, with whitish edges.

STALK: 2–4" (5–10 cm) long, 1/32–1/16" (1–2 mm) thick, hollow, base with strigose hairs; orangish near the apex, yellowish below.

SPORE PRINT COLOR: white.

MICROSCOPIC FEATURES: spores 7–9 × 4–5.5 μm, more or less oval with an eccentric apiculus, very slightly roughened, hyaline, lightly amyloid.

FRUITING: in small groups growing from decaying walnut or hickory nut shells; July–November; widely distributed in the Southeast; occasional to common.

EDIBILITY: inedible.

COMMENTS: This mushroom is easily identified if one notes the particular substrate, which is sometimes buried.

Mycena pura (Fries) Quélet   Illus. p. 39

COMMON NAME: Pink Mycena.

CAP: ¾–3" (2–7.5 cm) wide, convex to more or less flat; surface glabrous, moist, color quite variable, pink, red, purple, grayish lilac, or sometimes whitish at first, but invariably with some pink to purple tint; very hygrophanous but usually retaining some hint of the original pigments; margin translucent-striate.

FLESH: whitish to purplish or bluish, rather thick at the center, thinning abruptly about halfway toward the margin; distinct and strong radish odor and taste.

GILLS: variously attached, close to subdistant, very broad for the cap size, interveined, whitish to bluish or purplish but often with a strong grayish tint, edges whitish; lamellulae numerous.

STALK: 1–4" (2.5–10 cm) long, 1/16–1/4" (2–6 mm) thick, often flattened or twisted, sometimes enlarged near the base, hollow, tough, glabrous to scaly; colored more or less like the cap or paler, sometimes simply whitish.

SPORE PRINT COLOR: white.

MICROSCOPIC FEATURES: spores 6–10 × 3–3.5 μm, narrowly elliptic to subcylindric with an eccentric apiculus, smooth, hyaline, amyloid.

FRUITING: solitary, scattered, grouped, or gregarious on humus in coniferous or broadleaf woods; April–November; widely distributed in the Southeast; occasional to common.

EDIBILITY: reports vary; often considered to be poisonous.

COMMENTS: This mushroom's remarkable variability in color can complicate its identification; the radish-like odor and taste provide the most important clues.

Nolanea murraii (Berkeley and Curtis) Saccardo   Illus. p. 39

COMMON NAME: Yellow Unicorn Entoloma.

CAP: 3/8–1 1/4" (1–3 cm) wide, bell-shaped to conic at first, becoming more convex but retaining a small pointed umbo in most specimens; surface moist, silky smooth, yellow to orangish yellow, fading somewhat in age.

FLESH: yellowish; odor and taste pleasant or not distinctive.

GILLS: narrowly attached to the stalk, subdistant, broad, yellow, becoming pinkish tinged in age.

STALK: 2–4" (5–10 cm) long, 1/16–1/4" (2–5 mm) thick, dry, yellow, the base usually slightly coated with white mycelium.

SPORE PRINT COLOR: salmon pink.

MICROSCOPIC FEATURES: spores 9–12 × 8–10 μm, angular, typically 4-sided, appearing somewhat square, hyaline, inamyloid.

FRUITING: scattered or in groups on the ground in woods, especially in swamps or moist areas; June–November; widely distributed in the Southeast; occasional to fairly common.

EDIBILITY: unknown.

COMMENTS: Also known as *Entoloma murraii* and sometimes spelled *murrayi*. Formerly known as *Entoloma cuspidatum*.

Nolanea quadrata Berkeley and Curtis   Illus. p. 39

COMMON NAME: Salmon Unicorn Entoloma.

CAP: 5/8–2" (1.5–5 cm) wide, convex, bell-shaped to acutely conic, with a small pointed umbo in most specimens; surface smooth, moist, salmon orange, fading somewhat in age.

FLESH: orangish; odor and taste not distinctive.

GILLS: slightly attached, subdistant, broad, salmon orange.

STALK: 2–4" (5–10 cm) long, 1/16–1/4" (2–6 mm) thick, fragile, salmon orange, sometimes developing a greenish tinge in age, hollow, fragile, the base coated with white mycelium.

SPORE PRINT COLOR: salmon pink.

MICROSCOPIC FEATURES: spores 10–12 × 10–12 μm, angular, typically 4-sided, appearing somewhat square, hyaline, inamyloid.

FRUITING: scattered on humus and among mosses in woods; June–November; widely distributed in the Southeast; occasional to fairly common.

EDIBILITY: unknown, but believed to be toxic.

COMMENTS: Formerly known as *Entoloma salmoneum*. Some *Hygrocybe* species are similar in stature, but they have white gills and spores.

Nolanea strictia (Peck) Largent   Illus. p. 39
COMMON NAME: Straight-stalked Entoloma.
CAP: ¾–3" (2–7.5 cm) wide, conical, becoming bell-shaped to broadly convex or nearly plane in age, with a central broad umbo; surface smooth, moist, glabrous, pale pinkish brown to pale yellowish brown or pale grayish brown; margin striate when fresh.
FLESH: whitish to buff; odor slightly farinaceous or not distinctive; taste farinaceous or not distinctive.
GILLS: attached or nearly free, moderately close, whitish at first, becoming pinkish to dingy salmon at maturity.
STALK: 2–4" (5–10 cm) long, ⅛–⅜" (3–10 mm) thick, nearly equal, longitudinally twisted-striate, whitish, with a white basal mycelium.
SPORE PRINT COLOR: salmon pink.
MICROSCOPIC FEATURES: 10–13 × 7–9 μm, elliptic, angular, 5- to 6-sided, hyaline to slightly pinkish.
FRUITING: solitary, scattered, or in groups on the ground and on decaying wood, usually in wet areas; April–October; widely distributed in the Southeast; occasional to fairly common.
EDIBILITY: poisonous.
COMMENTS: Formerly known as *Nolanea strictior* and *Entoloma strictius*. Compare with *Nolanea verna* (p. 179).

Nolanea verna (Lundell) Kotlaba and Pouzar   Illus. p. 39
COMMON NAME: Early Spring Entoloma.
CAP: 1–2½" (2.5–6.5 cm) wide, conical with an acute umbo at first, becoming bell-shaped to broadly convex with an umbo; surface satiny, often streaked, glabrous, smooth to slightly wrinkled, dark brown to grayish brown, fading to pale brown or pale tan as moisture is lost; margin incurved at first, becoming expanded and somewhat uplifted and wavy in age.
FLESH: thin, fragile, watery brown; odor and taste not distinctive.
GILLS: attached and notched to nearly free, moderately close, broad, whitish at first, soon becoming pinkish to dull rose.
STALK: 1–3½" (2.5–9 cm) long, ⅛–⅜" (3–10 mm) thick, nearly equal or tapered slightly upward, often twisted-striate, slightly scurfy at the apex, moist or dry, soon hollow and easily splitting, gray to grayish brown with a white basal mycelium.
SPORE PRINT COLOR: salmon pink.
MICROSCOPIC FEATURES: spores 8–11 × 7–8 μm, elliptical, angular, smooth, inamyloid.
FRUITING: solitary to gregarious, sometimes in clusters on the ground in woods, along roadsides, and on disturbed ground; April–May; widely distributed in the Southeast; common.
EDIBILITY: poisonous.
COMMENTS: Formerly known as *Entoloma vernum*. *Nolanea strictia* (p. 179), poisonous, is similar but much larger overall, has larger spores that measure 10–13 × 7.5–9 μm, and typically appears later in the spring, although the fruiting periods for these two species may overlap.

Omphalina ectypoides (Peck) Bigelow   Illus. p. 39
  COMMON NAME: Wood Clitocybe.
  CAP: 1–2⅜" (2.5–6 cm) wide, broadly convex when young, soon becoming either flat with a sunken center or funnel-shaped; surface moist, brownish yellow to yellow-brown, covered with minute blackish brown to reddish brown matted fibers and scales that often disappear in age; margin uplifted at maturity.
  FLESH: yellowish; odor and taste not distinctive.
  GILLS: strongly decurrent, subdistant, narrow, occasionally forked, yellowish, sometimes with reddish brown stains in age.
  STALK: 1–2½" (2.5–6.5 cm) long, ⅛–⅜" (3–9 mm) thick, solid, smooth to slightly scurfy, enlarged downward, honey yellow, staining brownish when handled.
  SPORE PRINT COLOR: white.
  MICROSCOPIC FEATURES: spores 6.5–8 × 3.5–5 μm, elliptic, smooth, hyaline, amyloid.
  FRUITING: scattered or in groups on decaying conifer wood; June–October; widely distributed in the Southeast; occasional.
  EDIBILITY: unknown.
  COMMENTS: Formerly known as *Clitocybe ectypoides*. Compare with *Gerronema strombodes* (p. 132).

Omphalotus olearius (De Candolle : Fries) Singer   Illus. p. 40
  COMMON NAME: Jack O'Lantern.
  CAP: 2 ¾–7" (7–18 cm) wide, convex when young, becoming nearly flat and shallowly depressed at the disc to nearly funnel-shaped in age, often with a small umbo; surface dry, smooth, streaked with tiny fibrils, bright orange to yellow-orange, often stained reddish brown in age; margin uplifted and wavy in age.
  FLESH: white with an orange tint; odor not distinctive or somewhat unpleasant; taste not distinctive.
  GILLS: decurrent, close, narrow, thin, yellow-orange.
  STALK: 2–7" (5–18 cm) long, ¼–⅞" 5–22 mm) thick, nearly equal, tapered at the base, solid, dry, smooth, becoming scurfy in age, yellow-orange.
  SPORE PRINT COLOR: whitish cream.
  MICROSCOPIC FEATURES: spores 3–5 μm, round, smooth, hyaline, inamyloid.
  FRUITING: in clusters at the base of broadleaf trees and stumps, especially oak, and on the ground attached to buried wood; June–January; widely distributed in the Southeast; fairly common.
  EDIBILITY: poisonous, causing gastrointestinal upset.
  COMMENTS: Fresh specimens often glow green in the dark. The Golden Chanterelle, *Cantharellus cibarius* (p. 93), is similar, but its fertile surface has forked, blunt, gill-like ridges, and it grows on the ground. *Omphalotus olearius* is also known as *Omphalotus illudens* and *Clitocybe illudens*.

Panaeolus semiovatus (Sowerby : Fries) Lundell and Nannfeldt   Illus. p. 40
  COMMON NAME: Semi-ovate Panaeolus.
  CAP: 1¼–3½" (3–9 cm) wide, ovoid at first, then more or less bell-shaped and not further expanding; surface viscid when moist, shiny, smooth or somewhat wrinkled, at times cracked when drying, white to ivory or sometimes pale brownish, especially near the center; margin with a narrow sterile band, incurved, usually smooth but may have remnants of partial veil when young.

FLESH: thin, fragile, whitish, possibly yellowish in the lower stalk; odor and taste not distinctive.
GILLS: attached and notched, close, broad, grayish white at first, becoming mottled with blackish spots then black nearly overall; edges whitish.
STALK: 3–6" (8–15 cm) long, 3/8–5/8" (5–15 mm) thick, nearly equal or sometimes with an enlarged base, hollow, colored like the cap; partial veil white, leaving a thin, white, evanescent ring or ring zone near the middle of the stalk; ring or ring zone often black from the accumulation of spores.
SPORE PRINT COLOR: black.
MICROSCOPIC FEATURES: spores 15–20 × 8–11 μm, elliptic with a large germ pore, smooth dark brown.
FRUITING: solitary or in small groups on the dung of horses and other herbivores; June–December; widely distributed in the Southeast; occasional.
EDIBILITY: not recommended, although edible by some accounts.
COMMENTS: Also known as *Anellaria semiovatus* and *Panaeolus* (= *Anellaria*) *separatus*. The stalk ring is often inconspicuous, or it may disappear with age. This latter characteristic can lead to confusion with *Panaeolus* (= *Anellaria*) *phalaenarum* (not illustrated), edibility unknown, which is very similar but lacks a ring and has a solid stalk.

Panellus stipticus (Bulliard : Fries) Karsten   Illus. p. 40
COMMON NAME: Luminescent Panellus.
CAP: 1/4–1 1/4" (5–30 mm) wide, semicircular to kidney-shaped; surface minutely scaly to hairy or fuzzy, dingy white to pale brownish.
FLESH: thin, whitish to pale brown; odor not distinctive; taste rather acrid.
GILLS: attached to decurrent, close to crowded, narrow, pinkish to pale brownish.
STALK: short, stubby, hairy, lateral, whitish to brownish.
SPORE PRINT COLOR: white.
MICROSCOPIC FEATURES: spores 3–6 × 2–3 μm, oblong to sausage-shaped, smooth, hyaline, amyloid.
FRUITING: usually gregarious or in clusters on decaying broadleaf logs and sticks; May–December, sometimes overwintering; widely distributed in the Southeast; common.
EDIBILITY: inedible.
COMMENTS: When fresh, this mushroom will give off a whitish or greenish glow if viewed in total darkness for several minutes. The species name *stipticus* refers to its reputed value in stopping bleeding.

Paxillus atrotomentosus (Batsch : Fries) Fries   Illus. p. 40
COMMON NAME: Velvet-footed Pax.
CAP: 1 1/2–5 7/8" (4–15 cm) wide, convex, becoming flat, sometimes depressed at the center; surface dry, felty to smooth, covered with matted hairs, dull olive brown to rusty brown or yellowish brown to blackish brown; margin inrolled when young.
FLESH: whitish, thick, tough, not staining when cut or bruised; odor and taste not distinctive.
GILLS: decurrent, close, often forked or pore-like near the stalk, tan to yellow-brown or dull yellow, not staining when cut or bruised.
STALK: 3/4–4" (2–10 cm) long, 3/8–1 1/4" (1–3 cm) thick, equal, eccentric to nearly lateral, solid, velvety with a covering of densely matted dark brown or blackish brown hairs, apex often lighter; partial veil and ring absent.
SPORE PRINT COLOR: dull yellow to pale brownish yellow.

MICROSCOPIC FEATURES: spores 5–7 × 3–4 μm, elliptic, smooth, hyaline to pale brown.
FRUITING: solitary or in groups or clusters on decaying conifer stumps and logs and on partially buried wood; June–December; widely distributed in the Southeast; occasional to fairly common.
EDIBILITY: unknown.
COMMENTS: The conspicuous densely matted, dark brown to black hairs on the stalk make this mushroom easy to identify. *Paxillus panuoides*, edibility unknown, also grows in clusters on decaying conifer wood, but it lacks an obvious stalk (see *MNE*, p. 304).

Paxillus involutus (Batsch : Fries) Fries Illus. p. 40
COMMON NAME: Poison Paxillus.
CAP: 1⅝–4¾" (4–12 cm) wide, convex, becoming flat, often depressed on the disc; surface dry, smooth, somewhat sticky when moist, occasionally finely cracked in age, covered with matted hairs, dull brown or yellow-brown or red-brown, sometimes with olive tints; margin inrolled until maturity.
FLESH: dull yellow to pale tan, bruising reddish brown when cut or bruised; odor and taste not distinctive.
GILLS: decurrent, crowded, broad, forked and often pore-like near the stalk, tan to dull yellow or olive yellow, staining reddish brown when bruised or in age.
STALK: ¾–4" (2–10 cm) long, ¼–¾" (6–20 mm) thick, nearly equal or enlarged downward, solid, usually central, yellow-brown, often with darker brown stains, smooth; partial veil and ring absent.
SPORE PRINT COLOR: pale to dark yellow-brown.
MICROSCOPIC FEATURES: spores 7–9 × 4–6 μm, elliptic, smooth, pale brown.
FRUITING: solitary, scattered, or in groups on the ground and on decaying wood in conifer or mixed woods; June–December; widely distributed in the Southeast; fairly common.
EDIBILITY: poisonous (see Comments).
COMMENTS: Despite this mushroom's reputation, we have encountered several Europeans who collect it for the table. Their preparation technique includes boiling for several minutes in two or more changes of water. However, we strongly advise against this practice. According to mycologist Gary Lincoff, there are reports that this mushroom may produce a gradually acquired hypersensitivity that causes kidney failure. *Paxillus panuoides*, edibility unknown, is stalkless or has a rudimentary stalk; its gills are typically wavy and corrugated; and it grows scattered or in overlapping clusters on decaying conifer wood (see *MNE*, p. 304).

Pholiota highlandensis (Peck) A. H. Smith and Hesler Illus. p. 41
COMMON NAME: Burnsite Pholiota, Charred Pholiota.
CAP: ¾–2¼" (2–6 cm) wide, hemispherical at first, becoming convex to nearly flat or slightly depressed, with or without a small umbo; surface viscid to glutinous when moist, shiny when dry, smooth, yellowish brown with an olive tint to reddish brown; margin incurved and remaining so well into maturity.
FLESH: thin, pale yellowish to brownish; odor not distinctive; taste mild or astringent.
GILLS: attached, close to crowded, broad, white to yellowish at first, becoming grayish cinnamon at maturity; edges finely toothed.
STALK: 1–2" (2.5–5 cm) long, ¼–⅜" (5–10 mm) thick, cylindrical, moist or dry, solid when young, becoming hollow in age, fibrillose-scaly, pale yellow, covered with yellowish

brown scales; partial veil web-like, yellowish, leaving a fibrillose annular zone on the upper stalk.

SPORE PRINT COLOR: dark brown.

MICROSCOPIC FEATURES: spores 6–8 × 4–5 μm, elliptic to oval with a distinct apiculus and apical pore, smooth, grayish yellow.

FRUITING: solitary or more often in clusters on charred ground and on more localized substrates such as campfire pits and charred wood; March–February; widely distributed in the Southeast; fairly common.

EDIBILITY: unknown.

COMMENTS: *Pholiota carbonaria* (not illustrated), edibility unknown, which some mycologists consider to be synonymous, is very similar but has a rusty red to orangish veil and is common in the western United States. *Gymnopilus* species also produce rusty brown to orange-brown spore prints, but their spores are warted and lack a germ pore.

Pholiota polychroa (Berk.) Smith and Brodie   Illus. p. 41

COMMON NAME: Variable Pholiota.

CAP: ⅝–4" (1.5–10 cm) wide, obtuse to convex, becoming broadly convex to nearly flat, often with a low umbo; surface glutinous to viscid, at first with delicate vinaceous scales, especially on the margin, but soon glabrous, color variable, greenish to blue-green or dark olive, usually with olivaceous or purplish tones, often with dull orange to yellow hues, especially on the disc; margin incurved at first, usually adorned with hanging triangular flaps of veil tissue.

FLESH: soft, thick on the disc, thin at the margin, whitish to greenish; odor and taste not distinctive.

GILLS: attached, sometimes notched or somewhat decurrent as the cap margin becomes elevated, close to crowded, moderately broad, pale cream to lilaceous at first, becoming grayish to brown or dark purplish brown with an olivaceous tone; edges whitish.

STALK: ¾–3" (2–7.5 cm) long, up to ⁵⁄₁₆" (8 mm) thick, nearly equal or tapering toward the base, often curved, fibrillose-scaly from the base up to a thin evanescent ring, becoming glabrous in age, more or less striate above the ring, solid or hollow in age, yellowish to pale blue-green at the apex, soon becoming reddish brown on the lower portion, at times with a mat of tawny hairs at the base; partial veil leaving a thin, evanescent ring on the upper stalk.

SPORE PRINT COLOR: brown to cinnamon brown, tinged purplish when fresh.

MICROSCOPIC FEATURES: spores 6–7.5 × 3.5–4.5 μm, oblong to elliptical or bean-shaped with a minute apical pore, smooth, brownish.

FRUITING: solitary or more often in clusters on decaying wood of broadleaf trees and on sawdust, rarely occurring on conifer wood; June–December; widely distributed in the Southeast; fairly common.

EDIBILITY: unknown.

COMMENTS: *Pholiota lenta*, inedible, is similar but has a whitish to grayish cap with small white scales, and it grows on humus or soil as far south as South Carolina (see *MNE*, p. 305).

Phyllotopsis nidulans (Fries) Singer   Illus. p. 41

COMMON NAME: Orange Mock Oyster.

CAP: 1–3" (2.5–7.5 cm) wide, sometimes wider; surface yellowish orange, sometimes becoming somewhat brownish or fading to orangish yellow, densely fuzzy.

FLESH: orangish; odor unpleasant, often compared to rotting cabbage, but infrequently odorless; taste moderately to distinctly disagreeable.

GILLS: close to crowded, fairly narrow, yellow to orange.

STALK: absent.

SPORE PRINT COLOR: pale pink.

MICROSCOPIC FEATURES: spores 6–8 × 3–4 μm, sausage-shaped, smooth, hyaline, inamyloid.

FRUITING: solitary to clustered on dead wood of broadleaf and conifer trees; typically July–March; widely distributed in the Southeast; fairly common.

EDIBILITY: inedible.

COMMENTS: This mushroom is easy to identify because of its fuzzy orange stalkless cap, unpleasant odor, and distinctive pink spores.

Pleurotus dryinus (Persoon : Fries) Kummer   Illus. p. 41

COMMON NAME: Veiled Oyster.

CAP: 2–5" (5–12.5 cm) wide, convex, becoming broadly convex; surface dry, coated with tiny, matted, grayish fibrils on a whitish ground color, becoming slightly scurfy and whitish to dull yellowish tan overall in age; margin inrolled when young, becoming decurved at maturity, often rimmed with flaps of white veil.

FLESH: thick, firm, white; odor fragrant to slightly pungent; taste not distinctive.

GILLS: decurrent, close to subdistant, sometimes forked near the stalk or crossveined, white, discoloring yellowish.

STALK: 1½–4" (4–10 cm) long, ¾–1⅛" (1–3 cm) thick, eccentric to central, nearly equal or tapered downward, fibrous-tough, solid, whitish; partial veil white, membranous, sometimes leaving a white, superior ring.

SPORE PRINT COLOR: white.

MICROSCOPIC FEATURES: spores 9–14 × 3.5–5 μm, elliptic, smooth, hyaline, inamyloid.

FRUITING: solitary or in clusters on decaying broadleaf trees; July–October; widely distributed in the Southeast; occasional.

EDIBILITY: edible but tough; requires thorough cooking.

COMMENTS: Previously known as *Armillaria dryina*, this species is the only member of the genus *Pleurotus* that has a partial veil.

Pleurotus ostreatus complex (Jacquin : Fries) Kummer   Illus. p. 41

COMMON NAME: Oyster Mushroom.

CAP: 1½–7⅛" (4–18 cm) wide, convex, oyster-shell-shaped to fan-shaped; surface smooth, moist or dry but not viscid; color variable, dark brown, yellowish brown to grayish brown, pale gray, tan to yellowish buff, creamy white, or white.

FLESH: white; odor anise-like, fragrant, fruity, or not distinctive; taste not distinctive.

GILLS: decurrent, close to subdistant, white, grayish white or pale cream, typically with two or more tiers of lamellulae.

STALK: eccentric, lateral, rudimentary or absent; when present up to 1½" (4 cm) long and up to 1⅜" (3.5 cm) thick, dry, solid, enlarged in either direction or nearly equal; often coated, at least near the base, with downy white hairs; white to dingy yellow.

SPORE PRINT COLOR: white, buff, cream, or grayish lilac (see Comments).

MICROSCOPIC FEATURES: variable (see Comments).

FRUITING: typically growing in overlapping clusters, sometimes scattered, on logs, stumps, and standing trees, usually broadleaf wood; April–November or year-round when conditions allow; widely distributed in the Southeast; common.

EDIBILITY: edible and often rated as choice.

COMMENTS: Species in this complex are highly variable and often difficult to separate, even when microscopic features are studied. The information given here may assist those who have access to a microscope. The spores of *Pleurotus ostreatus* = *Pleurotus sapidus* measure 7–12 × 3.5–5 μm, and the spore print is white to grayish lilac. *Pleurotus populinus* (not illustrated), edible, grows on various broadleaf trees, especially cottonwood, and has a buff spore print and larger spores (9–15 × 3–5 μm). *Pleurotus cystidiosus* (not illustrated), edible, has a white spore print, larger spores (11–17 × 4–5 μm), and abundant clavate pileocystidia and short pyriform cheilocystidia. Compare with *Hypsizygus tessulatus* (p. 144).

Pluteus cervinus (Schaeffer : Fries) Kummer Illus. p. 41

COMMON NAME: Fawn Mushroom, Deer Mushroom.

CAP: 1⅛–4¾" (3–12 cm) wide, convex, becoming broadly convex to nearly flat in age; surface smooth, sometimes streaked with tiny fibers, often wrinkled when young, dull brown to grayish brown or pale cinnamon brown.

FLESH: soft, white, thick; odor and taste sometimes radish-like or not distinctive.

GILLS: free from the stalk, close to crowded, broad; white when young, becoming pale pink to salmon in age.

STALK: 2–4" (5–10 cm) long, ¼–¾" (6–20 mm) thick, nearly equal or enlarged downward, solid, dry, white, often with dull brown to grayish fibers; partial veil and ring absent.

SPORE PRINT COLOR: pink to salmon to brownish pink.

MICROSCOPIC FEATURES: spores 5.5–7 × 4–6 μm, elliptic, smooth, hyaline.

FRUITING: solitary or in groups on decaying wood of broadleaf trees and conifers, on the ground coming up from buried decaying wood, and on sawdust piles; May–November; widely distributed in the Southeast; fairly common.

EDIBILITY: edible.

COMMENTS: Also known as *Pluteus atricapillus*. Some authors recognize specimens with a blackish brown wrinkled cap when young and a thicker stalk, up to ¾" (2 cm) wide, as a separate species called *Pluteus magnus* (not illustrated), edible.

Pluteus petasatus (Fries) Gillet Illus. p. 42

COMMON NAME: none.

CAP: 1½–4" (4–10 cm) wide, bell-shaped at first, becoming convex or broadly convex with an obtuse umbo, then nearly flat in age; surface smooth, becoming cracked and scaly on the disc, whitish to cream, usually streaked with brownish fibrils or fine scales on the disc; margin smooth.

FLESH: thin, whitish; odor somewhat radish-like; taste not distinctive.

GILLS: free from the stalk, close, broad, white at first, soon becoming dull pink.

STALK: 1½–4" (4–10 cm) long, ¼–½" (6–12 mm) thick, nearly equal or enlarged downward, smooth or lightly streaked with fibrils, solid, whitish, becoming brownish toward the base.

SPORE PRINT COLOR: dull pink.

MICROSCOPIC FEATURES: spores 6–7.5 × 4.5–5 μm, elliptic, smooth, hyaline to pale pink.

FRUITING: solitary or in groups or clusters on wood mulch, sawdust, and decaying wood; June–February; widely distributed in the Southeast; common.

EDIBILITY: edible.

COMMENTS: This mushroom is often found on the ground near old tree stumps, arising from buried roots. *Pluteus pellitus* (not illustrated), edible, has a completely smooth white cap that lacks brownish streaks and cracks. The photograph labeled *Pluteus pellitus* on page 201 of *A Field Guide to Southern Mushrooms* (1985) by Weber and Smith, is actually *Pluteus petasatus*. Compare with *Pluteus cervinus* = *Pluteus atricapillus* (p. 185), edible, which has a brown to pale brown or grayish brown cap. *Lentinus lepideus,* edible but tough, can be somewhat similar, but it is whitish only when very young, then develops coarse brown scales, and has attached gills with serrated edges and a white spore print (see *MNE,* p. 297). *Volvariella* species also have free gills and pink spore prints, and they, too, grow on decaying wood, but they differ in having a distinct sac-like membranous volva at the stalk base.

Pouzarella nodospora (Atk.) Mazzer

Illus. p. 42

COMMON NAME: Hairy-stalked Entoloma.

CAP: ⅜–1¾" (1–4.5 cm) wide, conical to broadly conical when young, becoming bell-shaped to nearly flat with an umbo in age; surface dry, fibrillose-scaly to scaly, grayish brown to reddish brown; margin incurved, becoming even or wavy at maturity.

FLESH: grayish brown, thin, soft; odor and taste not distinctive.

GILLS: attached and deeply notched to nearly free, sinuate, pale gray when young, becoming grayish brown to reddish brown in age, subdistant, moderately broad.

STALK: 1½–4" (3.8–10 cm) long, ⅛–¼" (3–7 mm) thick, nearly equal; covered overall with fine, short, reddish brown to grayish brown hairs; solid; typically with coarse, stiff, dark reddish brown hairs at the base.

SPORE PRINT COLOR: pinkish cinnamon.

MICROSCOPIC FEATURES: spores 13–16 × 7–9 μm, elliptic, angular, 6- to 8-sided, smooth, moderately thick-walled, hyaline.

FRUITING: solitary, scattered, or in groups on leaf litter and decaying wood in broadleaf forests; June–December; widely distributed in the Southeast; occasional.

EDIBILITY: unknown.

COMMENTS: Formerly known as *Nolanea nodospora*. It is the largest species of the genus *Pouzarella*. The grayish brown to reddish brown scaly cap, grayish brown to reddish brown hairs on the stalk, and brownish gills are the distinctive characters of this species. *Inocybe calamistrata* (not illustrated), poisonous, is similar but has brown spores and a more scaly stalk that is bluish green at the base.

Psathyrella candolleana (Fries) Maire

Illus. p. 42

COMMON NAME: Common Psathyrella, Suburban Psathyrella.

CAP: 1⅛–4" (3–10 cm) wide, obtusely conic to convex when young, becoming nearly plane in age, sometimes with an umbo; surface smooth, pale to dark honey yellow, sometimes darkening to purplish brown near the margin, fading to whitish in age, coated with silky white fibrils when young; margin striate, typically rimmed with flaps of partial veil tissue.

FLESH: thin, pale yellow to whitish; odor and taste not distinctive.

GILLS: attached, close to crowded, white when young, becoming grayish brown, sometimes tinged with violet; edges whitish and very finely scalloped.

STALK: 2–4⅜" (5–11 cm) long, ⅛–⅜" (3–10 mm) thick, equal, hollow, white, coated with white fibrils; partial veil thin, fragile, fibrillose-membranous, not forming a ring or sometimes leaving an evanescent superior ring.

SPORE PRINT COLOR: purplish brown.

MICROSCOPIC FEATURES: spores 7–10 × 4–5 μm, elliptic, truncate, smooth, with an apical pore, pale brown.

FRUITING: scattered or in groups around decaying broadleaf stumps, among wood chips, and on lawns; May–October; widely distributed in the Southeast; fairly common.

EDIBILITY: edible.

COMMENTS: Compare with *Psathyrella velutina*, edible, which commonly grows in grassy areas. It has an orange-brown to dark yellow-brown cap with a dense layer of tiny flattened fibers, flaps of partial veil tissue on the cap margin, a ring on its stalk, and mottled brown gills with white edges (see *MNE*, p. 307).

Psathyrella hydrophila (Fries) Maire   Illus. p. 42

COMMON NAME: Clustered Psathyrella.

CAP: ¾–2¾" (2–7 cm) wide, obtusely conic to convex, becoming broadly convex to nearly plane, sometimes obscurely umbonate; surface moist when fresh, hygrophanous, glabrous, smooth or sometimes wrinkled, dark reddish brown fading to various shades of tan; margin typically adorned when young with whitish fibrillose patches of the torn partial veil, sometimes translucent-striate.

FLESH: fragile, watery brown to tan; odor and taste not distinctive.

GILLS: attached, crowded, narrow to moderately broad, pale brownish when young, becoming dark reddish brown at maturity; edges even, occasionally beaded with drops of moisture.

STALK: 1⅛–6" (3–15 cm) long, ⅛–¼" (2–7 mm) thick, nearly equal, hollow, moist or dry, fibrous, nearly smooth except for a pruinose apex, whitish to grayish at first, becoming brownish at least near the base in age; partial veil fibrillose to submembranous, whitish, leaving patches on the cap margin or sometimes a thin annular zone near the stalk apex.

SPORE PRINT COLOR: dark brown.

MICROSCOPIC FEATURES: spores 4–6 × 3–3.5 μm, elliptic, brown, smooth, apical pore present but inconspicuous.

FRUITING: in dense clusters on decaying broadleaf wood; July–November; widely distributed in the Southeast; fairly common.

EDIBILITY: unknown.

COMMENTS: Also known as *Psathyrella piluliformis*. This species at times may appear terrestrial, but really arises from buried wood.

Psathyrella umbonata (Peck) A. H. Smith   Illus. p. 42

COMMON NAME: none.

CAP: ¾–2" (2–5 cm) wide, bell-shaped to broadly conic, conspicuously umbonate at maturity; surface moist and purplish brown when fresh, fading to pale brown and eventually grayish white when dry, smooth or slightly wrinkled, glabrous, the umbo commonly paler than the marginal area; margin typically adorned with sparse, tiny grayish fibrils from the torn partial veil, weakly striate when moist.

FLESH: thin, fragile, pale brown; odor and taste not distinctive.

STALK: 2–4" (5–10 cm) long, about ⅛" (2–3 mm) thick, nearly equal, easily split, dry, hollow, white to pale gray, often slightly scurfy near the apex and smooth below, usually with a white tomentose base; partial veil thin, fibrillose, grayish, leaving scant remnants on the cap margin.

SPORE PRINT COLOR: dark brown.

MICROSCOPIC FEATURES: spores 12–15 × 6–8 μm, elliptic, brown, smooth, somewhat truncate, with an apical pore.

FRUITING: scattered or in dense clusters on mixtures of wood chips and soil; August–

November; Florida and Mississippi, also reported from New York and California; uncommon.

EDIBILITY: unknown.

COMMENTS: The combination of a darker marginal area and paler conspicuous umbo as the caps lose moisture is quite distinctive.

Psilocybe cubensis (Earle) Singer   Illus. p. 42

COMMON NAME: Magic Mushroom, Giggle Mushroom.

CAP: 1–3½" (2.5–9 cm) wide, broadly conical to bell-shaped at first, becoming convex to nearly flat with an umbo at maturity; surface smooth, viscid when moist, white to cream with a yellowish or ochraceous to brownish disc, becoming buff to pale brownish in age; margin often with whitish veil fragments.

FLESH: white, firm, bruising blue or greenish blue; odor and taste not distinctive.

GILLS: attached to the stalk, sometimes notched, close to crowded, narrow, gray at first, becoming purple-gray to nearly black; edges whitish.

STALK: 1½–6" (4–15.5 cm) long, 3/16–½" (4–12 mm) thick, usually tapering upward from a swollen or bulbous base, smooth or slightly grooved at the apex, whitish, bruising bluish or greenish blue, partial veil leaving a thin membranous ring on the upper stalk and sometimes fragments on the cap margin; upper surface of the ring soon blackened from the spore deposit.

SPORE PRINT COLOR: purple-brown.

MICROSCOPIC FEATURES: spores 11–17 × 7–12 μm, broadly elliptic, smooth, purplish brown (yellowish in KOH).

FRUITING: solitary or in groups on herbivore dung, especially cow dung, and on manure-enriched soil in pastures and grasslands; year-round; along the Gulf Coast from Florida to Texas; fairly common.

EDIBILITY: poisonous, hallucinogenic.

COMMENTS: This mushroom is sometimes sold to tourists in Mexico for "recreational purposes."

Rhodocollybia butyracea (Bulliard : Fries) Lennox   Illus. p. 42

COMMON NAME: none.

CAP: ¾–2¾" (2–7.5 cm) wide, obtusely convex, becoming broadly convex to nearly plane, sometimes with a small umbo; surface glabrous, lubricous when fresh and moist, hygrophanous, reddish brown to bay brown, fading to light brown or cinnamon brown; margin incurved at first, becoming decurved to uplifted, typically translucent-striate.

FLESH: moderately thick, white to pale watery gray; odor and taste not distinctive.

GILLS: attached and notched to nearly free, close to crowded, white, becoming slightly pinkish at maturity; edges wavy or straight, becoming eroded or torn in age.

STALK: 1½–2¾" (4–7.5 cm) long, ⅛–⅜" (3–10 mm) thick, enlarged downward, sometimes with a pinched base, moist or dry, stuffed then hollow in age, striate to finely sulcate, buff to pinkish buff overall when young, becoming pale reddish brown to pale bay brown in age, often with a white basal mycelium.

SPORE PRINT COLOR: pale pinkish buff.

MICROSCOPIC FEATURES: spores 6–10.5 × 3.5–5 μm, lacrymoid to ellipsoid or amygdaliform, smooth, hyaline.

FRUITING: scattered or in groups or clusters on the ground or among needle litter in conifer woods; July–November; widely distributed in the Southeast; fairly common.

EDIBILITY: edible.

COMMENTS: Also known as *Collybia butyracea*. *Gymnopus dryophilus* (p. 136), is similar, but it has a white spore print. *Gymnopus acervatus = Collybia acervata*, edibility unknown, has a reddish brown to chestnut brown cap that fades to tan or buff and a white to cream spore print, and it grows in dense clusters or groups among mosses in bogs and wet woodlands or on decaying wood (see *MNE*, p. 278). *Gymnopus luxurians = Collybia luxurians,* edibility unknown, has a dark reddish brown cap that fades to light brown or buff toward the margin and a cream spore print, and it grows in dense clusters on decaying wood, among wood chips, and in lawns on buried wood (see *MNE*, p. 279).

Rhodocollybia maculata (Fries) Singer   Illus. p. 43

COMMON NAME: Spotted Collybia.

CAP: 1 3/8–4" (3.5–10 cm) wide, convex when young, becoming broadly convex to nearly flat in age, sometimes with a low broad umbo; surface smooth, dry to moist, pinkish buff when young, becoming whitish in age, developing reddish brown to rusty streaks and spots overall, but especially toward the center.

FLESH: white; odor mildly unpleasant or not distinctive; taste bitter.

GILLS: attached, close to crowded, whitish to buff, developing brown to rusty spots in age.

STALK: 2–4 3/4" (5–12 cm) long, 3/8–1/2" (1–1.3 cm) thick, nearly equal, slightly rooting at the base, hollow, dry, longitudinally striate, whitish, developing rusty brown spots in age.

SPORE PRINT COLOR: pinkish buff to yellowish buff.

MICROSCOPIC FEATURES: spores 5.5–7 × 5–6 μm, globose to subglobose, smooth, hyaline, often dextrinoid.

FRUITING: solitary, scattered, or in groups or fairy rings on humus and buried wood under conifers and mixed woods; July–October; widely distributed in the Southeast; fairly common.

EDIBILITY: inedible owing to the bitter taste.

COMMENTS: Also known as *Collybia maculata*. *Rhodocollybia maculata* var. *scorzonerea*, (not illustrated), edibility unknown, is nearly identical, but it has distinctly yellow gills, and its stalk may be white or yellow.

Rhodocybe roseoavellanea (Murrill) Singer   Illus. p. 43

COMMON NAME: none.

CAP: 1 3/8–2 3/4" (3.5–7 cm) wide, convex at first, becoming broadly convex to nearly plane in age, often slightly depressed on the disc at maturity; surface moist or dry, dull, glabrous to finely matted-fibrillose, sordid pinkish buff with darker rosy tints; margin inrolled when young, becoming decurved, entire or slightly lobed.

FLESH: moderately thick, white; odor not distinctive; taste fragrant-farinaceous or not distinctive.

GILLS: attached and short-decurrent to decurrent, broad, close to subdistant; edges white when young, sometimes slowly bruising brownish.

STALK: 1 1/8–2 3/8" (3–6 cm) long, 3/8–1" (1–2.5 cm) thick, nearly equal, sometimes with a slightly bulbous base, dry, solid, glabrous, colored like the cap or paler, slowly staining brownish when bruised.

SPORE PRINT COLOR: rosy buff to light vinaceous cinnamon.

MICROSCOPIC FEATURES: spores 6.5–10 × 4–7 μm, broadly ellipsoid to somewhat amygdaliform, angular with 7–10 facets in polar view, slightly roughened to nearly smooth, hyaline.

FRUITING: scattered or in groups on sandy soil under oak, pine, and citrus trees; July–November; widely distributed in the Southeast; occasional.

EDIBILITY: unknown.

COMMENTS: Formerly known as *Clitopilus roseoavellaneus*. *Rhodocybe mundula*, edibility unknown, has a dingy grayish white cap that soon becomes cracked or lined in a concentric pattern, and whitish flesh with a farinaceous odor and bitter taste (see *MNE*, p. 308). *Clitopilus prunulus*, edible, has a whitish to grayish and somewhat felty cap, white fragile flesh with a farinaceous odor and taste, white decurrent gills that become pinkish in age, and longitudinally striate spores (see *MNE*, p. 278).

Ripartitella brasiliensis (Speg.) Singer   Illus. p. 43

COMMON NAME: none.

CAP: ⅜–2¾" (1–7 cm) wide, hemispheric to convex, becoming broadly convex to nearly plane; surface dry, matted-wooly, vinaceous reddish brown when young, breaking up and forming small concentric to scattered reddish brown scales on a whitish to buff ground color, sometimes whitish overall in age or following heavy rains that wash away the scales; margin incurved when young, becoming uplifted in age, often bearing patches of torn partial veil on young specimens, sometimes translucent-striate.

FLESH: up to ⅛" (3 mm) thick, whitish to buff; odor and taste not distinctive.

GILLS: attached, white, close, not staining when bruised or in age; edges entire.

STALK: ⅝–2" (1.5–5 cm) long, ⅛–¼" (2–6 mm) thick, nearly equal, dry, solid, silky-fibrillose with scattered tiny brownish scales on a whitish to buff ground color; often with a copious, white, cottony, basal mycelium and white mycelial cords; partial veil fibrillose-membranous, whitish to brownish, tearing and sometimes leaving remnants on the cap margin or a faint annular zone.

SPORE PRINT COLOR: white.

MICROSCOPIC FEATURES: spores 4.5–6 × 3.5–4.5 μm, broadly elliptic to subglobose, finely echinulate, hyaline, inamyloid; hymenial cystidia flask-shaped, usually with a narrow base, with crystalline-like ornamentation on the upper portion that dissolves quickly in 3 percent KOH.

FRUITING: in dense groups or clusters on decaying broadleaf wood, especially oak; July–December; along the Gulf Coast from Florida to Texas; occasional.

EDIBILITY: unknown.

COMMENTS: This species is fairly common in Central and South America and has also been reported from Africa and the Bonin Islands of the western Pacific.

Russula amoenolens Romagnesi   Illus. p. 43

COMMON NAME: none.

CAP: 1–4" (2.5–10 cm) wide, convex, becoming nearly flat or depressed in the center; surface viscid when moist, soon dry, dingy grayish yellow to yellow-brown, sometimes with dark reddish brown spots; margin prominently tuberculate-striate.

FLESH: white to pale yellowish; odor spermatic or rancid, sometimes likened to Jerusalem artichokes; taste oily and unpleasant or acrid.

GILLS: attached, close to subdistant, narrow, cream, sometimes staining brown, taste slowly acrid.

STALK: 1–2¾" (2.5–7 cm) long, ⅜–1" (1–2.5 cm) thick, nearly equal or tapered toward the base, dry, hollow at maturity, smooth, white to yellowish white with brown stains at the base.

SPORE PRINT COLOR: cream.

MICROSCOPIC FEATURES: spores 6–8.5 × 4.5–7 μm, elliptic, with isolated warts and a few connective lines, hyaline, amyloid.
FRUITING: usually in groups in woods, grassy woodland margins, parklands, and landscaped areas; June–December; widely distributed in the Southeast; fairly common.
EDIBILITY: inedible.
COMMENTS: *Russula pectinatoides* (not illustrated), inedible, is similar but has a paler, yellowish tan cap and a darker spore print, and it does not taste as acrid.

Russula ballouii Peck
Illus. p. 43

COMMON NAME: Ballou's Russula.
CAP: 1–3½" (2.5–9 cm) wide, broadly convex, becoming nearly flat or depressed in the center; surface dry, yellowish ochre to brownish orange, with the cuticle breaking up into small scale-like patches and revealing a cream to yellowish ground color; margin not striate.
FLESH: moderately thick, white; odor not distinctive; taste typically acrid.
GILLS: attached to subdecurrent, close to crowded, moderately broad, white to pale yellow, not staining when bruised.
STALK: 1–2½" (2.5–6.5 cm) long, ⅜–¾" (1–2 cm) thick, nearly equal, dry, solid, smooth, breaking into scale-like patches at the base, creamy white on the upper portion, colored like the cap on the lower portion.
SPORE PRINT COLOR: white to very pale yellow.
MICROSCOPIC FEATURES: spores 7.5–9 × 7–8 μm, oval, with warts and ridges forming a partial reticulum, hyaline, amyloid.
FRUITING: solitary or in small groups on the ground in broadleaf woods and grassy woodland margins; June–December; widely distributed in the Southeast; common.
EDIBILITY: unknown.
COMMENTS: Compare with *Russula compacta* (p. 191), which may be similar in color, but its gills quickly stain reddish brown when rubbed.

Russula compacta Frost
Illus. p. 43

COMMON NAME: Firm Russula.
CAP: 2¾–5⅞" (7–15 cm) wide, convex when young, becoming broadly convex to nearly flat in age, often depressed over the disc or funnel-shaped; surface dry, sticky when wet, dull, often cracked toward the center in age, whitish to cream, staining cinnamon to yellowish to orangish brown.
FLESH: very firm, white, with yellowish tones in age; odor fishy; taste not distinctive or slightly acrid.
GILLS: attached, close to subdistant, white, staining reddish brown when bruised.
STALK: ¾–4¾" (2–12 cm) long, ⅜–1½" (1–4 cm) thick, equal, dry, smooth, solid, becoming hollow in age, white, becoming brownish in age, staining reddish brown when bruised.
SPORE PRINT COLOR: white.
MACROCHEMICAL TESTS: flesh stains gray-green with the application of $FeSO_4$.
MICROSCOPIC FEATURES: spores 7–10 × 6–8 μm, elliptic to oval, with warts and reticulation, hyaline, amyloid.
FRUITING: solitary, scattered, or in groups on the ground in mixed woods; May–October; widely distributed in the Southeast; fairly common.
EDIBILITY: edible.

COMMENTS: The very firm flesh and fishy odor are the distinctive features of this mushroom.

Russula crustosa Peck   Illus. p. 44
COMMON NAME: Green Quilt Russula.
CAP: 2–5" (5–12.5 cm) wide, convex, becoming nearly flat or depressed on the center in age; surface dry, dull, often with a whitish bloom, cracking into quilt-like patches, color variable, ochraceous to dull green or yellowish green, often a fusion of these colors; margin striate.
FLESH: firm and compact when young, brittle, whitish; odor not distinctive, taste mild to slightly acrid.
GILLS: attached, close to subdistant, moderately broad, white to cream or pale yellow, some forking.
STALK: 1¼–3½" (3–9 cm) long, ½–1" (1.2–2.5 cm) thick, nearly equal, dry, hollow at maturity, smooth or wrinkled, white to cream or pale yellow.
SPORE PRINT COLOR: pale buff.
MICROSCOPIC FEATURES: spores 6–9 × 5.5–7 μm, elliptic, with isolated warts and sometimes connecting lines, hyaline, amyloid.
FRUITING: solitary, scattered, or in groups on the ground in mixed woods, often beneath oak trees; June–December; widely distributed in the Southeast; common.
EDIBILITY: edible.
COMMENTS: This mushroom is often confused with *Russula virescens* (not illustrated), edible, which has more fragile flesh and a white spore print. *Russula subgraminicolor*, edible, is similar, but it has a smooth, noncracking cap that is uniformly blue-grass green (see *MNE*, p. 309). *Russula aeruginea* (not illustrated), edible, is also similar, but it has a smooth, noncracking cap that is grayish olive green to moderate yellow-green. Some forms of *Russula variata* (not illustrated), edible, can also have a smooth green cap, but its gills are distinctly elastic rather than brittle.

Russula densifolia Secretan   Illus. p. 44
COMMON NAME: Dense-gilled Russula.
CAP: 1¾–5¾" (4.5–14.5 cm) wide, convex, becoming broadly convex to nearly plane with a depressed disc; surface viscid when fresh, smooth, often shiny, white at first, becoming brownish gray to smoky brown and finally blackish in age; margin even, incurved at first and remaining so well into maturity; cuticle peels up to one-half the distance to the center.
FLESH: white when exposed, soon staining reddish then grayish black; odor not distinctive; taste acrid.
GILLS: attached to decurrent, crowded, narrow, white to creamy white, slowly staining reddish then blackish when bruised; with numerous tiers of lamellulae.
STALK: ¾–3" (2–7.5 cm) long, ½–1" (1.2–2.5 cm) thick, nearly equal, dry, solid, smooth to slightly scurfy, white at first, becoming brownish black in age.
SPORE PRINT COLOR: white.
MICROSCOPIC FEATURES: spores 7–10 × 6–8 μm, oval to broadly elliptic, roughened with low warts and ridges, hyaline, amyloid.
FRUITING: solitary, scattered, or in groups on the ground in mixed conifer and broadleaf forests; June–November; widely distributed in the Southeast; occasional.
EDIBILITY: unknown.
COMMENTS: The Blackening Russula, *Russula adusta* (not illustrated), edible, has broad

gills and mild, white flesh that very slowly stains pink then fuscous gray when bruised, and it grows under conifers. *Russula nigricans,* edibility unknown, is very similar, but it has broad, distant gills, and its cuticle is matt when dry (see *MNE,* p. 309).

Russula foetentula Peck Illus. p. 44
COMMON NAME: none.
CAP: 1½–4" (4–10 cm) wide, convex to cushion-shaped at first, later becoming flattened, often with a central depression or nearly funnel-shaped in age; surface viscid when wet, color variable from pale yellow to orange-brown or yellow-brown or dark brown; margin prominently tuberculate-striate.
FLESH: moderately thick, firm and hard at first, becoming brittle with age; odor almond-like or oily, becoming unpleasant; taste acrid.
GILLS: attached, notched, or nearly free, close to subdistant, moderately broad, creamy yellow to yellow-orange, often developing brownish stains, often forked near the stalk.
STALK: 1½–4½" (4–11.5 cm) long, ½–1¼" (1.5–3 cm) thick, nearly equal, dry, becoming hollow at maturity, smooth to somewhat wrinkled, glabrous to minutely scurfy near the apex, colored like the cap but paler, often with reddish brown stains at the base.
SPORE PRINT COLOR: pale orange-yellow.
MICROSCOPIC FEATURES: spores $7–9 \times 5.5–7$ μm, broadly elliptic to subglobose, with isolated warts $0.7–1.2$ μm and some connective lines, pale yellow, amyloid.
FRUITING: solitary, scattered, or in groups on the ground in broadleaf and mixed woods; June–December; widely distributed in the Southeast; common.
EDIBILITY: unknown.
COMMENTS: This species is very close to the European *Russula subfoetens* and may be synonymous. *Russula laurocerasi,* edibility unknown, has a more pronounced almond or fetid odor and flesh that tastes mild to slightly acrid, and it has slightly larger spores that measure $7.5–10.5 \times 7.5–9$ μm with large warts and prominent ridges up to 2.5 μm (see *MNE,* p. 309). *Russula fragrantissima* (not illustrated), edibility unknown, has a larger cap, up to 8" (21 cm) wide, flesh that is moderately to strongly acrid, and spores with warts and ridges up to 1.5 μm high.

Russula perlactea Murrill Illus. p. 44
COMMON NAME: none.
CAP: 1¼–2½" (3–6.5 cm) wide, convex at first, becoming flat to somewhat depressed in the center; surface smooth, glabrous, dry to somewhat viscid, white to cream or pale yellow, sometimes pale brown over the disc; margin smooth, incurved at first.
FLESH: moderately thick, firm when young, becoming soft with age, white; odor not distinctive; taste sharply acrid.
GILLS: attached, close to subdistant, broad, white to pale cream, with few forking.
STALK: ¾–2½" (2–6.5 cm) long, ⅜–½" (1–1.5 cm) thick, nearly equal or slightly bulbous at the base, smooth, dry, solid, white.
SPORE PRINT COLOR: white.
MICROSCOPIC FEATURES: spores $9–11.5 \times 7.5–9$ μm, broadly elliptic, with warts up to 1.8 μm high and fine connecting lines, hyaline, amyloid.
FRUITING: solitary or more often in scattered groups on the ground in mixed woods, especially in sandy soil beneath oak and pines; August–February; along the Gulf Coast from Texas to Florida and northward to New Jersey and New York; common.
EDIBILITY: inedible owing to the extremely acrid flesh.
COMMENTS: Compare with *Russula subalbidula* (p. 196), which is larger overall and has

smaller spores. The relatively small size of the fruit bodies, with white gills, white spores, and acrid flesh, set *Russula perlactea* apart from other southeastern white Russulas.

Russula rubripurpurea Murrill — Not Illustrated
COMMON NAME: none.
CAP: 1¼–2" (3–5 cm) wide, convex at first, becoming depressed over the center; surface dry, smooth, glabrous, moist, red with purple tints, especially on the disc; margin even.
FLESH: white; odor and taste not distinctive.
GILLS: attached, narrow, close to crowded, white, not staining when bruised.
STALK: 1–1½" (2.5–4 cm) long, ⅜–½" (1–1.3 cm) thick, nearly equal, smooth, white.
SPORE PRINT COLOR: white to pale cream.
MICROSCOPIC FEATURES: spores 6.3–8 × 5–7.5 μm, subglobose to broadly elliptic, roughened with warts and fine lines that may form a reticulum, hyaline, amyloid.
FRUITING: solitary or in groups on the ground in oak woods.
EDIBILITY: unknown.
COMMENTS: The type was collected by Murrill on the ground in oak woods in Gainesville, Florida, on October 3, 1932.

Russula sanguinea Fries — Illus. p. 44
COMMON NAME: Rose-red Russula.
CAP: 1½–4" (4–10 cm) wide, convex at first, becoming flattened and often deeply depressed in age; surface viscid when moist, blood red to purplish red or bright rose red, paler in age; margin even at first, becoming tuberculate-striate.
FLESH: firm, thin, white or tinged red beneath the cuticle; odor faintly fruity; taste weakly to moderately acrid.
GILLS: attached or slightly decurrent, close to subdistant, creamy white when young, becoming pale ochre in age.
STALK: 1½–4" (4–10 cm) long, ⅜–1¼" (1–3 cm) thick, nearly equal or tapered toward the apex, smooth, glabrous or minutely pruinose at the apex, stuffed when young, becoming hollow in age, white, often flushed with pink or red.
SPORE PRINT COLOR: creamy yellow.
MICROSCOPIC FEATURES: spores 7–9 × 6–8 μm, subglobose, ornamented with isolated warts up to 1 μm high, sometimes with a few fine connecting lines, hyaline, amyloid.
FRUITING: usually gregarious on the ground beneath pines, also in mixed woods; June–January; widely distributed in the Southeast; common.
EDIBILITY: inedible.
COMMENTS: Also known as *Russula rosacea* in older literature.

Russula sericeonitans Kauffman — Illus. p. 44
COMMON NAME: Silky-shining Russula.
CAP: 1½–3½" (4–9 cm) wide, convex when young, becoming more or less flat to depressed in the center in age; surface smooth, glabrous with a silky sheen when fresh, lilac purple or paler, with a darker purple to nearly black center, at times with pale brownish or yellowish tones; margin even, becoming tuberculate-striate in age.
FLESH: brittle, white to grayish, sometimes purplish beneath the cuticle; odor not distinctive; taste mild.
GILLS: attached, close to crowded, with some forking near the stalk, intervenose, with few

lamellulae present, moderately broad and narrowed toward the stalk, white or slightly grayish.

STALK: 1¼–2¾" (3–7 cm) long, up to ⅝" (1.5 cm) thick, equal to subclavate; surface glabrous, smooth or somewhat longitudinally wrinkled, firm at first, becoming spongy or stuffed in age, white, becoming grayish in age.

SPORE PRINT COLOR: white.

MICROSCOPIC FEATURES: spores 7–8.5 × 5.5–7 μm, globose, spiny, hyaline, amyloid.

FRUITING: scattered or in small groups on the ground in broadleaf and mixed broadleaf and conifer woods, often with oak; July–November and during early winter along the Gulf Coast; widely distributed; occasional to locally common.

EDIBILITY: unknown.

COMMENTS: As with many red-capped species of *Russula,* the underside of *Russula sericeonitans* is frequently parasitized by the ascomycete *Hypomyces luteovirens* (p. 332), which forms an olive yellowish crust over the gill surface as shown in the photo. Some color variants of *Russula vinacea* (p. 196) are similar, but they have markedly acrid-tasting flesh. *Russula ornaticeps* (not illustrated), edibility unknown, has a more variegated cap, often including bluish tones, a pruinose bloom on the cap surface, and pinkish cream-colored spores. *Russula cyanoxantha* (not illustrated), edibility unknown, has prominently forked, flexible gills.

Russula silvicola Shaffer Illus. p. 44

COMMON NAME: none.

CAP: ¾–3" (2–7.5 cm) wide, convex, becoming broadly convex to nearly flat or depressed in the center in age; surface viscid and shiny or dry, at times pruinose when very young, typically smooth but sometimes slightly wrinkled, medium to dark red or deep pink or reddish orange or pale yellowish pink, at times with paler blotches; margin incurved at first, more or less tuberculate-striate.

FLESH: brittle, white to yellowish white, sometimes tinged red beneath the cuticle; odor fruity or spicy; taste acrid.

GILLS: attached and sometimes notched or subdecurrent, occasionally appearing free, close, occasionally forked near the stalk or not forked, white to yellowish white.

STALK: ¾–3" (2–7.5 cm) long, ¼–¾" (5–20 mm) thick, nearly equal or enlarged toward the base, dry, dull, glabrous, with or without longitudinal wrinkles, stuffed at first, becoming hollow in age, white or becoming yellowish in age.

SPORE PRINT COLOR: white.

MICROSCOPIC FEATURES: spores 6–10.7 × 5.3–9 μm, broadly elliptical, ornamented with warts and ridges up to 1.2 μm high, reticulate, amyloid.

FRUITING: solitary or gregarious on soil, humus, and well-decayed wood in broadleaf and conifer woods; July–January; widely distributed in the Southeast; common.

EDIBILITY: inedible.

COMMENTS: *Russula fragilis* (not illustrated), inedible, is similar but has a grayish purple to grayish red or yellowish pink cap, often with an olive to greenish tinge toward the center. *Russula emetica* (not illustrated), poisonous, is also similar but has a more uniformly red cap and typically grows among sphagnum mosses.

Russula subalbidula Murrill Illus. p. 45

COMMON NAME: none.

CAP: 2–6" (5–15.5 cm) wide, shallowly convex, often depressed in the center; surface dry, smooth, glabrous, white to creamy yellow, especially on the disc; margin even.

FLESH: moderately thick, brittle, white; odor not distinctive; taste acrid.

GILLS: attached and notched, close to subdistant, moderately broad, pale cream.

STALK: 1¼–2½" (3–6.5 cm) long, ½–1½" (1–4 cm) thick, nearly equal or slightly enlarged toward the base, smooth, hollow at maturity, white.

SPORE PRINT COLOR: cream.

MICROSCOPIC FEATURES: spores 7–9 × 6–7.5 μm, broadly elliptic to subglobose, with warts 0.6–0.8 μm high and some connecting lines, hyaline, amyloid.

FRUITING: solitary or in groups on the ground beneath oaks; June–February; widely distributed in the southeast; occasional.

EDIBILITY: unknown.

COMMENTS: *Russula anomala* (not illustrated), edibility unknown, is very similar but has a tuberculate-striate cap margin. Some mycologists place these two species in synonymy. Compare with *Russula perlactea* (p. 193), which is generally smaller in size and has larger spores.

Russula vinacea Burlingham   Illus. p. 45

COMMON NAME: Early Blackish-red Russula.

CAP: 1½–4½" (4–11.5 cm) wide, convex to broadly convex, becoming nearly flat and often depressed over the center in age; surface tacky to viscid and shiny when moist, dull when dry, smooth or slightly wrinkled, color variable from purplish red to blackish red, especially on the disc, paler pinkish red toward the margin, often with ochraceous to olivaceous yellow or pinkish yellow areas, or more rarely pale yellow overall; margin incurved at first, tuberculate-striate in age.

FLESH: white, brittle; odor not distinctive; taste mild or acrid.

GILLS: attached, sometimes notched, close to subdistant, sometimes forked near the stalk, moderately broad, white to creamy yellow, slowly developing rusty brown spots in age.

STALK: 1½–3½" (4–9 cm) long, ⅜–¾" (1–2 cm) thick, nearly equal or enlarged downward, dry, solid, dull, smooth or with longitudinal wrinkles, glabrous to minutely felted near the apex, white, bruising and becoming ashy gray in age, often developing rusty brown spots.

SPORE PRINT COLOR: white to yellowish cream.

MICROSCOPIC FEATURES: spores 7–10 × 5–8 μm, broadly oval to subglobose, with warts 0.4–1.4 μm high and fine connecting lines, hyaline, amyloid.

FRUITING: solitary or in scattered groups on the ground beneath broadleaf trees, especially oak, also in mixed broadleaf and conifer woods; April–November; widely distributed in the Southeast; common.

EDIBILITY: mild-tasting specimens reported to be edible.

COMMENTS: Although quite variable in color and taste, this common mushroom can often be identified because it is the first *Russula* of the season to appear (late spring) at least in the southern part of its range. It is also known as *Russula krombholzii* and *Russula atropurpurea*.

Squamanita umbonata (Sumstine) Bas   Illus. p. 45

COMMON NAME: Knobbed Squamanita.

CAP: 1¼–2¼" (3–6 cm) wide, conic at first, becoming nearly flat but retaining a conic umbo; surface orangish buff to orangish brown, dry, fibrous to scaly, the cuticle often splitting near the edge to reveal its flesh.

FLESH: thick, white; odor and taste not distinctive.

GILLS: attached, close to crowded, fairly broad, white; partial and universal veils evident on young specimens.

STALK: 1–3" (2.5–7.5 cm) long, ¼–¾" (5–20 mm) thick, whitish and cottony above, usually with fibers or scales like the cap on the lower portion; with rings of coarse brownish scales near the ground; arising from an enlarged cylindric tuber.

SPORE PRINT COLOR: white.

MICROSCOPIC FEATURES: spores $6–9 \times 3.5–5.5 \mu m$, elliptic, smooth, hyaline, inamyloid; chlamydospores sometimes present on lower stalk.

FRUITING: solitary, scattered, or in groups on the ground in mixed woods; July–November; reported from Mississippi and Texas, but most reports from more northern areas, including Pennsylvania, New York, and New England; rare.

EDIBILITY: unknown.

COMMENTS: Formerly known as *Armillaria umbonata*.

Strobilurus conigenoides (Ellis) Singer   Illus. p. 45

COMMON NAME: Magnolia-cone Mushroom.

CAP: ¼–¾" (5–20 mm) wide, convex to flat, white, dry, covered with a dense layer of minute hairs; margin incurved at first.

FLESH: very thin; odor and taste not distinctive.

GILLS: attached, close to almost crowded, broad, white.

STALK: 1–2" (2.5–5 cm) long, $\frac{1}{32}$–$\frac{1}{16}$" (0.75–2 mm) thick, dry, covered with a dense layer of minute hairs.

SPORE PRINT COLOR: white.

MICROSCOPIC FEATURES: spores $6–7 \times 3–3.5 \ \mu m$, elliptic, smooth, hyaline, inamyloid.

FRUITING: in groups on magnolia cones and sweet gum fruits; August–January; widely distributed in the Southeast; common.

EDIBILITY: unknown.

COMMENTS: This mushroom's range coincides with its host trees' natural range.

Stropharia bilamellata Peck   Illus. p. 45

COMMON NAME: Double-gilled Stropharia.

CAP: ¾–2¼" (2–6 cm) wide, convex, becoming flattened in age; surface smooth, somewhat viscid when moist, white to yellowish; margin even.

FLESH: white; odor and taste not distinctive.

GILLS: attached, close, moderately broad, pale purple-brown at first, becoming dark purple-brown at maturity.

STALK: 1¼–2¼" (3–6 cm) long, up to ¼" (6 mm) thick, nearly equal, dry, solid at first, becoming hollow in age, white, often dusted with purple-brown spores; partial veil leaving a superior membranous ring; upper surface of the ring with radiating lines that resemble gills and usually dusted with a purple-brown spore deposit.

SPORE PRINT COLOR: dark purple-brown.

MICROSCOPIC FEATURES: spores $11 \times 7 \ \mu m$, elliptic with an apical pore, smooth, pale purplish brown.

FRUITING: solitary or in small groups on the ground in grassy places and on disturbed ground; June–January; widely distributed in the Southeast; uncommon.

EDIBILITY: unknown; other related species poisonous.

COMMENTS: According to mycologist Scott Redhead, Peck may have named this species *bilamellata* because the gill-like surface on the upper portion of the ring resembles a second row of gills. The stalk ring may disappear on older specimens. *Stropharia coro-*

nilla (not illustrated), poisonous, is very similar but generally smaller overall and with smaller spores that measure 7–10 × 4–6 μm. Some smaller *Agaricus* species are also similar and can be found growing in the same habitat, but they have gills that are free from the stalk and produce chocolate brown spore prints.

Tricholoma caligatum (Viviani) Ricken complex Illus. p. 45
  COMMON NAME: none.
  CAP: 2⅜–4¾" (6–12 cm) wide, hemispheric when young, becoming broadly convex to nearly flat in age, sometimes shallowly depressed; surface dry, reddish brown to pale cinnamon brown, breaking up into coarse, flattened scales and fibers on a whitish to pinkish buff ground color; margin inrolled and cottony-membranous when young, becoming expanded to uplifted in age, sometimes rimmed with veil remnants.
  FLESH: white to cream; odor fragrant to spicy or not distinctive; taste nutty, bitter or not distinctive.
  GILLS: attached, close, broad, white.
  STALK: 1⅝–4" (4–10 cm) long, ⅝–1⅛" (1.5–3 cm) thick, nearly equal or enlarged in the middle and tapered in both directions, solid, dry, sheathed from the base up to the ring by a cinnamon brown veil that breaks into patches on a white ground color, white above the ring, sometimes pruinose; ring white, cottony-membranous, often flaring upward, median to superior.
  SPORE PRINT COLOR: white.
  MICROSCOPIC FEATURES: spores 6–8 × 4.5–5.5 μm, elliptic, smooth, hyaline, inamyloid.
  FRUITING: solitary, scattered, or in groups on the ground under conifers, especially hemlock, and under broadleaf trees, especially oak; June–December; widely distributed in the Southeast; occasional.
  EDIBILITY: mild-tasting forms edible, others too bitter.
  COMMENTS: *Tricholoma caligatum* is a complex of several forms, which explains the wide range of odors and tastes.

Tricholoma flavovirens (Persoon : Fries) Lundell Illus. p. 45
  COMMON NAME: Canary Trich.
  CAP: 2–4" (5–10 cm) wide, convex, becoming broadly convex to nearly flat in age; surface sticky when fresh, pale yellow to sulfur yellow, becoming reddish brown on the disc.
  FLESH: white; odor farinaceous; taste farinaceous or not distinctive.
  GILLS: notched, close, broad, bright sulfur yellow.
  STALK: 1⅛–2¾" (3–7 cm) long, ⅜–¾" (1–2 cm) thick, nearly equal, occasionally enlarged at the base, solid, dry, pale yellow to sulfur yellow.
  SPORE PRINT COLOR: white.
  MICROSCOPIC FEATURES: spores 6–7 × 4–5 μm, elliptic, smooth, hyaline, inamyloid.
  FRUITING: solitary, scattered, or in groups on the ground under conifers and in mixed woods; July–December; widely distributed in the Southeast; fairly common.
  EDIBILITY: not recommended. Traditionally considered to be edible and quite good, but recent reports of poisonings in France implicate *Tricholoma equestre,* which some authorities believe to be synonymous. Therefore, we can no longer recommend eating *T. flavovirens* until these concerns have been resolved.
  COMMENTS: *Tricholoma sejunctum,* edibility unknown, has a golden yellow cap with conspicuous radiating blackish fibrils and white to pale yellow gills (see *MNE,* p. 313).

Tricholoma odorum Peck   Illus. p. 46
  COMMON NAME: none.
  CAP: ¾–3½" (2–9 cm) wide, convex with a broad umbo, becoming flat or shallowly depressed in age; surface dry, dull or shiny, smooth, greenish yellow when young, becoming yellowish buff to yellowish tan in age, often with a brownish disc; margin incurved when young.
  FLESH: moderately thick, whitish; odor and taste farinaceous or disagreeable, like coal tar.
  GILLS: attached, close, broad, yellow to yellowish buff.
  STALK: 1½–4½" (4–12 cm) long, ¼–½" (6–12 mm) thick, nearly equal or enlarged toward the base, sometimes bulbous, dry, smooth, solid, sometimes longitudinally twisted, pale yellow to greenish yellow.
  SPORE PRINT COLOR: white.
  MICROSCOPIC FEATURES: spores 7–9 × 5–6 μm, elliptic or spindle-shaped, smooth, hyaline, inamyloid.
  FRUITING: solitary or in groups or clusters on the ground in broadleaf and mixed woods; July–November; widely distributed in the Southeast, fairly common.
  EDIBILITY: inedible.
  COMMENTS: *Tricholoma sulphureum* (p. 199) also has a strong coal tar odor but has a sulfur yellow cap and stalk, subdistant gills, and somewhat smaller spores. *Tricholoma sulphurescens,* edibility unknown, also has a strong odor of coal tar, but it has a whitish cap that bruises or ages dull yellow, and it grows in association with oaks (see *MNE,* p. 313).

Tricholoma sulphureum (Buller : Fries) Kummer   Illus. p. 46
  COMMON NAME: none.
  CAP: ¾–3" (2–7.5 cm) wide, hemispheric to conic at first, becoming convex, then broadly convex at maturity; surface smooth, dull, sulfur yellow with red-brown streaks and spots; margin incurved at first and remaining so well into maturity, acute.
  FLESH: thin, sulfur yellow; odor very strong of coal tar; taste unpleasant or not distinctive.
  GILLS: attached and notched, broad, subdistant; edges even.
  STALK: 1–3" (2.5–7.5 cm) long, ¼–¾" (7–20 mm) thick, nearly equal or enlarged downward, dry, solid, sulfur yellow with red-brown longitudinal fibrils, usually with a white tomentose base.
  SPORE PRINT COLOR: white.
  MICROSCOPIC FEATURES: spores 8–11 × 5–7 μm, broadly elliptic to almond-shaped, smooth, hyaline.
  FRUITING: solitary, scattered, or in groups on the ground in broadleaf or conifer woodlands; July–November; widely distributed in the Southeast; occasional.
  EDIBILITY: unknown.
  COMMENTS: *Tricholoma odorum* (p. 199), edibility unknown, also has a coal tar odor, but it has a cap that is greenish yellow when young and fades to yellowish buff to yellowish tan in age; pale yellow flesh that tastes farinaceous to disagreeable; a pale yellow stalk; and smaller spores that measure 7–9 × 5–6 μm. *Tricholoma inamoenum,* edibility unknown, which grows in association with conifers, also has a coal tar odor, but its cap is pale buff to whitish (see *MNE,* p. 312).

Tricholomopsis decora (Fries) Singer Illus. p. 46
  COMMON NAME: none.
  CAP: 1–2¼" (2.5–6 cm) wide, convex to nearly flat, sometimes with a slight central depression; surface yellowish to greenish yellow, with blackish fibrous scales.
  FLESH: yellowish; odor and taste not distinctive.
  GILLS: attached, crowded, narrow, yellow.
  STALK: 1–2¼" (2.5–6 cm) long, ⅛–⅜" (3–10 mm) thick, yellow with tiny blackish scales.
  SPORE PRINT COLOR: white.
  MICROSCOPIC FEATURES: spores 6–7.5 × 4–5.5 μm, elliptic, smooth, hyaline, inamyloid.
  FRUITING: solitary or in small groups on conifer wood; June–December; widely distributed in the Southeast; fairly common.
  EDIBILITY: inedible.
  COMMENTS: This mushroom resembles a *Tricholoma* species but grows on decaying wood. *Tricholomopsis sulfureoides* (not illustrated), edibility unknown, is similar but more brownish yellow, and its cap hairs are yellowish, not black.

Tricholomopsis formosa (Murrill) Singer Illus. p. 46
  COMMON NAME: none.
  CAP: 1⅛–3⅛" (3–8 cm) wide, convex to nearly plane; surface covered with ascending to recurved rusty brown to tawny scales over a cinnamon buff ground color, dry; margin incurved when young, often wavy in age.
  FLESH: whitish; odor not distinctive or slightly disagreeable; taste slightly disagreeable or not distinctive.
  GILLS: attached, close to crowded, whitish to pinkish cream.
  STALK: 1½–3" (4–7.5 cm) long, ¼–⅜" (5–10 mm) thick, nearly equal or tapered slightly downward, fibrillose-scaly, dry, colored like the cap or paler; partial veil and ring absent.
  SPORE PRINT COLOR: white.
  MICROSCOPIC FEATURES: spores 5–7 × 5–6 μm, ovoid, smooth, hyaline, inamyloid.
  FRUITING: solitary, scattered, or in groups on rich humus attached to buried wood and roots, on sawdust and litter debris, and on decaying wood; June–December; widely distributed in the Southeast; occasional to fairly common.
  EDIBILITY: unknown.
  COMMENTS: Plums and Custard, *Tricholomopsis rutilans,* typically too bitter to be considered edible, has red to purplish red cap scales and fibers, yellow gills, and red to purplish red scales and fibers on its stalk (see *MNE,* p. 314).

Volvariella gloiocephala (DC : Fr.) Singer Illus. p. 46
  COMMON NAME: none.
  CAP: 2–6 inches (5–15 cm) wide, conical at first, becoming bell-shaped to convex or flat to depressed in the center, with an umbo in age; surface smooth, viscid when moist, shiny when dry, white to grayish brown; margin incurved well into maturity, smooth.
  FLESH: white to brownish; odor somewhat radish-like; taste radish-like or not distinctive.
  GILLS: free from the stalk, close, broad, white at first, soon becoming pink to brownish pink.
  STALK: 4–6" (10–15.5 cm) long, ⅜–¾" (1–2 cm) thick, nearly equal or tapered toward the apex, base somewhat enlarged within a prominent sac-like volva, smooth, longitudinally fibrillose, pruinose at the apex, solid, white to brownish; universal veil white, membranous, leaving a large volva at the stalk base.

SPORE PRINT COLOR: pinkish brown
MICROSCOPIC FEATURES: spores 11–20 × 7–12 μm, elliptical, smooth, pale yellow.
FRUITING: solitary or scattered in groups on dung-enriched ground in woods, grassy areas, and pastures, as well as on straw and mulch; September–April; widely distributed in the Southeast; occasional to locally common.
EDIBILITY: edible with caution (see Comments).
COMMENTS: Also known as *Volvariella speciosa*. This mushroom has two color forms, white and grayish brown. Do not confuse white forms with members of the *Amanita virosa* complex (p. 113), which have white gills, white spores, and a skirt-like ring on the stalk. *Volvariella volvacea* (not illustrated), edible, is similar but less common. It has a streaked brownish cap and brownish tones on the volva. *Volvariella bombycina*, edible, has a silky-fibrillose white cap and grows on decaying wood (see *MNE*, p. 314).

Xeromphalina campanella (Bataille : Fries) Kühner and Maire Illus. p. 46
COMMON NAME: Fuzzy Foot, Golden Trumpets.
CAP: ⅛–1" (3–25 mm) wide, convex to broadly convex with a depressed center; surface smooth, moist, yellowish orange to orangish brown; margin striate.
FLESH: thin, yellowish to brownish yellow; odor and taste not distinctive.
GILLS: decurrent, subdistant to distant, pale yellow to pale orange.
STALK: ⅜–2" (1–5 cm) long, ¹⁄₃₂–⅛" (0.5–3 mm) thick, dry, yellow at the apex, shading below to dark reddish brown, base with a dense tuft of long orangish hairs.
SPORE PRINT COLOR: pale buff.
MICROSCOPIC FEATURES: spores 5–7 × 3–4 μm, elliptic to cylindric, smooth, hyaline, amyloid.
FRUITING: in dense clusters, typically recurved, on well-decayed conifer wood; May–November; widely distributed in the Southeast; common.
EDIBILITY: inedible.
COMMENTS: The densely hairy base and growth on conifer wood are the important field characters. *Xeromphalina kauffmanii* (not illustrated), inedible, is nearly identical but grows on decaying wood of broadleaf trees.

Tooth Fungi

Members of this group have downward-pointing, spine-like teeth on which they produce spores. Most tooth fungi grow on the ground and form teeth on the underside of their caps. Other members grow on wood and have teeth along branches or at the tips of branches. One species, usually found on standing trees, resembles a satyr's beard or goatee and has long spines hanging from a stalkless, solid mass of tissue. Species in this group differ from similar tooth-like polypores by forming conic teeth instead of elongated, flattened, irregular tubes. The spine-like tips of similar branched and clustered coral fungi do not point downward. Several species are excellent edibles, but many are much too tough to be eaten.

Hericium coralloides (Scopoli : Fries) S. F. Gray Illus. p. 47
  COMMON NAME: Comb Tooth.
  FRUIT BODY: 2¾–10" (7–25 cm) wide, 2¾–8" (7–20 cm) high, a cluster of spreading branches with spines arranged in rows along the branches like teeth on a comb, arising from a common base; white to salmon or pinkish.
  FLESH: thick, white, soft; odor and taste not distinctive.
  SPINES: up to 1" (2.5 cm) long, white, rather evenly distributed along the branches.
  SPORE PRINT COLOR: white.
  MICROSCOPIC FEATURES: spores 3–5 × 3–4 μm, oval to round, slightly roughened, hyaline.
  FRUITING: solitary or in groups hanging on decaying broadleaf logs and stumps; June–November; widely distributed in the Southeast; fairly common.
  EDIBILITY: edible.
  COMMENTS: Previously known as *Hericium ramosum*. The Bear's-head Tooth, *Hericium americanum* (not illustrated), edible, formerly known in North America as *Hericium coralloides*, is similar but has spines arranged in clusters at the branch tips.

Hericium erinaceus (Fries) Persoon Illus. p. 47
  COMMON NAME: Bearded Tooth, Satyr's Beard.
  FRUIT BODY: 2¾–8" (7–20.5 cm) wide, 3¾–8" (9.5–20 cm) high, a whitish to yellowish cushion-shaped mass giving rise to long spines, resembling a beard.
  FLESH: thick, white, soft; odor and taste not distinctive when young, sour and unpleasant in age.
  SPINES: up to 3½" (9 cm) long, white.
  SPORE PRINT COLOR: white.
  MICROSCOPIC FEATURES: spores 5–6.5 × 4–5.6 μm, oval to round, smooth to slightly roughened, hyaline.

FRUITING: solitary on standing trunks and fallen logs of broadleaf trees; July–December; widely distributed in the Southeast; fairly common.
EDIBILITY: edible.
COMMENTS: *Hericium erinaceus* ssp. *erinaceo-abietis* (not illustrated), edible, is similar and also grows on decaying logs and stumps of broadleaf trees, but it has small, short, white spines that cover the entire surface of the fruit body.

Hydnellum aurantiacum (Batsch) Karsten Illus. p. 47

COMMON NAME: Orange Rough-cap Tooth.
CAP: 2–7" (5–18 cm) wide, convex, becoming broadly convex to nearly plane, sometimes depressed; surface dry, nearly uniform at first but soon breaking into irregular projections, cavities, and channels; uneven, tomentose, becoming matted in age, sometimes zonate, orange-buff to whitish at first, soon becoming salmon orange then darkening to rusty brown or bay brown with a rusty orange to whitish or sometimes tan margin; undersurface covered with spines.
FLESH: in cap buff; in stalk rusty orange; fibrous-tough, zonate; odor and taste pungent and disagreeable.
SPINES: decurrent, 5–7 mm long, dark brown with grayish buff tips.
STALK: ¾–2¾" (2–7 cm) long, ³⁄₁₆–¾" (5–20 mm) thick, enlarged downward to a bulbous base, dry, solid, covered with a matted orange to brownish tomentum, with an orange basal mycelium.
SPORE PRINT COLOR: brown.
MACROCHEMICAL TESTS: cap stains black with the application of KOH; flesh stains dingy olive with the application of KOH.
MICROSCOPIC FEATURES: spores 5–8 × 5–6 μm, subglobose, distinctly tuberculate, brownish.
FRUITING: solitary, scattered, or in groups, commonly confluent and forming large rosettes, frequently concrescent, on the ground under conifers; July–May; widely distributed in the Southeast; occasional to fairly common.
EDIBILITY: inedible.
COMMENTS: The Red-juice Tooth, *Hydnellum peckii* (not illustrated), inedible, has a whitish to pinkish cap that exudes drops of red juice when young and fresh and becomes dark brown in age, with brownish flesh that has a peppery taste, and it grows on the ground under conifers throughout the Southeast.

Hydnellum spongiosipes (Peck) Pouzar Illus. p. 47

COMMON NAME: none.
CAP: ¾–4" (2–10 cm) wide, convex, becoming broadly convex to irregularly plane, occasionally depressed; surface dry, uneven, azonate, finely tomentose, cinnamon brown to reddish brown, sometimes with a grayish brown bloom, darkening when bruised, margin entire or sometimes with a concentric ridge of secondary growth, often misshapen because of fusing; undersurface covered with spines.
FLESH: in two layers; upper layer thick, spongy, dark brown; lower layer thin, fibrous-tough, cinnamon brown; odor and taste not distinctive.
SPINES: decurrent, up to ¼" (6 mm) long, dark to pale brown with slightly paler tips when young, darkening when bruised.
STALK: 1⅛–4" (3–10 cm) long, ¼–¾" (5–20 mm) thick, enlarged downward to a very broad bulbous base, up to 2" (5 cm) wide, often fused and arising with several others

from a thick, tomentose pad of mycelium, dark reddish brown to dull dark brown or grayish brown.

SPORE PRINT COLOR: cocoa brown.

MACROCHEMICAL TESTS: sections of dried material in KOH produce a violet flash, then stain black, and a dark olivaceous brown color leaches into the mounting medium (Harrison 1968).

MICROSCOPIC FEATURES: spores 5.5–7 × 5–6 μm, subglobose, moderately to coarsely tuberculate, pale brown.

FRUITING: solitary, scattered, or in groups or fused clusters on the ground in broadleaf woods, especially with oak; June–December, sometimes overwintering; North Carolina west to Tennessee and northward, southern distribution limits yet to be determined; fairly common.

COMMENTS: *Hydnellum pineticola* (not illustrated), inedible, is very similar but grows under conifers, especially pine.

Hydnum repandum Linnaeus : Fries   Illus. p. 47

COMMON NAME: Sweet Tooth, Hedgehog.

CAP: ¾–6" (2–15.5 cm) wide, convex to nearly plane, sometimes slightly depressed; surface dry, felty, becoming somewhat wrinkled and pitted in age, pale buff or yellow-orange or pale brownish orange to apricot orange or reddish orange, staining dark orange when bruised; margin wavy, sometimes deeply indented or lobed; undersurface covered with spines.

FLESH: thick, firm, brittle, white, staining orange-yellow when cut and rubbed; odor pleasant, somewhat nutty to sweet, or not distinctive; taste mild or sometimes peppery.

SPINES: attached to subdecurrent, ⅛–⅜" (3–10 mm) long; creamy white to orange-yellow, darkening when bruised.

STALK: 1–4" (2.5–10 cm) long, ⅜–1⅜" (1–3.5 cm) thick, nearly equal, solid, white with orange tints or colored like the cap, bruising orange-yellow.

SPORE PRINT COLOR: white.

MICROSCOPIC FEATURES: spores 7–8.5 × 6–7 μm nearly round, smooth, hyaline.

FRUITING: solitary, scattered, or in groups on the ground under conifer and broadleaf trees; June–November; widely distributed in the Southeast; common.

EDIBILITY: edible.

COMMENTS: Formerly known as *Dentinum repandum*. The name *Hydnum rufescens* is a synonym. *Hydnum repandum* var. *album* (see photo, p. 47), is nearly identical, but it has a whitish to pale tan cap that stains orange when bruised, a whitish stalk, and whitish spines that darken when bruised.

Mycorrhaphium adustum (Schw.) M. Geesteranus   Illus. p. 48

COMMON NAME: Kidney-shaped Tooth.

CAP: 1–3" (2.5–7.5 cm) wide, kidney- to fan-shaped or sometimes circular, broadly convex to nearly flat; surface finely roughened to somewhat velvety, whitish to tan, staining smoky gray when bruised; margin wavy, thin, faintly zoned, sometimes blackish in age.

FLESH: thin, fibrous-tough, white.

SPINES: up to ⅛" (3 mm) long, somewhat flattened, typically fused and appearing forked at their tips, white at first, becoming pinkish brown to cinnamon brown at maturity.

STALK: ¾–1⅛" (2–3 cm) long, ⅜–¾" (1–2 cm) thick, lateral, somewhat velvety, whitish to dull cream, sometimes rudimentary or absent.

SPORE PRINT COLOR: white.

MICROSCOPIC FEATURES: spores 2.5–4 × 1–1.5 μm, cylindric, smooth, hyaline.
FRUITING: solitary, scattered, or in groups, often fused or overlapping on decaying broadleaf logs and fallen branches, especially oak; July–December, sometimes overwintering; widely distributed in the Southeast; fairly common.
EDIBILITY: inedible.
COMMENTS: Formerly known as *Hydnum adustum*.

Phellodon niger (Fries) Karsten   Illus. p. 48

COMMON NAME: none.
CAP: 1⅛–2¾" (3–7 cm) wide, broadly convex to nearly plane or depressed to funnel-shaped; surface dry, tomentose or rarely smooth, sometimes roughened and irregular with pits, horn-like projections, and ridges over the disc, usually with dark-colored zones present, whitish to grayish white or pale grayish at first, becoming smoky brown to olive brown to blackish olive or dark purple toward the center, staining brownish black when bruised.
FLESH: duplex, in both the cap and stalk; upper layer spongy, colored like the cap surface; lower layer hard, bluish black to black; odor fragrant and stronger when dry; taste not distinctive.
SPINES: subdecurrent, up to ⅛" (4 mm) long, gray, staining blackish gray when bruised.
STALK: ¾–2" (2–5 cm) long, ³⁄₁₆–⅜" (4–10 mm) thick, nearly equal to tapered downward to a bulbous base and a mycelial pad, velvety to felted, blackish gray to dark purple.
SPORE PRINT COLOR: white.
MACROCHEMICAL TESTS: flesh stains blue-green with the application of KOH or NH_4OH.
MICROSCOPIC FEATURES: spores 3.5–5 × 3.5–5 μm, subglobose to globose, echinulate, hyaline.
FRUITING: solitary, scattered, or concrescent in sandy soil under conifers, especially pine and hemlock, under broadleaf trees, especially oak, and under mixed woods; June–November; reported from North Carolina west to Tennessee and northward, distribution limits yet to be determined; occasional.
EDIBILITY: inedible.
COMMENTS: *Phellodon alboniger* (not illustrated), inedible, is very similar, but its cap retains some white coloration well into maturity, and it has a smooth cap surface.

Boletes

Boletes, also known as fleshy pored fungi, are among the most fascinating and highly prized mushrooms. Their beautiful colors, distinctive features, and relative abundance make them one of the most popular groups collected. Boletes are relatively safe to collect for the table and are immensely popular among mycophagists. Most boletes grow on the ground and are soft and fleshy. They have a cap, a stalk, and a sponge-like layer of tubes on the undersurface of the cap. Except for species in the genus *Gastroboletus,* which have enclosed and irregularly arranged tubes, boletes have vertically arranged tubes, each of which terminates in a pore. The tube layer is easily detached and typically separates cleanly from the cap flesh. Polypores also have tubes on the underside, but can be differentiated from boletes because they grow on wood or sometimes on the ground arising from buried wood. Their fruit bodies are typically tough and leathery to woody, and their tube layers usually do not separate cleanly from the cap flesh.

The majority of boletes are mycorrhizal with trees and only rarely are found in open fields or grassy areas. One of the most important steps in bolete identification is obtaining a spore print. Although some boletes can be very difficult to identify even with the aid of chemical tests and the microscope, many are distinctive and easily identified using only macroscopic features.

For additional information about boletes, see *North American Boletes—A Color Guide to the Fleshy Pored Mushrooms* (Bessette, Roody, and Bessette 2000), listed in the Nontechnical Publications section of "Recommended Reading."

Austroboletus betula (Schweinitz) Horak Illus. p. 48
  COMMON NAME: Shaggy-stalked Bolete.
  CAP: 1⅛–3½" (3–9 cm) wide, convex, becoming broadly convex; surface smooth, viscid when moist, dark red, reddish, orange, bright yellow, or yellow-brown to red-brown, sometimes with a yellow margin; margin even.
  FLESH: greenish yellow to orange-yellow, not blueing when cut or bruised; odor and taste not distinctive.
  PORE SURFACE: yellow to greenish yellow, becoming olive brown in age, not staining when bruised; pores circular, 1 mm wide; tubes 1–1.5 cm deep.
  STALK: 4–8" (10–20 cm) long, ¼–¾" (6–20 mm) thick, nearly equal, solid, yellow to dark red or dull red, coarsely reticulate to shaggy with raised yellow ribs that may redden in age, often with white mycelium at the base; partial veil and ring absent.
  SPORE PRINT COLOR: dark olive to olive brown.

MACROCHEMICAL TESTS: unknown.
MICROSCOPIC FEATURES: spores 15–19 × 6–10 μm; narrowly elliptic, ornamented with a loose reticulum and scattered minute pits, typically with a distinct apical pore, pale brown.
FRUITING: solitary to scattered on the ground in mixed oak-pine and beech forests; July–September; occasional to locally common; widely distributed in the Southeast.
EDIBILITY: edible.
COMMENTS: Also known as *Boletellus betula*. This exceptionally beautiful and distinctive bolete is not likely to be mistaken for any other.

Austroboletus gracilis var. gracilis (Peck) Wolfe

Illus. p. 48

COMMON NAME: Graceful Bolete.
CAP: 1⅛–4" (3–10 cm) wide, convex to broadly convex; surface dry, finely velvety when young, sometimes rimose in age, maroon to reddish brown or cinnamon, at times tawny to yellow-brown; margin even.
FLESH: white or tinged pink, not staining blue or brown; odor not distinctive; taste mild or slightly tart.
PORE SURFACE: white when young, becoming pinkish to pinkish brown or burgundy-tinged in age, darkening or staining brownish when bruised; pores circular, 1–2 per mm; tubes 1–2 cm deep.
STALK: long and slender in relation to the cap diameter, 3–7" (7.5–18 cm) long, ¼–⅜" (6–10 mm) thick at the apex, enlarging downward or nearly equal, colored like the cap or paler, whitish at the base, solid, with elevated, anastomosing lines that sometimes form an obscure, narrow reticulation overall or at least on the upper half; partial veil and ring absent.
SPORE PRINT COLOR: pinkish brown to reddish brown.
MACROCHEMICAL TESTS: cap surface stains mahogany red to reddish brown with the application of KOH, stains brilliant blue-green then amber orange to dull orange with NH_4OH, and is negative with $FeSO_4$; flesh stains pale greenish gray with the application of $FeSO_4$ and is negative with KOH or NH_4OH.
MICROSCOPIC FEATURES: spores 10–17 × 5–8 μm, narrowly ovoid to subelliptic, pitted, pale brown.
FRUITING: solitary, scattered, or in groups on the ground and on decaying wood in conifer and broadleaf forests; June–October; widely distributed in the Southeast; frequent.
EDIBILITY: edible.
COMMENTS: Also known as *Tylopilus gracilis* and *Porphyrellus gracilis*. The species epithet means "slender." *Gyroporus* species have a hollow stalk at maturity, a white to yellow pore surface, and a pale to bright yellow spore print color. *Tylopilus* species have smooth (not pitted) spores.

Austroboletus subflavidus (Murrill) Wolfe

Illus. p. 48

COMMON NAME: none.
CAP: 1⅛–4" (3–10 cm) wide, convex, becoming nearly plane and usually slightly depressed in age; surface dry, finely velvety at first, soon becoming areolate, white to buff or pale yellowish to grayish yellow, often with a pale ochraceous salmon tinge; margin even.
FLESH: white, unchanging when cut; odor somewhat fruity; taste bitter.
PORE SURFACE: white to grayish at first, becoming pinkish at maturity, not staining when bruised; pores angular to nearly circular, about 1 mm wide on mature specimens, sometimes beaded with clear fluid on young specimens; tubes 1–2 cm deep.

STALK: 1¾–5¾" (4.5–14.5 cm) long, ¼–1⅛" (7–30 mm) thick, nearly equal or tapered downward, rarely enlarged at the base, colored like the cap, coarsely reticulate with conspicuous raised ribs that give a pitted appearance, solid; flesh yellow in the base and white above; partial veil and ring absent.

SPORE PRINT COLOR: reddish brown.

MACROCHEMICAL TESTS: cap surface stains cinnamon buff with the application of KOH or NH_4OH.

MICROSCOPIC FEATURES: spores 15–20 × 6–9 μm, fusoid, minutely pitted, pale brown.

FRUITING: scattered or in groups on the ground under oak and pine; June–October; fairly common to infrequent; New Jersey Pine Barrens south to Florida, west to Texas.

EDIBILITY: inedible because of the bitter taste.

COMMENTS: Previously known as *Tylopilus subflavidus* and *Porphyrellus subflavidus*.

Boletellus ananas (Curtis) Murrill   Illus. p. 48

COMMON NAME: Pineapple Bolete.

CAP: 1⅛–4" (3–10 cm) wide, obtuse at first, becoming convex to broadly convex; surface dry, coated with coarse and overlapping purplish red to dark red scales that become dull pinkish tan to dingy yellow in age and extend beyond the margin.

FLESH: whitish to yellowish, typically blueing when cut; odor and taste not distinctive.

PORE SURFACE: yellow when fresh, sometimes tinged reddish brown in age, blueing when bruised; pores irregular to angular, 1–2 mm wide; tubes 9–16 mm deep.

STALK: 2⅜–6" (6–15.5 cm) long, ¼–⅝" (7–16 mm) thick, enlarged downward or nearly equal, solid, dry, glabrous to slightly fibrillose, white to pale tan, sometimes with a reddish zone near the apex; partial veil whitish, usually leaving remnants on the cap margin and typically not forming a ring on the stalk.

SPORE PRINT COLOR: dark rusty brown to dark brown.

MACROCHEMICAL TESTS: cap surface stains olive yellow then maroon with the application of KOH and olive yellow with NH_4OH.

MICROSCOPIC FEATURES: spores 15–24 × 7–11 μm, fusoid, with conspicuous longitudinally ridged thin wings spirally arranged, often with an indistinct apical pore, pale brown.

FRUITING: scattered or in groups under oaks and pines, often on the bases of trees; May–November; North Carolina to Florida, west to Texas; occasional.

EDIBILITY: inedible.

COMMENTS: Previously known as *Boletus ananus* and *Boletus coccineus*. *Boletellus pictiformis* = *Suillellus pictiformis* (not illustrated), edibility unknown, is a rare species known only from Florida. It has a dry cap coated with appressed to erect, shaggy, chestnut brown to dark brown scales, a stalk sheathed with shaggy to fibrillose chestnut brown to brown scales, and a reddish to orange-brown pore surface that stains blue when bruised. *Boletellus fallax* (not illustrated), edibility unknown, another Florida species, is nearly identical to *Boletellus pictiformis*, but it has a yellow to olive yellow pore surface that also stains blue when bruised.

Boletus abruptibulbus W. C. Roody and Both   Illus. p. 49

COMMON NAME: none.

CAP: 1–2¼" (2.5–5.7 cm) wide, convex, becoming broadly convex to flattened with a narrow band of sterile tissue at the margin; surface dull to shiny, glabrous, cracking and becoming more or less areolate in dry conditions, chestnut brown to reddish brown or cinnamon; margin incurved at first, at times becoming decurved in age.

FLESH: whitish to pale yellow, unchanging when exposed; odor pleasant; taste mild.
PORE SURFACE: pale golden yellow, becoming dingy yellowish olive, not changing when bruised; pores angular, 1–2 per mm.
STALK: 1¼–2" (3–5 cm) long, 3⁄16–1" (1–2.5 cm) thick, typically enlarged to abruptly bulbous at the base, often with a narrow root-like projection, solid, glabrous to minutely pruinose on the lower half and basal area, pallid or yellowish at the apex, gradually darker and becoming colored like the cap toward the base; partial veil and ring absent.
SPORE PRINT COLOR: not recorded.
MACROCHEMICAL TESTS: unknown.
MICROSCOPIC FEATURES: spores 14–21 × 5.5–7.5 μm, spindle-shaped, smooth, pale amber in Melzer's.
FRUITING: solitary or in scattered groups in sand in coastal oak-pine woods, especially in old colonized dunes; December–March; distribution yet to be established; fairly common locally.
EDIBILITY: edible.
COMMENTS: This smallish bolete seems to be limited to coastal woods and barrier islands of the Gulf Coast and is currently only recorded from Florida. The abruptly bulbous stalk base is very unusual for boletes and is distinctive.

Boletus albisulphureus (Murrill) Murrill   Illus. p. 49
COMMON NAME: Chalky-white Bolete.
CAP: 1½–4¾" (4–12 cm) wide, obtuse to convex, becoming broadly convex at maturity; surface dry, glabrous, sometimes finely areolate in age, milk white to grayish white, sometimes with yellowish or brownish tinges, especially near the margin; margin even.
FLESH: white, unchanging when exposed; odor somewhat pungent and medicinal or not distinctive; taste not distinctive.
PORE SURFACE: white to buff at first, becoming yellow at maturity then dingy olive yellow in age, not blueing when bruised, adnate to subdecurrent; pores circular to angular, mostly less than 1 mm wide; tubes 5–15 mm deep.
STALK: 2–3½" (5–9.5 cm) long, 5⁄8–1⅜" (1.6–3.5 cm) thick, nearly equal or enlarged downward, sometimes abruptly narrowed at the base, dry, solid, milk white to whitish with a yellowish apex, white to yellowish reticulation on the upper portion; partial veil and ring absent.
SPORE PRINT COLOR: olive brown.
MACROCHEMICAL TESTS: cap surface weakly stains pale pinkish with the application of KOH.
MICROSCOPIC FEATURES: spores 11–15 × 3.5–4.5 μm, subfusiform, smooth, nearly hyaline to pale yellowish, some dextrinoid in Melzer's.
FRUITING: solitary, scattered, or in groups in sandy soil under oaks; July–September; widely distributed in the Southeast; fairly common.
EDIBILITY: edible and sometimes regarded as choice.
COMMENTS: *Xanthoconium stramineum* = *Boletus stramineus* (p. 245), edible, is similar, but lacks reticulation on the stalk and has a white to buff pore surface. *Tylopilus peralbidus* (not illustrated), edibility unknown, also has a white cap and stalk, but its pore surface does not become yellow, and its stalk lacks reticulation.

Boletus auriflammeus Berkeley and Curtis   Illus. p. 49
COMMON NAME: none.
CAP: 1⅛–3½" (3–9.5 cm) wide, convex, becoming broadly convex to nearly plane in age;

surface dry, distinctly pulverulent at first, becoming tomentose, and sometimes rimose in age, bright orange-yellow, sometimes with olive tints; margin even.

FLESH: yellow, not blueing when exposed; odor not distinctive; taste acidic or not distinctive.

PORE SURFACE: yellow to yellow-orange at first, becoming olive yellow to greenish yellow at maturity, then becoming bright crimson to crimson-orange in age, not blueing when bruised, adnate to subdecurrent and often depressed near the stipe at maturity; pores angular, radially elongated near the stipe, typically more than 1 mm at maturity; tubes up to 1.5 cm deep.

STALK: 2–3½" (5–9 cm) long, ¼–½" (5–12 mm) thick, nearly equal or enlarged in either direction, narrowed at the base, solid, pulverulent, becoming nearly glabrous, orange-yellow, typically reticulate at least on the upper portion of mature specimens but often absent or indistinct on young ones; mycelium white; partial veil and ring absent.

SPORE PRINT COLOR: olive brown to ochre-brown.

MACROCHEMICAL TESTS: pileipellis stains vinaceous brown or darker orange with a faint slate blue flash with the application of NH_4OH, stains dull brown or dark amber after dissolving the yellow with KOH, and is negative with $FeSO_4$.

MICROSCOPIC FEATURES: spores 8–12 × 3–5 μm, subellipsoid to subfusiform, smooth, nearly hyaline.

FRUITING: solitary, scattered, or in groups on the ground in woods, usually associated with oaks; July–November; New York south to Florida, west to Ohio and Tennessee; occasional.

EDIBILITY: unknown.

COMMENTS: Handling this bolete stains fingers yellow. *Auriflammeus* means "flaming gold," which is a good description for the deep, rich color of this striking bolete. *Boletus aurantiosplendens* (not illustrated), edibility unknown, is similar but has an orange to brownish orange or brownish yellow cap, a yellow to apricot or orange stalk with tawny to reddish brown streaks that do not stain fingers when handled, and yellow flesh that darkens when exposed. *Boletus roxanae* (not illustrated), edibility unknown, is somewhat similar but is not as brilliantly colored, typically has brownish tints on its cap, and lacks reticulation on the stalk.

Boletus auriporus Peck   Illus. p. 49

COMMON NAME: none.

CAP: ¾–3⅛" (2–8 cm) wide, convex, becoming broadly convex to nearly plane; surface coated with tiny appressed fibrils, smooth, moist, and viscid when fresh, becoming dull when dry, pinkish cinnamon to pinkish brown or vinaceous brown when fresh, often fading in age or when dry; taste of the cap surface usually acidic; margin with a narrow band of sterile tissue that is whitish when young, brownish at maturity.

FLESH: white to pale yellow, except vinaceous under the cap surface, not blueing when cut or bruised; odor and taste not distinctive.

PORE SURFACE: brilliant golden yellow when young and fresh, becoming dull yellow in age, usually slowly staining dull brick red when bruised; pores angular, 1–2 per mm; tubes 6–15 mm deep.

STALK: 1½–4½" (4–11.5 cm) long, ¼–⅝" (6–17 mm) thick, solid, slightly enlarged downward or nearly equal, typically narrowed abruptly at the base, viscid when fresh, pale yellow at the apex, streaked and flushed pale pinkish brown downward, with copious white mycelium at the base; partial veil and ring absent.

SPORE PRINT COLOR: olive brown.

MACROCHEMICAL TESTS: cap surface stains burgundy red to dull blood red or dark reddish vinaceous or bleaches reddish tones to pale rusty orange with the application of NH_4OH; stains reddish, amber orange, or yellowish with KOH; flesh is negative with the application of KOH, NH_4OH, or $FeSO_4$.

MICROSCOPIC FEATURES: spores 11–16 × 4–6 μm, fusiform-elliptic, smooth, pale brown.

FRUITING: scattered or in groups on the ground under oak; July–October; widely distributed in the Southeast; fairly common.

EDIBILITY: edible.

COMMENTS: The cap surface and tube layer of this bolete are easily removed. *Boletus viridiflavus* is a synonym of *Boletus auriporus*. Compare with *Boletus innixus* (p. 217), which is similar but has a yellowish brown cap, a stalk that is swollen above a tapered base, and flesh that has an odor similar to *Scleroderma citrinum*.

Boletus bicolor var. bicolor Peck Illus. p. 49

COMMON NAME: Two-colored Bolete, Red-and-Yellow Bolete.

CAP: 2–6" (5–15.5 cm) wide, convex, becoming broadly convex to nearly plane, or irregular; surface dry, velvety-subtomentose when young, sometimes rimose in age, dark red to purple-red, red to rose red or rose pink when fresh, turning yellow to ochraceous brown in age; margin even.

FLESH: pale yellow, slowly staining blue when bruised or cut; odor and taste not distinctive.

PORE SURFACE: yellow when fresh, dingy yellow or olive-tinged at maturity, sometimes with reddish tints in age, staining greenish blue (sometimes weakly), when bruised; pores angular, 1–2 per mm; tubes 3–10 mm deep.

STALK: 2–4" (5–10 cm) long, 3/8–1 1/8" (1–3 cm) thick, nearly equal or club-shaped, solid, yellow at the apex, red or rosy red on the lower two-thirds or more, unchanging or slowly staining blue when bruised or cut; partial veil and ring absent.

SPORE PRINT COLOR: olive brown.

MACROCHEMICAL TESTS: cap surface stains blackish with the application of $FeSO_4$ and is negative with KOH or NH_4OH; flesh stains bluish gray to olive green with the application of $FeSO_4$, stains pale orange to pale yellow with KOH, and is negative with NH_4OH.

MICROSCOPIC FEATURES: spores 8–12 × 3.5–5 μm, oblong to slightly ventricose, smooth, pale brown, inamyloid; pileotrama inamyloid.

FRUITING: solitary or in groups on the ground under oaks; June–October; widely distributed in the Southeast; fairly common.

EDIBILITY: edible and choice; be careful not to confuse this bolete with similar poisonous species such as *Boletus sensibilis*.

COMMENTS: *Boletus bicolor* var. *subreticulatus* (not illustrated), edible, is nearly identical, but its stalk has conspicuous reticulation over the upper 1–3 cm. *Boletus sensibilis* (p. 226), poisonous, has a dark to pale brick red cap that fades to dull rose or dingy cinnamon in age; a predominantly yellow stalk lightly flushed with red; pale yellow flesh that has an odor of curry, licorice, or fenugreek; and a yellow pore surface; both the flesh and pore surface stain blue quickly. *Boletus pallidoroseus* (not illustrated), edibility unknown, is similar, but its cap and stalk are paler, pink to reddish pink, and the odor of its cut flesh is like beef bouillon at first and like rotten meat in age.

Boletus carminiporus A. E. Bessette, Both, and Dunaway　　　　　　　　　Illus. p. 49
　COMMON NAME: none.
　CAP: 1⅛–5½" (3–14 cm) wide, convex, becoming broadly convex to nearly plane in age; surface dry to subviscid, glabrous or nearly so, dull red at first, becoming pinkish red to orange-red at maturity, fading to reddish orange to dull golden orange in age; margin incurved to inrolled at first, becoming decurved at maturity, with a narrow band of sterile tissue.
　FLESH: whitish to pale yellow, becoming darker yellow when exposed or in age, not blueing at all when exposed; odor and taste not distinctive.
　PORE SURFACE: yellow when very young, soon becoming dark red to brownish red, fading to dull red or orange-red in age, staining bluish green then dull olive when bruised, depressed near the stalk at maturity; pores angular to irregular, 2–3 per mm; tubes 3–12 mm deep, yellow, staining bluish green when cut.
　STALK: 2–4½" (5–11.5 cm) long, ⅜–1⅛" (1–3 cm) thick, enlarged downward to a pinched base, rarely tapered downward or nearly equal, dry, solid, distinctly reticulate overall or at least on the upper portion, rose pink at first, soon becoming dark red at the apex and paler red below, staining brownish red or slowly olive green to olive yellow when bruised; flesh pale yellow to yellow, darker than in the cap, unchanging when exposed, becoming dull red around larval tunnels, lacking reddish hairs at the base; partial veil and ring absent.
　SPORE PRINT COLOR: olive brown.
　MACROCHEMICAL TESTS: cap surface stains dull golden yellow to pale amber with the application of KOH, olive to yellowish olive with NH_4OH, and olive gray with $FeSO_4$; flesh stains orange-buff with the application of KOH, bluish gray with NH_4OH, and gray to bluish gray with $FeSO_4$.
　MICROSCOPIC FEATURES: spores 8–11 × 3–4 μm, subfusoid, smooth, ochraceous.
　FRUITING: solitary, scattered, or in groups on the ground in mixed broadleaf forests, especially with beech, hickory, and oak, and in mixed woods with oak and pine; June–September; fairly common; from North Carolina south to Florida, west to Arkansas and Louisiana; northern distribution limits yet to be established.
　EDIBILITY: unknown.
　COMMENTS: *Boletus rubroflammeus,* poisonous (see *MNE,* p. 366), and *Boletus flammans* (p. 215), edibility unknown, are similar to *B. carminiporus,* but all parts of these mushrooms stain blue when cut or bruised.

Boletus curtisii Berkeley　　　　　　　　　Illus. p. 49
　COMMON NAME: none.
　CAP: 1⅛–3½" (3–9.5 cm) wide, obtuse to convex, becoming broadly convex to nearly plane in age; surface viscid to glutinous when fresh, glabrous, bright yellow to orange-yellow, sometimes with brownish tints or whitish areas in age; margin with a narrow band of sterile tissue, incurved when young.
　FLESH: whitish, unchanging when exposed; odor and taste not distinctive.
　PORE SURFACE: whitish to buff or pale yellow at first, duller and darker at maturity, often depressed near the stalk in age, not blueing when bruised; pores circular to angular, 2–3 per mm; tubes 6–12 mm deep.
　STALK: 2⅜–4¾" (6–12 cm) long, ¼–½" (6–13 mm) thick, nearly equal, viscid to glutinous when fresh, solid or often hollow, somewhat scurfy near the apex, nearly smooth below, pale yellow to yellow down to a base that is sheathed with cottony white mycelium, not reticulate; partial veil and ring absent.

SPORE PRINT COLOR: olive brown.

MACROCHEMICAL TESTS: unknown.

MICROSCOPIC FEATURES: spores 9.5–17 × 4–6 μm, elliptic to subventricose, smooth, yellowish.

FRUITING: solitary, scattered, or in groups on the ground in conifer and mixed woods, often with pines; August–November; widely distributed in the Southeast; frequent.

EDIBILITY: unknown.

COMMENTS: The gluten has an acidic taste and stains fingers yellow. The overall aspect of this brightly colored bolete has a superficial resemblance to members of the genus *Suillus*.

Boletus dupainii Boudier   Illus. p. 49

COMMON NAME: none.

CAP: 1–4⅜" (2.5–11 cm) wide, hemispherical at first, becoming convex and finally broadly convex to nearly plane in age; surface smooth, slimy-viscid when fresh, becoming shiny when dry, purplish red to pinkish red or bright red, sometimes with yellowish spots; margin incurved at first, becoming decurved and extending as a thin band of sterile tissue on mature specimens.

FLESH: whitish to pale yellow, quickly blueing when exposed, becoming blackish blue; odor pleasant and musky or not distinctive; taste not distinctive.

PORE SURFACE: dark red at first, becoming carmine red to orange-red and finally fading to yellow toward the margin in age, blueing when bruised; pores circular to angular, 1–3 per mm; tubes 4–10 mm deep, yellow to olive, blueing slightly when bruised.

STALK: 1⅛–4" (3–10 cm) long, ⅜–1⅜" (1–3.5 cm) thick, enlarged downward and typically clavate with a somewhat pointed base or nearly equal, dry, solid, coated with fine reddish punctae that are more abundant toward the base, ground color yellow, typically with a white basal mycelium; flesh yellowish, becoming reddish toward the base, blueing when exposed.

SPORE PRINT COLOR: olive brown.

MACROCHEMICAL TESTS: unknown.

MICROSCOPIC FEATURES: spores 10–17 × 3.5–6 μm, fusiform-ellipsoid to elliptic, with prominent guttules, smooth, light yellow.

FRUITING: solitary, scattered, or in groups on sandy soil under broadleaf trees, especially oak, and in mixed woods; August; reported only from North Carolina; rare.

EDIBILITY: unknown.

COMMENTS: The combination of a reddish pore surface and a slimy-viscid cap is most unusual. The color photograph of this mushroom was taken by mycologist Owen McConnell. His collection is the first reported from North America.

Boletus edulis Bulliard   Illus. p. 50

COMMON NAME: King Bolete, Cep, Steinpilz, Porcini.

CAP: 1¾–15" (4.5–38 cm) wide, convex to nearly plane; surface smooth to slightly wrinkled, dry to somewhat viscid when moist, brown to reddish brown, pale cinnamon brown, rusty red, or yellowish tan; margin even.

FLESH: white, not blueing when bruised; odor and taste not distinctive.

PORE SURFACE: white when young, becoming yellow to olive yellow then brownish yellow to brown in age, staining yellowish olive to dull orange cinnamon or pale yellowish brown when bruised; pores small, circular, 2–3 per mm; tubes 1–3 cm deep.

STALK: 2–12" (5–31 cm) long, ¾–3" (2–7.5 cm) thick, enlarging downward or nearly equal, sometimes bulbous, white or pale brown, with a distinct whitish reticulum on the upper one-third or more, solid; partial veil and ring absent.

SPORE PRINT COLOR: olive brown.

MACROCHEMICAL TESTS: cap surface stains orange with the application of KOH or NH_4OH and slowly pale grayish green with $FeSO_4$; flesh stains pale grayish green with the application of $FeSO_4$ and is negative with KOH or NH_4OH.

MICROSCOPIC FEATURES: spores 13–19 × 4–6.5 μm, elliptic, smooth, pale yellowish brown.

FRUITING: solitary, scattered, or in groups on the ground in woods, under conifers, especially spruce; June–October; widely distributed in the Southeast; occasional.

EDIBILITY: edible and choice; one of the most highly prized edible mushrooms.

COMMENTS: *Boletus edulis* is variable in color and stature. Several forms, varieties, and subspecies have been recognized. The Bitter Bolete, *Tylopilus felleus*, inedible, has a brown stalk with coarse brown reticulation, a vinaceous spore print color, and very bitter-tasting flesh (see *MNE*, p. 371). *Boletus pinophilus* (not illustrated), edible and choice, is very similar, but it grows on the ground under pines, and its stalk has a reticulum that is whitish near the apex and brownish below and that darkens when handled. *Boletus edulis* may be easily confused with several similar species. Compare with *Boletus variipes* (p. 229), *Boletus nobilis* (p. 220), and *Xanthoconium separans* (p. 245).

Boletus firmus Frost   Illus. p. 50

COMMON NAME: none.

CAP: 2⅜–6" (6–15.5 cm) wide, convex, becoming broadly convex in age; surface dry, glabrous but appearing finely velvety, whitish to grayish, tan to pinkish tan or pale grayish olive; margin even.

FLESH: whitish to pale yellow, blueing when cut, sometimes slowly or slightly; odor not distinctive; taste mild to slightly bitter.

PORE SURFACE: red to red-orange, blueing when bruised; pores angular, 1–2 per mm; tubes 8–20 mm deep.

STALK: 2–4¾" (5–12 cm) long, ⅜–1⅛" (1–3 cm) thick, nearly equal or slightly tapered in either direction, dry, solid, nearly smooth, colored like the cap, sometimes with reddish tints, staining olive to brownish when handled or bruised, with a slight red reticulation near the apex or nearly overall, occasionally lacking reticulation; partial veil and ring absent.

SPORE PRINT COLOR: olive brown.

MACROCHEMICAL TESTS: cap surface stains pale blue or produces a pale slate flash that becomes dingy gray or pale brownish with the application of NH_4OH, dingy yellow or pale brownish to blackish brown with KOH, and pale slate blue with $FeSO_4$; flesh stains dull brownish with the application of NH_4OH, reddish brown to dull amber with KOH, and dark gray to grayish olive with $FeSO_4$.

MICROSCOPIC FEATURES: spores 9–15 × 3.5–5 μm, subellipsoid, smooth yellowish.

FRUITING: solitary, scattered, or in groups on the ground in broadleaf forests and mixed woods with oak and beech; July–September; Georgia west to Mississippi and northward; occasional.

EDIBILITY: inedible because of the bitter taste.

COMMENTS: *Boletus piedmontensis* and *Boletus satanas* var. *americanus* are synonyms of *Boletus firmus*.

Boletus flammans Dick and Snell Illus. p. 50
  COMMON NAME: none.
  CAP: 1½–4¾" (4–12 cm) wide, convex, becoming broadly convex to nearly plane in age; surface dry to slightly viscid when moist, minutely subtomentose to glabrous, dark red to brick red, rosy red or brownish red, sometimes with tan areas near the margin in age, blueing when bruised; margin even or nearly so.
  FLESH: yellowish, blueing when bruised; odor and taste not distinctive.
  PORE SURFACE: bright red to orange-red without a yellow or orange-yellow marginal zone, depressed near the stalk at maturity, blueing when bruised; pores circular to angular, 1–3 per mm; tubes 8–20 mm deep.
  STALK: 1½–4" (4–10 cm) long, ⅜–¾" (1–2 cm) thick, nearly equal or slightly enlarged downward, dry, solid, colored like the cap, typically yellow near the base, finely reticulate on the upper half or at least near the apex, blueing when bruised; partial veil and ring absent.
  SPORE PRINT COLOR: olive brown.
  MACROCHEMICAL TESTS: unknown.
  MICROSCOPIC FEATURES: spores 10–16 × 4–5 μm, cylindric to ellipsoid, smooth, yellowish.
  FRUITING: solitary, scattered, or in groups on the ground under conifers, especially spruce, hemlock, and pine; July–October; widely distributed in the Southeast; occasional.
  EDIBILITY: unknown.
  COMMENTS: *Boletus rubroflammeus*, poisonous, is very similar but has a darker red to deep wine red cap and a more prominently reticulate stalk, and it grows under broadleaf trees (see *MNE*, p. 366). *Boletus subluridellus* (see photo, p. 53), edibility unknown, is also similar but grows on the ground under broadleaf trees, especially oak, and in mixed woods. It has a pale yellow stalk that darkens from the base upward in age or where handled and is faintly scurfy overall but that lacks reticulation. *Boletus carminiporus* (p. 212), edibility unknown, is also similar, but it does not stain blue when cut or bruised.

Boletus floridanus (Singer) Singer Illus. p. 50
  COMMON NAME: none.
  CAP: 1½–4" (4–10 cm) wide, convex, becoming broadly convex to nearly plane; surface dry but viscid when wet, subtomentose, pinkish red, rose red, purplish red, or brownish red; typically yellowish or whitish along the margin; margin incurved at first, even or nearly so.
  FLESH: pale to bright yellow, quickly blueing when cut; odor and taste not distinctive.
  PORE SURFACE: reddish orange to pinkish red, sometimes with yellow tints, often beaded with yellow drops when young and fresh, usually depressed near the stalk at maturity; pores circular to angular, 1–3 per mm; tubes 6–12 mm deep, blueing when bruised.
  STALK: 1½–4" (4–10 cm) long, ⅜–1" (1–2.5 cm) thick, often bulbous when young, becoming nearly equal or enlarged downward at maturity, dry, solid, typically yellow near the apex and red below or sometimes red overall, with a conspicuous red reticulation at least on the upper half; reticulation venose but not lacerated; partial veil and ring absent.
  SPORE PRINT COLOR: olive brown.
  MACROCHEMICAL TESTS: unknown.

MICROSCOPIC FEATURES: spores 13–18 × 4–5 μm, ellipsoid-fusoid, smooth, pale yellow-brown.

FRUITING: scattered or in groups in sandy soil under oaks; July–December; Tennessee and the coastal plain of North Carolina south to Florida, west to Texas; occasional.

EDIBILITY: edible with a somewhat acidic taste.

COMMENTS: Formerly known as *Boletus frostii* ssp. *floridanus*. Compare with *Boletus frostii*, edible, which has a dark red cap and much coarser raised reticulation (see *MNE*, p. 365).

Boletus griseus Frost   Illus. p. 50

COMMON NAME: none.

CAP: 2–5¾" (5–14 cm) wide, convex, becoming broadly convex to nearly plane, sometimes slightly depressed; surface dry, appressed-fibrillose, fibrils grayish, often somewhat scaly in age, pale to dark gray or brownish gray when young, sometimes developing yellowish or ochre tints in age; margin even.

FLESH: whitish with dark yellow-brown around larval tunnels, unchanging or staining dingy red or brown when cut or bruised; odor and taste not distinctive.

PORE SURFACE: whitish to grayish or dingy gray-brown, not yellow, unchanging or staining brownish or gray when bruised; pores circular, 1–2 per mm; tubes 8–20 mm deep.

STALK: 1⅝–5¾" (4–14.5 cm) long, ⅜–1⅜" (1–3.5 cm) thick, equal or tapered downward, often curved near the base, solid, whitish or grayish when young, developing yellow tones from the base upward as it matures, sometimes with reddish stains; surface covered overall with a coarse pale to yellowish reticulum that becomes brownish to blackish in age; partial veil and ring absent.

SPORE PRINT COLOR: pinkish brown to olive brown.

MACROCHEMICAL TESTS: flesh typically stains bluish gray with the application of $FeSO_4$.

MICROSCOPIC FEATURES: spores 9–13 × 3–5 μm, oblong, smooth, pale brown.

FRUITING: solitary to scattered on the ground in mixed broadleaf forests, especially under oaks; June–September; widely distributed in the Southeast; occasional.

EDIBILITY: edible, but its flesh often riddled with insects.

COMMENTS: Also known as *Xerocomus griseus*.

Boletus hortonii A. H. Smith and Thiers   Illus. p. 50

COMMON NAME: none.

CAP: 1⅝–4¾" (4–14 cm) wide, convex, becoming broadly convex in age; surface dry or viscid when moist, conspicuously corrugated and pitted, nearly glabrous, color ranging from chestnut to ochre, cinnamon, or reddish brown, sometimes paler in age; margin even.

FLESH: whitish, not reddening when cut or bruised, sometimes blueing slowly and weakly when cut or not staining at all; odor and taste not distinctive.

PORE SURFACE: yellow when young, becoming duller yellow or olive yellow in age, sometimes slowly and weakly staining blue when bruised; pores circular to angular, 2–3 per mm; tubes 5–10 mm deep.

STALK: 2⅜–4" (6–10 cm) long, ⅜–¾" (1–2 cm) thick, nearly equal or enlarged toward the base, dry, solid, sometimes pruinose, pale yellow to tan, occasionally with reddish tones near the base; partial veil and ring absent.

SPORE PRINT COLOR: olive brown.

MACROCHEMICAL TESTS: cap surface produces a blue-green flash then stains olive brown with the application of NH_4OH, olive brown with KOH, and pale olive tan with

$FeSO_4$; flesh stains bluish gray with the application of $FeSO_4$ and is negative with KOH or NH_4OH.

MICROSCOPIC FEATURES: spores 12–15 × 3.5–4.5 μm, somewhat boat-shaped, smooth, yellow.

FRUITING: solitary to scattered on the ground in mixed woods and under broadleaf trees, often with oak, beech, and hemlock; June–October; widely distributed in the Southeast; frequent.

EDIBILITY: edible.

COMMENTS: Previously called *Boletus subglabripes* var. *corrugis*. The corrugated to pitted cap is distinctive. *Boletus subglabripes* (p. 227), edible, is similar but has a smooth cap.

Boletus innixus Frost

Illus. p. 50

COMMON NAME: none.

CAP: 1⅛–3⅛" (3–8 cm) wide, convex, becoming broadly convex; surface dry, becoming somewhat viscid when wet, often rimose in age; dull reddish brown, yellow-brown, or grayish brown, often dull purplish red on the margin, fading to dull cinnamon when dry; margin even.

FLESH: white to pale yellow, vinaceous under the cap surface, with pale brownish tinges in the stalk; odor musty, reminiscent of witch hazel or similar to the characteristic pungent odor of a sectioned fruiting body of fresh *Scleroderma citrinum*; taste not distinctive.

PORE SURFACE: brilliant golden yellow when young and fresh, becoming dull yellow in age; pores circular to angular, 1–3 per mm when young, up to 2 mm wide at maturity; tubes 3–10 mm deep.

STALK: short and stout, 1⅛–2⅜" (3–6 cm) long, ⅜–⅝" (1–1.6 cm) thick at the apex, often enlarged up to 1" (2.5 cm) downward, typically club-shaped and distinctly swollen above a tapered base, yellowish, streaked with dark brown tones, moist or dry on the upper portion, viscid near the base, solid, with yellow mycelium sometimes visible at the base; partial veil and ring absent.

SPORE PRINT COLOR: olive brown.

MACROCHEMICAL TESTS: cap surface stains mahogany red to reddish brown with the application of KOH, produces a green flash then stains dull orange-red with NH_4OH, and is pale olive with $FeSO_4$; flesh stains pale dull pinkish orange with the application of KOH or NH_4OH and pale gray with $FeSO_4$.

MICROSCOPIC FEATURES: spores 8–11 × 3–5 μm, elliptic, smooth, pale brown.

FRUITING: often growing in small clusters fused at the stalk bases; or solitary, scattered, or in groups on the ground under broadleaf trees, especially oak; June–October; widely distributed in the Southeast; occasional to fairly common.

EDIBILITY: edible.

COMMENTS: Also known as *Aureoboletus innixus*. It has frequently been misnamed *Boletus auriporus*. Carefully compare *Boletus innixus* with *Boletus auriporus* (p. 210), with which it is often confused.

Boletus longicurvipes Snell and A. H. Smith

Illus. p. 51

COMMON NAME: none.

CAP: ⅝–2⅜" (1.6–6 cm) wide, convex, becoming broadly convex; surface smooth to slightly wrinkled or pitted, viscid to glutinous when fresh, butterscotch yellow to yellow-brown, yellow-orange, orange, orange-brown, or reddish orange, often becoming olive green with an orange center in age; margin even.

FLESH: white, becoming pale yellowish, unchanging when cut or bruised; odor and taste not distinctive.

PORE SURFACE: pale yellow to dingy yellow or olive yellow, becoming greenish gray in age, not blueing when cut or bruised, but may stain brownish; pores circular, 2 per mm; tubes 8–12 mm deep.

STALK: 2–3½" (5–9 cm) long, ⅜–¾" (1–2 cm) thick, nearly equal or enlarging downward, often curved above the base, solid, whitish to yellowish or with reddish or pinkish tan tones, scurfy from tiny scabrous dots that are yellow when young but reddish to red-brown in age, basal mycelium white to whitish; partial veil and ring absent.

SPORE PRINT COLOR: olive brown.

MACROCHEMICAL TESTS: cap surface stains cherry red to reddish orange with the application of KOH or NH_4OH and is negative with $FeSO_4$; flesh is negative with the application of KOH, NH_4OH, or $FeSO_4$.

MICROSCOPIC FEATURES: spores 13–17 × 4–5 μm, narrowly subfusiform to oblong, smooth, pale brown.

FRUITING: solitary, scattered, or in groups on the ground in woods, especially oak-pine forests; July–September; widely distributed in the Southeast; fairly common.

EDIBILITY: edible.

COMMENTS: *Boletus rubropunctus* (not illustrated), edible, has a yellow stalk that is punctate with reddish scabrous dots; a pale yellow basal mycelium; and larger spores that measure 17–21 × 5.5–7.5 μm.

Boletus luridiformis Rostkovius   Illus. p. 51

COMMON NAME: Slender Red-pored Bolete.

CAP: 2–4¾" (5–12 cm) wide, convex, becoming broadly convex in age; surface dry, subtomentose to nearly smooth, dark brown to nearly blackish brown when young, becoming reddish brown to olive brown at maturity, staining dark blue to blackish brown when bruised; margin even.

FLESH: greenish yellow to yellow, quickly staining blue when cut; odor and taste not distinctive.

PORE SURFACE: orange-red to dull orange, sometimes yellow when very young, staining blue when bruised; pores circular to irregular, 1–3 per mm; tubes 8–14 mm deep.

STALK: 2⅜–4" (6–10 cm) long, ½–2" (1.3–5 cm) thick, nearly equal or enlarged in either direction, dry, solid, with reddish to orange cinnamon pruina on a yellow ground color, often reddish at the base, blueing when bruised, not reticulate, lacking reddish hairs at the base; flesh yellow except for reddish in the base, rapidly blueing when exposed; partial veil and ring absent.

SPORE PRINT COLOR: olive brown.

MACROCHEMICAL TESTS: cap surface stains dark amber with the application of KOH or NH_4OH and dark olive green with $FeSO_4$; flesh slowly stains olivaceous with the application of $FeSO_4$ and bleaches the blued surface yellow then white with KOH or NH_4OH.

MICROSCOPIC FEATURES: spores 12–16 × 4.5–6 μm, subfusoid, smooth, ochraceous.

FRUITING: solitary, scattered, or in groups on the ground under conifers and broadleaf trees; July–January; reported from Mississippi and Texas, more common northward; occasional.

EDIBILITY: not recommended; although consumed in Europe, American collections are thought to be poisonous, causing gastrointestinal distress.

COMMENTS: Also known as *Boletus erythropus*. *Boletus hypocarycinus* (see photo, p. 50), edibility unknown, is very similar but has smaller spores, 8–12 × 3–4 μm. *Boletus subvelutipes* (p. 227), poisonous, has reddish hairs on the stalk base and a brighter, more orange cap. *Boletus luridus* (p. 219), poisonous, has a reticulate stalk.

Boletus luridus Schaeffer Illus. p. 51

COMMON NAME: none.

CAP: 1½–4¾" (4–12 cm) wide, convex, becoming broadly convex to nearly plane in age, sometimes shallowly depressed at maturity; surface dry, fibrillose, color variable, yellow to olive yellow or olive brown, sometimes with pinkish or reddish tints, staining greenish blue when bruised; margin with a narrow band of sterile tissue, incurved at first.

FLESH: yellowish to reddish with a red line above the tubes of freshly sectioned specimens, quickly blueing when cut; odor and taste not distinctive.

PORE SURFACE: dark red at first, becoming orange-red and depressed near the stalk at maturity, quickly blueing when bruised; pores circular, 2–3 per mm; tubes 1–2 cm deep.

STALK: 2⅜–6" (6–15.5 cm) long, ⅜–1⅛" (1–3 cm) thick, nearly equal or enlarged downward, dry, solid, covered with red reticulation at least near the apex or nearly overall, somewhat pruinose to scurfy-punctate, yellow with a red base or apex, or variously streaked with red; partial veil and ring absent.

SPORE PRINT COLOR: olive brown.

MACROCHEMICAL TESTS: cap surface stains dark red to blackish with the application of KOH; flesh stains orange-yellow with the application of KOH and dark blue with Melzer's after the natural blueing has faded.

MICROSCOPIC FEATURES: spores 11–16 × 5–7 μm, ovoid to subellipsoid, smooth, yellow.

FRUITING: solitary, scattered, or in groups on the ground under broadleaf trees, especially oak, and under conifers; July–November; widely distributed in the Southeast; occasional.

EDIBILITY: poisonous, causing gastrointestinal distress.

COMMENTS: Similar species—including *Boletus subvelutipes* (p. 227), poisonous; *Boletus luridiformis* (p. 218), poisonous; and *Boletus hypocarycinus* (see photo, p. 50), edibility unknown—lack reticulation on their stalks.

Boletus mahagonicolor A. E. Bessette, Both, and Dunaway Illus. p. 51

COMMON NAME: none.

CAP: 1–3½" (2.5–9 cm) wide, convex, becoming broadly convex in age; surface dry, glabrous to silky and shiny or finely tomentose, mahogany to reddish brown or rose brown, paler toward the margin, blueing along the margin when bruised; margin incurved at first and remaining so well into maturity, with a narrow band of sterile tissue.

FLESH: yellow, rapidly blueing when exposed, then slowly changing to pale reddish brown; odor and taste not distinctive.

PORE SURFACE: adnate to slightly depressed, bright yellow at first, becoming greenish yellow then brownish yellow at maturity, staining dark blue to bluish black when bruised; pores angular, 1–3 per mm; tubes very shallow, 2–8 mm deep, concolorous with the pore surface, rapidly blueing.

STALK: 1⅛–2¾" (3–7 cm) long, ⅜–1" (1–2.5 cm) thick, enlarged downward, dry, solid, minutely scurfy to nearly glabrous, bright yellow on the upper portion, with orange-red to brownish red tones on the lower one-half to two-thirds or more, blueing when bruised, lacking reticulation; partial veil and ring absent; basal mycelium whitish to very pale yellow.

SPORE PRINT COLOR: olive brown.
MACROCHEMICAL TESTS: cap surface darkens with the application of KOH, stains yellowish with NH_4OH, and is negative with $FeSO_4$; flesh stains pale brownish orange with the application of KOH, pale orange with NH_4OH, and grayish with $FeSO_4$; pore surface stains burnt orange with the application of KOH, stains pale reddish brown with NH_4OH, and is negative with $FeSO_4$.
MICROSCOPIC FEATURES: spores 10–13 × 3.5–4 μm, elliptic, smooth, ochraceous.
FRUITING: scattered or in groups on the ground under oak and pine; May–July; known only from Mississippi, distribution limits yet to be established; uncommon.
EDIBILITY: unknown.
COMMENTS: The name *mahagonicolor* refers to the mahogany tree, *Swietenia mahagoni*, which has wood similar to the color of this species' cap. *Boletus rubricitrinus* (p. 225), edible, is very similar but has a more reddish cap, different macrochemical test reactions, longer tubes, and much larger spores, 13–19 × 5–8 μm. *Boletus oliveisporus* (not illustrated), edibility unknown, is also similar but has a dark fulvous cap tinged with bay that instantly stains blue-black when bruised; much longer tubes, up to 20 mm deep; and larger spores that measure 11–17 × 4–6 μm.

Boletus nobilis Peck

Illus. p. 51

COMMON NAME: none.
CAP: 2¾–8" (7–20 cm) wide, convex, becoming broadly convex to nearly plane in age; surface dry, smooth, glabrous, yellowish brown or reddish brown at first, becoming ochraceous to olive ochraceous or reddish ochraceous with age; margin even.
FLESH: white, becoming yellowish near the tubes, not staining when exposed; odor and taste not distinctive.
PORE SURFACE: white at first, becoming yellow to pale ochraceous or brownish yellow and often slightly depressed near the stalk in age; pores circular to angular, 1–3 per mm; tubes 8–25 mm deep.
STALK: 2¾–6" (7–16 cm) long, ½–1⅜" (1.2–3.5 cm) thick, nearly equal or sometimes ventricose to slightly enlarged near the base, dry, solid, white to pale ochraceous, delicately reticulate on the upper half or at least near the apex and glabrous below, or glabrous overall; partial veil and ring absent.
SPORE PRINT COLOR: dull ochre-brown to dull rusty brown.
MACROCHEMICAL TESTS: cap surface stains brown to amber brown with the application of KOH; flesh stains vinaceous pink to reddish with the application of NH_4OH and purplish red with KOH; stipitipellis stains reddish with the application of NH_4OH and amber to brown with KOH.
MICROSCOPIC FEATURES: spores 11–16 × 4–5 μm, subfusiform, smooth, pale olivaceous.
FRUITING: solitary, scattered, or in groups on the ground under broadleaf trees, especially oak and beech; July–September; uncommon; New England west to the Great Lakes region, south to West Virginia and Tennessee, southern distribution limits yet to be established.
EDIBILITY: edible.
COMMENTS: *Boletus atkinsonii* (not illustrated), edible, is similar, but it has more prominent reticulation on the stalk, a darker cap roughened with tufts of hyphae that are best observed with a hand lens, and smaller spores that measure 10–13 × 4–5 μm. Also compare with *Boletus variipes* (p. 229).

Boletus ornatipes Peck   Illus. p. 51
  COMMON NAME: Ornate-stalked Bolete.
  CAP: 1⅝–6¼" (4–16 cm) wide, convex, becoming nearly plane in age; surface dry, subtomentose to velvety-subtomentose, color variable, pale gray to purplish gray, yellow, olive, yellow-brown, or olive brown, often darkest at the center with yellow at the margin; margin even.
  FLESH: yellow, not blueing when cut or bruised; odor not distinctive; taste bitter or sometimes mild.
  PORE SURFACE: lemon yellow, staining deeper yellow, orange-yellow, or orange-brown when cut or bruised; pores circular, 1–2 per mm; tubes 4–15 mm deep.
  STALK: 3⅛–6" (8–15.5 cm) long, ⅜–1" (1–2.5 cm) thick, equal to ventricose or slightly enlarged downward, solid, bright yellow, often developing brown tones in age, with coarse and raised reticulation often extending to the base, staining darker orange-yellow when bruised, staining fingers yellow when handled; partial veil and ring absent.
  SPORE PRINT COLOR: olive brown to dark yellow-brown.
  MACROCHEMICAL TESTS: cap surface stains pale orange-brown with the application of KOH, stains pale brownish with NH_4OH, and is negative with $FeSO_4$; flesh stains pale orange-brown with the application of KOH, becomes pale brownish with NH_4OH, and bleaches to dull white with $FeSO_4$.
  MICROSCOPIC FEATURES: spores 9–13 × 3–4 μm, oblong to slightly ventricose with apex obtuse, smooth, pale brown.
  FRUITING: solitary or scattered, frequently caespitose on the ground along road banks and near or under oak, beech, and other broadleaf trees; June–September; widely distributed in the Southeast; fairly common.
  EDIBILITY: edible to inedible; most collections from the Southeast are very bitter.
  COMMENTS: Some authors use *Boletus retipes* as a synonym, whereas others reserve this name for smaller specimens with a powdery yellow cap. *Boletus auriflammeus* (p. 209), edibility unknown, has a bright orange-yellow cap that sometimes has olive tints, mild-tasting flesh, and an orange-yellow reticulate stalk; it also stains fingers yellow when handled.

Boletus pallidus Frost   Illus. p. 52
  COMMON NAME: none.
  CAP: 1¾–6" (4.5–15.5 cm) wide, convex when young, becoming broadly convex to nearly plane, sometimes slightly depressed in age; surface dry, often rimose in age, slightly viscid when moist, whitish to buff or pale gray-brown when young, dingy brown with occasional rose tints in age; margin even.
  FLESH: whitish or pale yellow, sometimes slowly and weakly staining bluish or pinkish when cut or bruised; odor not distinctive; taste mild or slightly bitter.
  PORE SURFACE: whitish to pale yellow when young, becoming yellow to greenish yellow in age, staining greenish blue then grayish brown when bruised; pores circular, becoming nearly angular in age, 1–2 per mm; tubes 1–2 cm deep.
  STALK: 2–4¾" (5–12 cm) long, ⅜–1" (1–2.5 cm) thick, nearly equal to enlarging downward, solid, whitish when young, sometimes yellow at the apex, with occasional reddish flushes near the base in age, smooth to slightly reticulate near the apex, often with white mycelium at the base, possibly blueing slightly when cut or bruised; partial veil and ring absent.
  SPORE PRINT COLOR: olive or olive brown.

MACROCHEMICAL TESTS: cap surface stains rusty orange with the application of KOH and pale rusty orange with NH_4OH or $FeSO_4$; flesh stains pale rusty orange with the application of KOH, becomes blue-green with NH_4OH, and is negative with $FeSO_4$; pore surface stains rusty orange with the application of KOH, blue-green with NH_4OH, and rusty orange with $FeSO_4$.

MICROSCOPIC FEATURES: spores 9–15 × 3–5 μm, narrowly oval to subfusoid, smooth, pale brown.

FRUITING: solitary or in groups or caespitose on the ground under mixed broadleaf trees, especially oaks; July–September; widely distributed in the Southeast; occasional to fairly common.

EDIBILITY: edible.

COMMENTS: Similarly colored species of *Boletus* typically have a distinctly reticulate stalk.

Boletus patrioticus Baroni, A. E. Bessette, and Roody — Illus. p. 52

COMMON NAME: none.

CAP: 1⅛–5⅛" (3–13 cm) wide, convex, becoming broadly convex to nearly plane in age; surface dry, velvety-tomentose, olive when very young, soon pinkish to brick red or dark red, often brownish red toward the margin, usually with olive to tarnished brass tints; margin incurved at first, becoming decurved, with a narrow band of sterile tissue.

FLESH: pale yellowish or pinkish red to purplish red under the cap surface or extending throughout the cap, slowly staining blue below the reddish area when exposed; odor not distinctive; taste tart to acidic.

PORE SURFACE: pale yellow at first, becoming olive yellow at maturity, depressed near the stalk in age, blueing when bruised; pores angular, 1–2 per mm; tubes 3–15 mm deep, blueing when cut.

STALK: 1–4" (2.5–10 cm) long, ⅜–¾" (1–2 cm) thick, usually enlarged downward, sometimes tapered downward or nearly equal, base often pinched, dry, solid, pruinose to scurfy, color variable, usually rosy red on the upper portion or nearly overall and olive toward the base, often a mixture of these colors, with a yellow ground color; basal mycelium whitish to pale yellow; partial veil and ring absent; flesh whitish to pale yellow, becoming dingy yellow to brownish toward the base, staining bluish to greenish and sometimes rosy red when exposed.

SPORE PRINT COLOR: olive brown.

MACROCHEMICAL TESTS: cap surface stains olive brown with the application of KOH, olive amber with NH_4OH, and olive gray with $FeSO_4$; flesh stains pale orange-yellow with the application of KOH, NH_4OH, or $FeSO_4$.

MICROSCOPIC FEATURES: spores 10–13 × 4–5.5 μm, subfusiform, smooth, deep golden brown.

FRUITING: solitary, scattered, or in groups on the ground, often in grassy areas, under oak, and in mixed woods with oak present; May–October; North Carolina south to Florida, west to Ohio and Texas; fairly common.

EDIBILITY: unknown.

COMMENTS: Formerly known as *Boletus communis*. The name *patrioticus* means "patriotic," a reference to the red, white, and blue colors displayed by the exposed flesh of this bolete.

Boletus pulverulentus Opatowski — Illus. p. 52

COMMON NAME: none.

CAP: 1½–5" (4–12.5 cm) wide, pulvinate to convex then broadly convex to nearly plane, margin even; surface dry, subtomentose, becoming glabrous and often somewhat shiny in age, dark yellow-brown to blackish brown or dark cinnamon brown and often developing reddish tints in age, instantly bruising blackish blue when handled.

FLESH: yellow, instantly blueing upon exposure; odor and taste not distinctive.

PORE SURFACE: yellow when young, darkening to golden yellow to brownish yellow when mature, instantly blueing then slowly staining dull brown when bruised; pores angular, 1–2 per mm; tubes 6–12 mm deep.

STALK: 1⅜–3½" (3.5–9 cm) long, ⅝–1⅜" (1.5–3.5 cm) thick, nearly equal or sometimes enlarged downward, dry, solid, pruinose at the apex, yellow above and darker yellow to orange-yellow downward, typically reddish brown and pruinose toward the base, quickly blueing then slowly staining dull brown when handled; flesh reddish brown in the base, yellow above, and instantly staining blackish blue when exposed; lacking reticulation but often with raised longitudinal ridges; partial veil and ring absent.

SPORE PRINT COLOR: dark olive to olive brown.

MACROCHEMICAL TESTS: pileipellis displays a green flash with the application of NH_4OH.

MICROSCOPIC FEATURES: spores 11–15 × 4–6 μm, fusoid to elliptical, smooth, yellowish.

FRUITING: scattered or in groups on soil under conifers or broadleaf trees; July–November; eastern Canada south to Florida, west to Michigan; occasional.

EDIBILITY: edible.

COMMENTS: This species is also known as *Xerocomus pulverulentus*.

Boletus purpureorubellus Baroni, Yetter, and Norarevian   Illus. p. 52

COMMON NAME: none.

CAP: 3–5" (7.5–12.5 cm) wide, convex, becoming broadly convex to nearly plane in age; surface viscid when wet, shiny when dry, glabrous or nearly so, color variable, dark red, purplish blood red, yellow on the margin; margin undulating, with a narrow band of sterile tissue.

FLESH: yellow with a reddish line under the cap surface, becoming pale vinaceous red above the tube layer; yellow areas stain blue then slowly whitish when exposed; odor and taste not distinctive; taste of the cap surface slightly acidic.

PORE SURFACE: bright yellow to golden yellow, staining dark bluish gray to blackish blue and eventually brown when bruised, slightly depressed with a decurrent tooth; pores angular, 0.5–2 mm wide, conspicuously lamellate near the stalk; tubes 8–10 mm deep, staining fuscous blue when cut.

STALK: 2–3⅛" (5–8 cm) long, ¼–⅝" (5–17 mm) thick, nearly equal or tapered downward, dry, solid, pseudoreticulate at the apex with decurrent lines, delicately pruinose in the red areas, glabrous elsewhere, apex yellow to pale creamy, streaked with pale red to red downward, becoming dull red to brown or golden yellow at the base; basal mycelium bright yellow; partial veil and ring absent; flesh yellow at the apex, vinaceous red and yellow below, yellow areas staining dark blue when exposed.

SPORE PRINT COLOR: olive to brownish olivaceous.

MACROCHEMICAL TESTS: cap surface stains yellow with the application of KOH.

MICROSCOPIC FEATURES: spores 5–7.3 × 3–4 μm, short-ellipsoid, smooth, dull yellow.

FRUITING: scattered or in groups in a dense swamp, growing in mats of moss with tree roots under loblolly, bay, Cyrilla, red maple, and cabbage palm; July–September; Florida and Georgia, distribution limits yet to be established; rare.

EDIBILITY: unknown.

COMMENTS: Formerly known as *Boletus rubellus* ssp. *purpureus*. The very small spores are unusual for a bolete. Another species, *Boletus rubellus,* edibility unknown, has a dry, tomentose cap and much larger spores (see *MNE*, p. 366).

Boletus roseolateritius A. E. Bessette, Both, and Dunaway Illus. p. 52
COMMON NAME: Rosy Brick-red Bolete.
CAP: 1½–6" (4–15.5 cm) wide, convex, becoming broadly convex to nearly plane in age; surface dry, velvety-subtomentose, with a faint grayish bloom when young, finely scurfy to nearly glabrous at maturity, dark reddish salmon or burnt orange at first, becoming rosy brick red to reddish brown at maturity, fading to brownish orange or brownish pink with dull yellow tints and finally to dull dingy yellow in age; margin incurved at first, becoming decurved and remaining so well into maturity, with a narrow band of sterile tissue, pale salmon pink to whitish, drying purplish rose to dull brick red, instantly staining dark blue where bruised.
FLESH: pale lemon yellow, instantly blueing on exposure then slowly fading to pale lemon yellow; odor and taste not distinctive.
PORE SURFACE: orange-red at first, becoming dull orange and finally orange-yellow in age, instantly blueing when bruised; pores circular to angular, 1–3 per mm; tubes 5–12 mm deep, pale lemon yellow, instantly blueing on exposure.
STALK: 2–4⅜" (5–11 cm) long, ½–1⅜" (1.2–3.5 cm) thick, nearly equal or tapered in either direction, often with a pinched base, dry, solid, faintly longitudinally striate, lacking reticulation, pale lemon yellow, instantly blueing where bruised, slowly developing rusty brown stains in age or when bruised.
SPORE PRINT COLOR: olive brown.
MACROCHEMICAL TESTS: cap surface stains black with the application of $FeSO_4$ and is negative with KOH or NH_4OH; flesh quickly stains yellow-orange with the application of KOH and bleaches blued areas a pale lemon yellow.
MICROSCOPIC FEATURES: spores 8.5–11 × 3.5–4.5 (5.5) μm, narrowly ellipsoid to subfusoid, smooth, pale ochraceous, lacking an apical pore.
FRUITING: solitary, scattered, or in groups on the ground in river bottomland under beech, with oak and hickory nearby; June–August; reported only from McComb, Mississippi, distribution range yet to be determined; occasional.
EDIBILITY: unknown.
COMMENTS: The name *roseolateritius* means "rosy brick red," a reference to the color of the mature cap. *Boletus fairchildianus* (not illustrated), edibility unknown, is very similar, but its cap stains olive with the application of NH_4OH and deep maroon or orange-yellow to brown with KOH, and it has much larger spores that measure 13–19 × 5–8 μm. *Boletus luridellus* (see photo, p. 51), edible, has a yellow-brown to amber brown cap that is sometimes streaked brown over a yellow ground color and that becomes ochraceous tawny to hazel in age; flesh that quickly stains blue when cut; a yellow pore surface that quickly bruises blue when bruised; and a yellow stalk that becomes dark red to brownish toward the base. The stalk is covered with brown reticulation on the upper portion, is often punctate below, and quickly stains blue when bruised.

Boletus roseopurpureus Both, A. E. Bessette, and Roody Illus. p. 52
COMMON NAME: none.
CAP: 2¾–6" (7–16 cm) wide, convex, becoming broadly convex to nearly plane in age; surface viscid-tacky in wet weather but drying quickly, velvety-tomentose at first, becoming appressed-tomentose to obscurely fibrillose, the fibrils darker, narrow; a striking pinkish purple when fresh, becoming darker purplish pink to dark purplish red

or brownish red, at times somewhat mottled in these colors, the marginal areas becoming grayish in age; margin incurved at first, later decurved, projecting as a sterile band of tissue.

FLESH: pale yellow, instantly dark blue when exposed but soon fading to pale slate color; odor fragrant, taste very sour, like lemon.

PORE SURFACE: at first lemon yellow or more golden, becoming yellowish olive to greenish olive at maturity, instantly dark blue to greenish blue when bruised, soon fading to pale greenish blue or greenish gray; pores circular, 1–2 per mm; tubes 5–10 mm deep, concolorous with the pore surface, blueing rapidly when injured.

STALK: 1¾–3" (4.5–8.5 cm) long, ¾–1½" (2–4 cm) thick, equal or slightly tapered downward, reticulated nearly the entire length or at least over the upper two-thirds, the lower third appressed-tomentose, the reticulation in part strongly raised; bright yellow over most of the length of the stalk, occasionally with burgundy red areas near the base; with a white base and white basal mycelium, instantly staining dark blue when bruised; flesh deep yellow, some specimens burgundy red in basal area, instantly blueing when exposed but soon fading to pale slate color.

SPORE PRINT COLOR: light brownish olive.

MACROCHEMICAL TESTS: cap surface stains dark vinaceous brown or dingy dull orange with the application of KOH and dingy pale yellowish or dull orange with NH_4OH; flesh and stalk surface stains dull orange with the application of KOH or NH_4OH.

MICROSCOPIC FEATURES: spores 9.4–13 × 2.7–3.5 μm, narrowly ellipsoid to subfusoid, smooth, pale ochraceous, lacking an apical pore.

FRUITING: solitary or in small groups in mixed woods of oak, beech, hemlock, and maple, apparently associated with red oak; July–August; along the Gulf Coast from Florida to Texas, north to New England and New York; occasional.

EDIBILITY: unknown.

COMMENTS: The name *roseopurpureus* refers to the pinkish purple color of the cap. *Boletus speciosus* (not illustrated), edible, has a bright red cap that lacks purplish tones, and it has longer and wider spores that measure 11–15 × 3–5.5 μm.

Boletus rubricitrinus (Murrill) Murrill   Illus. p. 52

COMMON NAME: none.

CAP: 1⅛–6" (3–15 cm) wide, pulvinate to convex, becoming broadly convex to nearly plane, sometimes slightly depressed in age; surface dry, glabrous to velvety-subtomentose, color variable, dull rose red to dull brick red or reddish brown to tawny cinnamon when fresh, fading to tawny olive or dull brown in age, sometimes with yellow tints; margin incurved at first, with a band of sterile tissue.

FLESH: pale yellow to yellow, quickly blueing when exposed; odor and taste not distinctive.

PORE SURFACE: yellow at first, becoming dull yellow to olive yellow and depressed near the stalk in age, blueing when bruised; pores irregular, 1–3 per mm; tubes 8–20 mm deep.

STALK: 2–4¾" (5–12 cm) long, ⅝–1⅜" (1.5–3.5 cm) thick, tapered in either direction or nearly equal, with a ventricose base, dry, solid, yellow with olive gray tints, typically with dull reddish to reddish brown streaks and dots, especially toward the base, lacking reticulation or weakly reticulate at the very apex, typically longitudinally striate nearly overall; partial veil and ring absent.

SPORE PRINT COLOR: olive brown.

MACROCHEMICAL TESTS: pileipellis stains olive with the application of NH_4OH and deep maroon or orange-yellow to brown with KOH.

MICROSCOPIC FEATURES: spores 13–19 × 5–8 μm, fusoid to subfusoid-ellipsoid, smooth, yellowish.

FRUITING: solitary, scattered, or in groups in sandy soil in oak or oak-pine woods; May–November; New Jersey south to Florida, west to Texas; fairly common.

EDIBILITY: edible.

COMMENTS: *Boletus fairchildianus* (not illustrated), edibility unknown, is very similar but has a red pore surface and more red on the stalk.

Boletus rufomaculatus Both

Illus. p. 53

COMMON NAME: none.

CAP: 2½–5½" (6.5–14 cm) wide, hemispheric, becoming broadly convex to nearly plane in age; surface subviscid when wet, tomentose to granular tomentose when young, becoming appressed-fibrillose in age, variable in color, at first dull rusty brown to ochre-brown with marginal areas rusty brown to reddish or pale brown or honey yellow; in age the lighter colors predominate and become dotted, splashed, or mottled with brick red to brownish red; margin incurved at first, with a narrow band of sterile tissue.

FLESH: pale yellow, slowly blueing when exposed especially above the tubes, at times blueing only weakly and erratically; odor and taste not distinctive.

PORE SURFACE: pale yellow at first, becoming darker yellow and finally yellowish olive and shallowly depressed near the stalk at maturity, sometimes with rusty spots or reddish tints in age, blueing when bruised; pores variable in shape, about 1.5 mm wide; tubes 6–14 mm deep.

STALK: 2⅜–3⅛" (6–8 cm) long, 5⁄16–1⅛" (8–30 mm) thick, nearly equal or subventricose, dry, solid, minutely appressed-tomentose to nearly glabrous, somewhat ribbed to pseudoreticulate, golden yellow near the apex, pale yellow below, strongly dotted or mottled with burgundy red in the lower half and sometimes nearly overall, staining blue when bruised or handled; basal mycelium white; partial veil and ring absent.

SPORE PRINT COLOR: dark olive.

MACROCHEMICAL TESTS: cap surface stains dull yellowish to yellow with the application of NH_4OH and dark amber with KOH; flesh stains dark yellow with the application of NH_4OH and pinkish amber to amber with KOH.

MICROSCOPIC FEATURES: spores 10–13 × 3–4.5 μm, subfusiform, smooth, yellowish in Melzer's; pileotrama inamyloid.

FRUITING: solitary, scattered, or in groups on the ground under beech in mixed woods; June–September; reported from western New York and Mississippi, distribution limits yet to be established; occasional.

EDIBILITY: unknown.

COMMENTS: The name *rufomaculatus* means "spotted reddish," a reference to the reddish spots and splashes on the cap. Compare with *Boletus bicolor* var. *bicolor* (p. 211), edible.

Boletus sensibilis Peck

Illus. p. 53

COMMON NAME: none.

CAP: 2–6¼" (5–16 cm) wide, convex, becoming broadly convex to nearly plane in age; surface velvety-subtomentose when young, dry, dark to pale brick red, fading to dull rose or sometimes dingy cinnamon in age, quickly staining blue when bruised; margin even.

FLESH: pale yellow, blueing instantly when cut or bruised; odor variously described as faintly fruity, like maple syrup, fenugreek, curry, or licorice; taste not distinctive.

PORE SURFACE: yellow when young, becoming duller or browner in age, blueing instantly when bruised; pores circular, 1–2 per mm; tubes 8–12 mm deep.

STALK: 3 1/8–4 3/4" (8–12 cm) long, 3/8–1 3/8" (1–3.5 cm) thick, equal or enlarging slightly downward, solid, mostly yellow but often tinged pink or red near the base, occasionally finely reticulate only at the very top, quickly staining blue when bruised; flesh bright yellow, quickly staining blue when exposed; partial veil and ring absent.

SPORE PRINT COLOR: olive brown.

MACROCHEMICAL TESTS: cap surface stains yellow with the application of KOH or NH_4OH and greenish gray with $FeSO_4$; flesh instantly stains yellow with the application of KOH or NH_4OH and yellow-orange to yellow-brown with $FeSO_4$.

MICROSCOPIC FEATURES: spores 10–13 × 3.5–4.5 μm, suboblong to slightly ventricose, smooth, pale brown; hymenial cystidia fusoid-ventricose with an elongated neck.

FRUITING: scattered or in groups on the ground in woods, usually under broadleaf trees; July–September; widely distributed in the Southeast; occasional to fairly common.

EDIBILITY: poisonous, causing gastrointestinal distress that may be severe in some individuals.

COMMENTS: *Boletus miniato-pallescens*, edibility unknown, has a red to brick red cap that fades to reddish orange or orange-yellow (see *MNE*, p 366). *Boletus miniato-olivaceus* (not illustrated), poisonous, has different microscopic features and a red to rosy red cap that becomes olivaceous to yellowish in age, and the odor and taste of its flesh are not distinctive.

Boletus subglabripes Peck
Illus. p. 53

COMMON NAME: none.

CAP: 1 3/4–4" (4.5–10 cm) wide, convex, becoming broadly convex to nearly plane in age, sometimes broadly umbonate; surface smooth to slightly wrinkled, deep auburn or chestnut or bay brown; margin even.

FLESH: pale yellow to whitish, rarely blueing slightly when cut or bruised; odor not distinctive; taste mild to slightly acidic.

PORE SURFACE: bright yellow when fresh, duller or slightly greenish yellow in age; pores circular, 2 per mm; tubes 5–16 mm deep.

STALK: 2–4" (5–10 cm) long, 3/8–3/4" (1–2 cm) thick, nearly equal, solid, yellow with occasional onion skin pink to reddish, possibly with reddish brown tinges on the lower portion, minutely scurfy with a thin coating of tiny yellow scales, dots, or points, not reticulate; partial veil and ring absent.

SPORE PRINT COLOR: olive brown.

MACROCHEMICAL TESTS: flesh stains grayish olive green with the application of $FeSO_4$.

MICROSCOPIC FEATURES: spores 13–21 × 5–7 μm, fusoid, smooth, pale brown.

FRUITING: scattered on the ground under broadleaf trees especially oak, occasionally under conifers; July–October; widely distributed in the Southeast; fairly common.

EDIBILITY: edible.

COMMENTS: *Boletus hortonii* (p. 216), is very similar, but its cap is conspicuously pitted.

Boletus subvelutipes Peck
Illus. p. 53

COMMON NAME: Red-mouth Bolete.

CAP: 2 3/8–5 1/8" (6–13 cm) wide, convex, becoming broadly convex to nearly plane in age; surface dry, velvety-subtomentose when young, occasionally rimose in age, color variable, cinnamon brown to yellow-brown, reddish brown, or reddish orange to orange-yellow, quickly staining blue to blue-black when bruised; margin even.

FLESH: bright yellow, quickly staining dark blue to blackish when cut or bruised; odor not distinctive, taste mild to slightly acidic.

PORE SURFACE: color variable, red, brownish red, dark maroon red, or red-orange to orange when fresh, often with a yellow rim, duller in age, quickly staining dark blue to blackish when cut or bruised; pores circular, 2 per mm; tubes 8–26 mm deep.

STALK: 1⅛–4" (3–10 cm) long, ⅜–¾" (1–2 cm) thick, nearly equal, solid, furfuraceous, flushed red and yellow, typically yellow at the apex, not reticulate, quickly staining blue to blackish when bruised; often with short, stiff, dark red hairs at the base on mature specimens (immature specimens may have yellow hairs at the base that become dark red in age); partial veil and ring absent.

SPORE PRINT COLOR: dark olive brown.

MACROCHEMICAL TESTS: cap surface stains mahogany red with the application of KOH or NH_4OH and grayish green to dark olive green with $FeSO_4$; blued flesh stains rusty orange with the application of KOH or NH_4OH and pale yellow-orange with $FeSO_4$.

MICROSCOPIC FEATURES: spores 13–18 × 5–6.5 μm, fusoid-subventricose, smooth, pale brown.

FRUITING: solitary to scattered on the ground under broadleaf trees, especially oak, and sometimes under conifers; June–September; widely distributed in the Southeast; fairly common.

EDIBILITY: poisonous, causing gastrointestinal distress.

COMMENTS: *Boletus luridiformis* (p. 218), poisonous, is similar, but it has a darker reddish brown cap, and it lacks hairs on its stalk base.

Boletus tenax A. H. Smith and Thiers

Illus. p. 53

COMMON NAME: none.

CAP: 1⅝–4" (4–10 cm) wide, convex when young, becoming broadly convex to nearly plane in age, sometimes slightly depressed; surface dry, finely velvety when young, sometimes rimose toward the margin in age, dull brick red to reddish brown with an olive tint; margin incurved at first, sometimes uplifted, lobed, and irregular in age, even.

FLESH: whitish, slowly staining reddish purple when exposed; odor not distinctive; taste acidic.

PORE SURFACE: yellow to dingy yellow, depressed near the stalk at maturity, staining slightly cinnamon red when bruised or in age; pores angular, 1–3 mm wide; tubes up to 12 mm deep.

STALK: 1⅛–3½" (3–9 cm) long, ⅜–1⅛" (1–3 cm) thick, dry, solid, acutely tapered downward, yellow to whitish; covered with a coarse, distinctly wide-meshed, brown to dark brown reticulum, the base bound firmly to a mass of substrate by a pale yellowish mycelium; partial veil and ring absent.

SPORE PRINT COLOR: olive brown.

MACROCHEMICAL TESTS: cap surface stains green, then blue, and finally fuscous with the application of NH_4OH and becomes fuscous with KOH.

MICROSCOPIC FEATURES: spores 9–12 × 4–6 μm, subfusoid to ovate, smooth, yellow.

FRUITING: scattered or in groups on the ground near or under oak and pine; July–October; widely distributed in the Southeast; occasional.

EDIBILITY: unknown.

COMMENTS: The stalk base of this bolete can be almost woody at times. *Boletus illudens,* edible, has a less acutely tapered stalk with longitudinal rib-like lines that may form a partial reticulum (see *MNE*, p. 365).

Boletus variipes Peck   Illus. p. 53
  COMMON NAME: none.
  CAP: 2⅜–8" (6–20 cm) wide, convex when young, becoming broadly convex to nearly plane in age; surface dry, somewhat velvety, color variable, creamy tan to yellowish tan, tan, grayish brown to yellow-brown or brown, often rimose to rimose-areolate at maturity; margin even.
  FLESH: white, unchanging; odor and taste not distinctive.
  PORE SURFACE: white at first, becoming yellowish to yellowish olive in age, unchanging when bruised; pores circular, 1–2 per mm; tubes 1–3 cm deep.
  STALK: 3⅛–6" (8–15 cm) long, ⅜–1⅜" (1–3.5 cm) thick, nearly equal or enlarged downward, dry, solid, whitish to yellow-brown or grayish brown, usually distinctly reticulate or sometimes inconspicuously so, reticulation white or brown; partial veil and ring absent.
  SPORE PRINT COLOR: olive brown.
  MACROCHEMICAL TESTS: cap surface stains dark amber with a purplish black zone with the application of KOH or NH_4OH and weakly olivaceous or negative with $FeSO_4$; flesh stains pale pinkish orange then grayish with the application of KOH, pale grayish or negative with NH_4OH, and slowly pale grayish or yellowish with $FeSO_4$.
  MICROSCOPIC FEATURES: spores 12–18 × 4–6 μm, subfusoid, smooth, yellow; hyphal walls of the cap surface smooth.
  FRUITING: scattered or in groups on the ground under mixed broadleaf trees, especially oak, beech, maple, and aspen, also sometimes hemlock, pine, and spruce; May–October; widely distributed in the Southeast; fairly common.
  EDIBILITY: edible and good.
  COMMENTS: *Boletus atkinsonii* (not illustrated), edible, has a darker cap that is roughened with tufts of hyphae that are best observed with a hand lens and smaller spores that measure 10–13 × 4–5 μm. *Boletus nobilis* (p. 220), edible, has a whitish to pale ochraceous stalk with delicate white reticulation. *Boletus edulis* (p. 213), edible, has fine whitish reticulation and a moist, smooth, reddish brown cap that does not become conspicuously rimose to rimose-areolate at maturity. *Xanthoconium separans* (p. 245), edible, has lilaceous tones on its cap and stalk.

Boletus weberi Singer   Illus. p. 53
  COMMON NAME: none.
  CAP: 1½–2¾" (4–7 cm) wide, convex, becoming broadly convex to nearly plane in age; surface dry, areolate-scaly with pale yellow flesh showing in the cracks, brownish olive at first, becoming olive brown at maturity, sometimes with reddish tints, not staining when bruised; margin incurved at first, becoming decurved at maturity, entire, even.
  FLESH: pale yellow, not blueing on exposure; odor not distinctive; taste tart, somewhat lemony.
  PORE SURFACE: dull red at first, becoming dull orange red and finally reddish orange and depressed near the stalk in age, not blueing when bruised; pores angular, 1–2 per mm; tubes 3–6 mm deep, yellow, not blueing on exposure.
  STALK: 1½–2¾" (4.5–7 cm) long, ⅜–¾" (1–2 cm) thick, enlarged downward to a conspicuously pinched base, dry, solid, fibrillose-punctate on the upper portion, becoming somewhat scaly-punctate on the lower portion; pinkish red to dull purplish red over a

pale yellow ground color on the upper portion, olive to brownish olive or olive brown over a yellow ground color on the lower portion; not blueing when bruised; punctae and squamules dull dark brown; flesh pale yellow, becoming olive yellow to dull mustard yellow at the base, with red or brown stains around larval tunnels.

SPORE PRINT COLOR: olive brown.

MACROCHEMICAL TESTS: cap surface stains reddish brown with the application of KOH or NH_4OH and is negative with $FeSO_4$; flesh quickly stains greenish blue with the application of NH_4OH and is negative with KOH or $FeSO_4$.

MICROSCOPIC FEATURES: spores 10–13 × 4–6 μm, narrowly ellipsoid to subfusoid, smooth, pale ochraceous.

FRUITING: solitary, scattered, or in groups on sandy soil under longleaf pine, often with blue-jack oak nearby; May–August; Florida west along the Gulf Coast into Texas; occasional.

EDIBILITY: unknown.

COMMENTS: *Boletus inedulis* (not illustrated), inedible, is similar, but its pore surface is yellow, and it has bitter-tasting flesh. The flesh or pore surface of other similar red-pored boletes stains blue when cut or bruised.

Chalciporus pseudorubinellus (A. H. Smith and Thiers) L. D. Gomez   Illus. p. 54

COMMON NAME: none.

CAP: ⅝–2⅜" (1.5–6 cm) wide, convex to broadly convex, sometimes broadly umbonate; surface glabrous or nearly so, dry, somewhat viscid when moist, color variable, yellowish or pinkish red, becoming pinkish cinnamon in age; margin even.

FLESH: pale yellow, not blueing when exposed; odor and taste not distinctive.

PORE SURFACE: bright rose red when fresh, fading to orangish or yellowish brown in age, not blueing when bruised; pores irregular, 1–2 mm wide; tubes 6–10 mm deep.

STALK: 1⅛–2¾" (3–7 cm) long, ⅛–½" (4–12 mm) thick, nearly equal or enlarged downward, solid, dry, finely pruinose, mostly rose pink, typically yellow near the base, with a yellow basal mycelium; partial veil and ring absent.

SPORE PRINT COLOR: brown.

MACROCHEMICAL TESTS: unknown.

MICROSCOPIC FEATURES: spores 9–13 × 3–4 μm, fusoid, smooth, distinctly olivaceous; pleurocystidia very rare, 32–45 × 7–12 μm, fusoid-ventricose.

FRUITING: scattered or in groups on the ground, often among mosses, typically in conifer woods; July–September; reported from Florida and Mississippi, distribution limits yet to be established; occasional.

EDIBILITY: unknown.

COMMENTS: Also known as *Boletus pseudorubinellus*. *Chalciporus rubinellus*, edibility unknown, has larger spores that measure 12–15 × 3–5 μm and are pale dull ochraceous in KOH, scattered to abundant pleurocystidia, and a redder cap, especially when immature (see *MNE*, p. 368).

Gyrodon merulioides (Schweinitz) Singer   Illus. p. 54

COMMON NAME: Ash-tree Bolete.

CAP: 2–4¾" (5–12 cm) wide, convex, becoming plane to concave in age; surface dry to slightly viscid, glabrous or with tiny fibrils, yellowish brown to reddish brown, bruising dull yellow-brown; margin incurved when young, with a band of sterile tissue, often wavy at maturity, even.

FLESH: yellow, unchanging or sometimes slowly staining blue-green when cut; odor and taste not distinctive.

PORE SURFACE: pale yellow to dull gold or olive, decurrent, usually blueing slowly when bruised, gradually discoloring reddish brown; pores boletinoid, sometimes sublamellate, 1 mm or more wide; tubes 3–6 mm deep.

STALK: ¾–1⅝" (2–4 cm) long, ¼–1" (6–25 mm) thick, nearly equal, often curved, eccentric, lacking glandular dots, not reticulate, solid, apex colored like the pore surface, lower portion colored like the cap surface, bruising red-brown; partial veil and ring absent.

SPORE PRINT COLOR: olive brown.

MACROCHEMICAL TESTS: cap surface stains blackish then dull red to orange with the application of KOH or NH_4OH.

MICROSCOPIC FEATURES: spores 7–10 × 6–7.5 μm, oval to nearly round, smooth, pale yellow.

FRUITING: scattered or in groups on the ground almost always near or under ash trees, but also reported as occurring under maple and white pine; July–October; widely distributed in the Southeast; fairly common.

EDIBILITY: edible.

COMMENTS: Also known as *Boletinellus merulioides*. *Gyrodon rompelii* (not illustrated), edible, is very similar, but it has a pinkish red to red zone near the base of its stalk, which becomes rusty red to reddish brown in age.

Gyroporus castaneus (Fries) Quélet   Illus. p. 54

COMMON NAME: Chestnut Bolete.

CAP: 1⅛–4" (3–10 cm) wide, rounded to broadly convex, becoming nearly plane, sometimes slightly depressed; surface velvety-subtomentose to nearly glabrous, dry, chestnut brown to yellow-brown or orange-brown; margin often split and flaring in age.

FLESH: brittle, white, not staining blue when cut or bruised; odor and taste not distinctive.

PORE SURFACE: whitish to buff or yellowish, never pinkish or reddish; pores circular, 1–3 per mm; tubes 5–8 mm deep.

STALK: 1⅛–3½" (3–9 cm) long, ¼–⅝" (6–16 mm) thick, equal or often swollen in the middle or below, often constricted at the apex and base, brittle, stuffed with a soft pith, developing several cavities or becoming hollow in age, surface uneven, not reticulate, colored like the cap or slightly paler toward the apex; partial veil and ring absent.

SPORE PRINT COLOR: pale yellow to buff.

MACROCHEMICAL TESTS: cap surface stains yellow then bleaches to whitish with the application of KOH, stains amber orange with NH_4OH, and is negative with $FeSO_4$; flesh stains very pale brownish or is negative with the application of KOH, NH_4OH, or $FeSO_4$.

MICROSCOPIC FEATURES: spores 8–13 × 5–6 μm, elliptic to ovoid, smooth, hyaline.

FRUITING: solitary, scattered, or in groups on the ground under mixed conifers and broadleaf trees; June–October; widely distributed in the Southeast; fairly common.

EDIBILITY: edible.

COMMENTS: Compare with *Boletus variipes* (p. 229), which has similar colors, and a reticulate stalk.

Gyroporus cyanescens var. **cyanescens** (Bull.) Quélet　　　　　　　　　　Illus. p. 54
  COMMON NAME: Bluing Bolete.
  CAP: 1½–4" (4–12 cm) wide, convex to broadly convex, sometimes nearly plane; surface dry, coarsely tomentose to floccose-scaly, buff or straw-colored to pale olive, tan or yellowish, sometimes with darker streaks or an olive tinge, staining greenish yellow then greenish blue to blue when bruised.
  FLESH: brittle, whitish to pale yellow, staining greenish yellow then greenish blue to blue when cut; odor and taste not distinctive.
  PORE SURFACE: white to yellowish, yellowish green or tan, staining greenish yellow then greenish blue to blue when bruised; pores circular, 1–2 per mm; tubes 5–10 mm deep.
  STALK: 1½–4" (4–10 cm) long, ⅜–1" (1–2.5 cm) thick, equal or swollen in the middle or below, brittle, stuffed with a soft pith, becoming hollow or developing several cavities in age; surface dry, coarsely tomentose to floccose-scaly when young, often smoother in age, not reticulate, colored like the cap or paler, staining greenish yellow then greenish blue to blue when bruised; partial veil and ring absent.
  SPORE PRINT COLOR: pale yellow.
  MACROCHEMICAL TESTS: unknown.
  MICROSCOPIC FEATURES: spores 8–10 × 5–6 μm, elliptic, smooth, hyaline.
  FRUITING: solitary to scattered or in groups on the ground, usually in sandy soil, in broadleaf forests and mixed woods, especially under birch and poplar, also along road cuts; July–September; eastern Canada south to Florida, west to Minnesota; fairly common.
  EDIBILITY: edible.
  COMMENTS: A variety of this bolete has flesh that does not stain blue when exposed. It has been repeatedly collected in North Carolina, but the distribution limits are yet to be established. *Gyroporus cyanescens* var. *violaceotinctus* (not illustrated), edible, is nearly identical, but all tissues instantly stain dark lilaceous to indigo when cut or bruised. *Gyroporus umbrinosquamosus* (not illustrated), edibility unknown, is found along the Gulf Coast and is similar but has white flesh that does not change color when exposed. *Gyroporus phaeocyanescens* (not illustrated), edibility unknown, has a much smaller, fulvous to yellow-brown cap, context that stains indigo-blue when exposed, a yellow pore surface that does not stain blue when bruised, and larger spores that measure 9–15 × 5–7 μm.

Gyroporus subalbellus Murrill　　　　　　　　　　Illus. p. 54
  COMMON NAME: none.
  CAP: 1–4¾" (2.5–12 cm) wide, convex, becoming nearly plane, often shallowly depressed in age; surface dry, subtomentose to nearly glabrous, color variable, apricot buff to pinkish buff or pale pinkish cinnamon to orange cinnamon, sometimes pale yellow to whitish, darkening to brownish in age or when handled; margin even.
  FLESH: white, unchanging when cut; odor and taste not distinctive.
  PORE SURFACE: whitish at first, becoming pale yellow then dull yellow in age, slowly staining pinkish cinnamon when bruised or in age, sometimes deeply depressed near the stalk; pores circular to angular, 1–3 per mm; tubes 3–8 mm deep.
  STALK: 1½–4" (4–10 cm) long, ⅜–1⅛" (1–3 cm) thick, typically enlarged downward to a swollen or sometimes tapered base, hollow at maturity, dry, smooth, whitish at first, soon flushed pinkish to salmon orange especially near the base, frequently stained cinnamon to brownish or olivaceous; partial veil and ring absent.
  SPORE PRINT COLOR: yellowish buff.

MACROCHEMICAL TESTS: unknown.
MICROSCOPIC FEATURES: spores 8–14 × 4–6 μm, ellipsoid to ovoid, smooth, hyaline.
FRUITING: scattered or in groups in sandy soil in oak and pine woods; July–October; occasional; along the Atlantic Seaboard from Cape Cod, Massachusetts, south to Florida, west to Texas.
EDIBILITY: edible.
COMMENTS: The tendency to darken to brownish in age or when handled is a useful field character for this species.

Leccinum albellum (Peck) Singer   Illus. p. 54

COMMON NAME: none.
CAP: ¾–2⅜" (2–6 cm) wide, obtuse to convex, becoming broadly convex in age; surface dry to moist, smooth to slightly pitted, subvelutinous, often rimose-areolate in age, color variable, whitish, pale tan, pale gray to pinkish gray, pinkish brown to medium brown, sometimes tinged yellow; margin even.
FLESH: white, not staining when cut; odor and taste not distinctive.
PORE SURFACE: whitish to pale tan or pale gray, unchanging when bruised; pores angular, less than 1 mm wide; tubes up to 10 mm deep.
STALK: 2–3½" (5–9 cm) long, ¼–⅜" (6–10 mm) thick, nearly equal to slightly enlarged downward, dry, solid, whitish to pale olive buff, with tiny white scabers that darken to grayish or brownish in age; partial veil and ring absent.
SPORE PRINT COLOR: brown to olive brown.
MACROCHEMICAL TESTS: cap surface stains olive with the application of KOH or NH_4OH and dark bluish gray with $FeSO_4$; flesh stains pale yellow with the application of KOH or NH_4OH and gray with $FeSO_4$.
MICROSCOPIC FEATURES: spores 14–22 × 4–6 μm, cylindric to subfusiform, smooth, pale yellow.
FRUITING: scattered or in groups on the ground under broadleaf trees, especially oak; July–September; widely distributed in the Southeast; fairly common.
EDIBILITY: edible.
COMMENTS: Although the species name *albellum* means "whitish," pinkish brown and whitish caps commonly grow beside each other in the same collection.

Leccinum nigrescens (Richon and Roze) Singer   Illus. p. 54

COMMON NAME: none.
CAP: 1½–6" (4–15 cm) wide, convex, becoming broadly convex; surface dry to moist but not viscid, wrinkled, uneven, pitted, often coarsely areolate in age, dark brown to blackish brown when young, fading to pale yellow-brown in age; margin incurved at first and remaining so well into maturity.
FLESH: pale yellow, typically staining pinkish gray to dull reddish or pale fuscous, sometimes slowly; odor and taste not distinctive.
PORE SURFACE: pale yellow to dingy yellow, often staining brownish when bruised, usually depressed near the stalk in age; pores circular to angular, 1–3 per mm; tubes 8–15 mm deep.
STALK: 2–3½" (5–9 cm) long, ⅜–⅝" (1–1.6 cm) thick, nearly equal or often swollen near the middle or toward the base and tapered in either direction, dry, solid, pale yellow, with brown scabers that darken to blackish brown in age, often dull reddish at the base; partial veil and ring absent.
SPORE PRINT COLOR: honey yellow.

MACROCHEMICAL TESTS: unknown.
MICROSCOPIC FEATURES: spores 14–20 × 6–9 μm, fusiform, smooth, yellowish; caulocystidia narrowly fusoid-ventricose with flexuous necks or broadly fusoid-ventricose to clavate-mucronate.
FRUITING: solitary, scattered, or in groups in sandy soil under broadleaf trees, especially oak; July–December; widely distributed in the Southeast; fairly common.
EDIBILITY: edible.
COMMENTS: Also known as *Leccinum crocipodium*. Compare with *Leccinum rugosiceps*, edible, which has a paler cap, a nearly equal stalk with paler scabers, and flesh that slowly stains reddish or burgundy when exposed (see *MNE*, p. 369).

Leccinum snellii A. H. Smith, Thiers, and Watling   Illus. p. 54

COMMON NAME: none.
CAP: 1⅛–3½" (3–9 cm) wide, rounded to convex, becoming broadly convex to nearly plane in age; surface dry, covered with tiny dark brown to black fibrils when young, often fading to dark yellowish brown in age as the fibrils erode; margin even.
FLESH: white, staining pinkish to reddish (sometimes slowly) when exposed, especially at the juncture of the cap and stalk, darkening to purple-gray or black after one hour or more; odor and taste not distinctive.
PORE SURFACE: whitish when young, becoming grayish to dingy grayish brown in age, unchanging or staining yellowish or brownish when cut or bruised; pores circular, 2–3 per mm.
STALK: 1⅝–4⅜" (4–11 cm) long, ⅜–¾" (1–2 cm) thick, nearly equal or enlarging slightly downward, solid, whitish beneath gray to black scabers, often with blue-green stains on the lower portion; flesh typically staining blue-green at least near the base and often reddish above; partial veil and ring absent.
SPORE PRINT COLOR: brown.
MACROCHEMICAL TESTS: cap surface stains yellow with the application of KOH; flesh stains blue with the application of $FeSO_4$.
MICROSCOPIC FEATURES: spores 16–22 × 5–7.5 μm, fusoid, smooth, pale brown; caulocystidia mostly ventricose with a long flexuous neck.
FRUITING: scattered or in groups on the ground in mixed broadleaf forests, especially under yellow birch and oak; June–October; widely distributed in the Southeast; fairly common.
EDIBILITY: edible.
COMMENTS: Named in honor of the American boletologist Walter H. Snell. In the eastern United States, this dark and inconspicuous species is one of the first boletes to appear during spring.

Phylloporus boletinoides A. H. Smith and Thiers   Illus. p. 54

COMMON NAME: none.
CAP: ¾–2" (2–5 cm) wide, broadly convex when young, becoming nearly plane and sometimes shallowly depressed in age; surface dry, velvety-subtomentose to minutely scaly, becoming nearly smooth at maturity, cinnamon to dark pinkish brown, fading to dull yellow-brown in age; margin strongly incurved when young and remaining so at maturity, with a narrow band of sterile tissue.
FLESH: white to whitish, slowly staining grayish when exposed, but not blueing; odor not distinctive; taste slightly acidic or not distinctive.
PORE SURFACE: pale olive buff when young, becoming dark olive buff at maturity, some-

times staining dark blue to bluish green when bruised, but usually not bluing at all; pores strongly boletinoid to sublamellate, strongly decurrent; tubes 3–5 mm deep.
- STALK: 1–2⅜" (2.5–6 cm) long, ¼–⅝" (7–16 mm) thick, tapered slightly downward, solid or hollow in the base, smooth, dry, pale yellow at the apex, pale cinnamon below, with a sparse layer of pale yellow basal mycelium.
- SPORE PRINT COLOR: olive brown.
- MACROCHEMICAL TESTS: unknown.
- MICROSCOPIC FEATURES: spores 11–16 × 5–6.5 μm, subcylindric to narrowly oval, smooth, pale brown.
- FRUITING: solitary or scattered on the ground in mixed pine and oak woods; July–December; widely distributed in the Southeast; infrequent to fairly common.
- EDIBILITY: unknown.
- COMMENTS: Both *Phylloporus leucomycelinus* (p. 235), edible, and *Phylloporus rhodoxanthus* (see *MNE*, p. 369), edible, have redder caps and strongly lamellate yellow pores.

Phylloporus leucomycelinus Singer Illus. p. 55
- COMMON NAME: none.
- CAP: 1½–3⅛" (4–8 cm) wide, obtuse to convex at first, becoming nearly plane and sometimes shallowly depressed; surface dry, subvelutinous, often rimose to rimose-areolate in age, dark red to reddish brown or chestnut, usually paler on the disc at maturity; margin incurved at first, typically with a narrow band of sterile tissue.
- FLESH: whitish to pale yellow; odor and taste not distinctive.
- PORE SURFACE: strongly lamellate, decurrent, subdistant to distant, yellow to golden yellow, sometimes forked, strongly intervenose, sometimes poroid near the stalk, not blueing when bruised, separating cleanly from the cap.
- STALK: 1½–3⅛" (4–8 cm) long, ¼–½" (5–13 mm) thick, nearly equal or ventricose at the base, often with distinct ribs near the apex, scurfy or punctate with small reddish brown dots and points, yellow with reddish tinges, a white to whitish basal mycelium; partial veil and ring absent.
- SPORE PRINT COLOR: yellowish ochraceous.
- MACROCHEMICAL TESTS: cap surface stains blue with the application of NH_4OH.
- MICROSCOPIC FEATURES: spores 8–14 × 3–5 μm, ellipsoid to fusoid, smooth, pale yellowish.
- FRUITING: scattered or in groups on the ground in broadleaf forests under beech and oak; July–October; widely distributed in the Southeast; fairly common.
- EDIBILITY: edible.
- COMMENTS: *Phylloporus rhodoxanthus*, edible, is very similar, but it usually has a darker red cap, and it has a yellow basal mycelium (see *MNE*, p. 369).

Pulveroboletus ravenelii (Berkeley and Curtis) Murrill Illus. p. 55
- COMMON NAME: Powdery Sulfur Bolete.
- CAP: ⅜–4" (1–10 cm) wide, bluntly rounded to convex, becoming nearly plane in age; surface dry and pulverulent at first, becoming appressed-fibrillose to fibrillose-scaly, often slightly wrinkled or rimose in age, bright sulfur yellow, becoming orange-red to brownish red from the center toward the margin; margin incurved when young, typically appendiculate.
- FLESH: white to pale yellow, slowly staining pale blue then dingy yellow to pale brown when cut; odor and taste not distinctive.
- PORE SURFACE: bright yellow, becoming dingy yellow to grayish brown at maturity,

staining greenish blue then grayish brown when bruised; pores angular to nearly circular, 1–3 per mm; tubes 5–8 mm deep.

STALK: 1½–5¾" (4–14.5 cm) long, ¼–⅝" (6–16 mm) thick, equal or enlarging downward, solid, sheathed from the base upward with tiny appressed fibrils, bright sulfur yellow, above the ring bright yellow and smooth; partial veil membranous and powdery, bright sulfur yellow, typically leaving a prominent or sometimes inconspicuous superior ring.

SPORE PRINT COLOR: olive gray to olive brown.

MACROCHEMICAL TESTS: unknown.

MICROSCOPIC FEATURES: spores 8–10.5 × 4–5 μm, elliptic to oval, smooth, pale brown.

FRUITING: solitary, scattered, or in groups on the ground in woods; July–October; widely distributed in the Southeast; fairly common.

EDIBILITY: edible.

COMMENTS: This distinctive and unique bolete is named for H. W. Ravenel, a nineteenth-century mycologist who worked in South Carolina and Texas. *Pulveroboletus melleoluteus* (not illustrated), edibility unknown, has a smaller cap that does not develop orange-red to brownish red tones at maturity; it also lacks a partial veil and ring and has slightly longer spores that measure 7–12 × 3.5–4.5 μm.

Strobilomyces dryophilus Cibula and Weber   Illus. p. 55

COMMON NAME: none.

CAP: 1⅛–4¾" (3–12 cm) wide, convex, becoming broadly convex and finally plane in age; surface dry, with a whitish ground color, covered with coarse, wooly or cottony, appressed or erect, grayish pink or pinkish tan or pinkish brown scales; margin appendiculate with cottony pieces of whitish to tan torn partial veil.

FLESH: whitish, quickly staining orange-red to orange when cut or bruised; odor and taste not distinctive.

PORE SURFACE: white when young, soon becoming gray, finally black, staining reddish orange or brick red then black when bruised; pores angular, 1–2 mm wide, tubes 1–1.7 cm deep.

STALK: 1⅝–3⅛" (4–8 cm) long, ⅜–¾" (1–2 cm) thick, nearly equal, sometimes enlarged at the base, dry, solid, pinkish tan to brownish, ridged or reticulate above the ring, shaggy below; partial veil cottony to wooly, whitish to pale pinkish tan, leaving a ring or shaggy zones on the stalk; flesh quickly staining reddish orange then blackening when exposed.

SPORE PRINT COLOR: blackish brown to black.

MACROCHEMICAL TESTS: cap surface stains pale yellow with the application of NH_4OH and reddish with KOH; flesh stains pale yellow with the application of NH_4OH and reddish with KOH.

MICROSCOPIC FEATURES: spores 9–12 × 7–9 μm, short-elliptic to subglobose, covered by a distinct and complete reticulum, grayish.

FRUITING: solitary, scattered, or in groups on the ground in sandy soil under oak; July–December; widely distributed in the Southeast; fairly common.

EDIBILITY: edible.

COMMENTS: Compare with *Strobilomyces floccopus*, edible, which has darker cap and stalk color (see *MNE*, p. 10), and *Strobilomyces confusus* (not illustrated), edible, which has darker cap and stalk color, smaller and more pointed scales on the cap, and spores with irregular projections and short ridges that sometimes resemble a partial reticulum.

Suillus decipiens (Berkeley and Curtis) Kuntze   Illus. p. 55
  COMMON NAME: none.
  CAP: 1½–2¾" (4–7 cm) wide, convex to broadly convex when young, becoming nearly plane in age; surface dry, distinctly fibrillose to scaly, orangish to dull yellow, tan or pale reddish brown; margin strongly incurved, usually appendiculate with whitish to yellowish or grayish veil tissue.
  FLESH: pale yellow to pinkish buff, unchanging or sometimes reddening when exposed; odor and taste not distinctive.
  PORE SURFACE: orange-yellow to yellow, becoming brownish yellow in age, slightly avellaneous when bruised; pores irregular, often compound, 0.5–1 mm wide at maturity; tubes up to 6 mm deep.
  STALK: 1½–2¾" (4–7 cm) long, ¼–⅝" (7–16 mm) thick, usually enlarged downward and often curved at the base, dry, solid, cottony-tomentose to appressed-fibrillose, orangish to dull yellow, often bright yellow-orange above the annular zone, lacking glandular dots; partial veil fibrillose, whitish to yellowish or grayish, evanescent or leaving a thin, fragile, superior annular zone.
  SPORE PRINT COLOR: pale brown.
  MACROCHEMICAL TESTS: cap surface stains green-black with the application of KOH and rusty brown then dull blackish lilac with NH_4OH.
  MICROSCOPIC FEATURES: spores 9–12 × 3.5–4 μm, cylindric to subellipsoid, smooth, hyaline to pale ochraceous.
  FRUITING: scattered or in groups on the ground and among sphagnum mosses in pine-oak woods; July–December; widely distributed in the Southeast; fairly common.
  EDIBILITY: edible.
  COMMENTS: *Suillus hirtellus* (p. 237), edible, is similar but lacks a veil, is less scaly, and has prominent glandular dots. *Suillus pictus*, edible, has much redder cap colors when fresh, but faded specimens strongly resemble *Suillus decipiens* (see *MNE*, p. 370).

Suillus hirtellus (Peck) Kuntze   Illus. p. 55
  COMMON NAME: none.
  CAP: 2–4¾" (5–12 cm) wide, convex, becoming nearly plane in age; surface viscid when moist, soon becoming dry, covered with scattered tufts of reddish, brownish, or grayish fibrils on a yellowish ground color, typically staining vinaceous brown from handling; margin incurved when young, with tiny cottony tufts of sterile tissue that disappear at maturity.
  FLESH: pale yellow, not blueing when exposed; odor and taste not distinctive.
  PORE SURFACE: pale yellow at first, becoming dull yellow to olive yellow or dingy orange-buff in age, slightly vinaceous brown when bruised, sometimes with whitish droplets when young; pores angular, elongated and compound at maturity, about 1 mm wide; tubes 3–8 mm deep.
  STALK: 1⅛–3⅛" (3–8 cm) long, ⅜–¾" (1–2 cm) wide, nearly equal or enlarged downward, dry, solid, pale yellow, sometimes with reddish tints, also showing glandular dots that are yellowish at first and brown or blackish brown in age; partial veil and ring absent.
  SPORE PRINT COLOR: dull cinnamon.
  MACROCHEMICAL TESTS: unknown.
  MICROSCOPIC FEATURES: spores 7–9 × 3–3.5 μm, nearly oblong, smooth, pale ochraceous.
  FRUITING: scattered or in groups on the ground under conifers, especially pine, spruce, and fir; July–September; widely distributed in the Southeast; occasional to fairly common.

EDIBILITY: edible.

COMMENTS: *Suillus hirtellus* ssp. *cheimonophilus* (not illustrated), edibility unknown, is nearly identical but has yellow tufts of fibrils on the cap, a deep olive spore print color, and longer spores, 8–13.5 × 3–3.3 μm. Also carefully compare with *Suillus subaureus*, edible, which often grows under oak in mixed conifer woods and under various broad-leaf trees and which has a pale yellowish orange, radially arranged pore surface (see *MNE*, p. 370). *Suillus granulatus*, edible, is similar but typically has a stouter stalk, and its cap lacks fibrils (see *MNE*, p. 369).

Suillus salmonicolor (Frost) Halling Illus. p. 56

COMMON NAME: Slippery Jill.

CAP: 1⅛–3¾" (3–9.5 cm) wide, bluntly rounded or convex to nearly plane; viscid to glutinous when moist, shiny when dry, color variable, dingy yellow, yellowish orange to ochraceous salmon, cinnamon brown or olive brown to yellow-brown.

FLESH: pale orange-yellow to orange-buff or orange, not staining when exposed; odor and taste not distinctive.

PORE SURFACE: yellow to dingy yellow or yellowish orange to salmon, darkening to brownish in age, not staining when bruised; pores circular to angular, 1–2 per mm; tubes 8–10 mm deep.

STALK: 1–4" (2.5–10 cm) long, ¼–⅝" (6–16 mm) thick, equal or enlarged downward, whitish to yellowish or pinkish ochraceous, with reddish brown to dark brown glandular dots and smears; flesh ochraceous to yellowish; partial veil at first thick and distinctly baggy, rubbery, often with a conspicuously thickened cottony roll on the base, sometimes flaring from the stalk on the lower portion, forming a gelatinous superior to median ring.

SPORE PRINT COLOR: cinnamon brown to brown.

MACROCHEMICAL TESTS: cap surface stains purplish red then purplish black with the application of KOH or NH_4OH and dull dark brown with $FeSO_4$; flesh stains dark purplish red with the application of KOH or NH_4OH and dull dark brown to blackish with $FeSO_4$.

MICROSCOPIC FEATURES: spores 6–11 × 2.5–4 μm, elliptic, smooth, pale brown; glandular dots on the upper stalk consisting of fascicles of clavate caulocystidia, up to 45 μm long, with brown content or incrustations.

FRUITING: scattered or in groups on the ground under pine; August–November; fairly common; eastern Canada south to Florida, the western range unknown.

EDIBILITY: edible, with a lemony flavor, but cap surface and partial veil must be removed.

COMMENTS: Also known as *Suillus subluteus* and *Suillus pinorigidus*. *Suillus cothurnatus* (not illustrated), edible, is nearly identical but has a thinner, less rubbery veil that typically lacks a conspicuously thickened cottony roll at the base, and its glandular dots have parallel, hyaline, multiseptate hyphae terminating in a palisade of sterile, large, basidium-like dermatocystidia and rather small, hyaline, ampullaceous dermatocystidia with narrowed bases.

Suillus subalutaceus (A. H. Smith and Thiers) A. H. Smith and Thiers Illus. p. 56

COMMON NAME: none.

CAP: 1⅛–3½" (3–9 cm) wide, obtuse to convex, becoming broadly convex to nearly plane in age; surface covered with viscid gluten that tastes mild, not acidic, when fresh, pinkish buff to vinaceous buff or pale pinkish cinnamon; margin often appendiculate.

FLESH: whitish near the cap surface, yellowish above the tubes, slowly staining vinaceous brown, odor and taste not distinctive.

PORE SURFACE: pale yellow at first, soon darkening to pale brownish, staining vinaceous cinnamon when bruised, sometimes decurrent; pores circular to irregular, about 2 per mm; tubes 3–6 mm deep, staining vinaceous cinnamon with the application of KOH.

STALK: 1½–4" (4–10 cm) long, ⅜–⅝" (1–1.5 cm) thick, nearly equal, moist or dry, often hollow at maturity, dull white to yellow, often yellow in both the interior and exterior at the apex, surface with vinaceous cinnamon glandular dots both above and below the ring; partial veil membranous and coated with a gelatinous layer, leaving a superior gelatinous band-like ring that is typically collapsed on mature specimens.

SPORE PRINT COLOR: tawny brown to brown.

MACROCHEMICAL TESTS: cap surface stains olive with the application of KOH and is negative for NH_4OH or $FeSO_4$; flesh stains gray with the application of KOH, pinkish then quickly lilac-gray with NH_4OH, and greenish gray with $FeSO_4$.

MICROSCOPIC FEATURES: spores 8–11 × 3–3.5 μm, subfusoid, smooth, yellowish.

FRUITING: scattered or in groups on the ground under mixed stands of red and white pine; July–October; reported from North Carolina northward, distribution limits yet to be established; occasional.

EDIBILITY: edible.

COMMENTS: *Suillus granulatus*, edible, is very similar, but it often has a streaked or checkered cap when fresh, has conspicuous pinkish tan to brownish glandular dots and smears, and lacks a partial veil and ring (see *MNE*, p. 369). *Suillus brevipes* (see photo, p. 55), edible, is also similar, but it usually has a very short white to pale yellowish stalk that lacks glandular dots and smears or has sparse glandular dots in age.

Tylopilus alboater (Schweinitz) Murrill   Illus. p. 56

COMMON NAME: Black Velvet Bolete.

CAP: 1⅛–5⅞" (3–15 cm) wide, convex, becoming broadly convex to nearly plane in age; surface dry, velvety-tomentose, occasionally finely rimose in age, black to dark grayish brown, often covered with a thin whitish bloom when young; margin often with a narrow band of sterile tissue.

FLESH: white or tinged gray, staining pinkish to reddish gray when cut or bruised, eventually blackening; odor and taste not distinctive.

PORE SURFACE: white or with a tinge of gray when young, becoming dull pinkish or reddish in age, not dark gray or black, usually staining reddish then slowly black when bruised; pores angular to irregular, about 2 per mm; tubes 5–10 mm deep.

STALK: 1½–4" (4–10 cm) long, ⅝–1½" (2–4 cm) thick, equal or enlarging downward, solid, colored like the cap or paler, especially near the apex, often covered with a thin whitish bloom, not reticulate or only slightly so at the apex.

SPORE PRINT COLOR: pinkish to deep reddish color.

MACROCHEMICAL TESTS: cap surface stains amber orange with the application of KOH and is negative with NH_4OH or $FeSO_4$; flesh stains pinkish orange with the application of KOH, olive then brownish orange with NH_4OH, and instantly grayish blue to greenish blue with $FeSO_4$.

MICROSCOPIC FEATURES: spores 7–11 × 3.5–5 μm, narrowly oval, smooth, hyaline.

FRUITING: solitary to scattered on the ground under broadleaf trees, especially oak; June–September; widely distributed in the Southeast; fairly common.

EDIBILITY: edible.

COMMENTS: This is one of the best edible species of *Tylopilus*. It has firm flesh and a pleasant mild flavor when cooked. It is often overlooked or shunned by mycophagists because of its somber colors. When young, it is very dense and heavy. *Tylopilus atronicotianus* (not illustrated), edibility unknown, is very similar, but its cap is ochraceous to olive brown and nearly glabrous. *Tylopilus atratus* (not illustrated), edibility unknown, has a smaller cap and stalk and white flesh that lacks a reddish phase when exposed, and it grows on the ground under conifers. *Tylopilus griseocarneus* (not illustrated), edibility unknown, is also similar but has a coarsely reticulated stalk.

Tylopilus ballouii (Peck) Singer Illus. p. 56
  COMMON NAME: Burnt-orange Bolete.
  CAP: 2–4¾" (5–12 cm) wide, convex, becoming nearly plane in age, often irregular; surface dry, bright orange to bright orange-red, fading to dull orange, cinnamon, or tan in age; margin incurved at first, even.
  FLESH: white, staining pinkish tan to violet-brown when cut or bruised; odor not distinctive; taste mild to bitter.
  PORE SURFACE: white to dingy white, becoming tan or slightly pinkish in age, not yellow, staining brown when bruised; pores somewhat angular, 1–2 per mm; tubes up to 8 mm deep.
  STALK: 1–4¾" (2.5–12 cm) long, ¼–1" (6–25 mm) thick, equal or swollen at or above the base, solid, not reticulate or only finely so at the apex, whitish or tinged yellow to orange, staining brownish when cut or bruised or in age, surface smooth or scurfy.
  SPORE PRINT COLOR: pale brown, tan, or reddish brown.
  MACROCHEMICAL TESTS: cap surface stains amber yellow with the application of KOH or NH_4OH and olive gray with $FeSO_4$; flesh stains yellow with the application of KOH, stains bluish gray with $FeSO_4$, and is negative with NH_4OH.
  MICROSCOPIC FEATURES: spores 5–11 × 3–5 μm, elliptic, smooth, hyaline to pale brown.
  FRUITING: solitary, scattered, or in groups on the ground on lawns under trees and in woods, especially near oak, beech, and pine; June–November; widely distributed in the Southeast; fairly common.
  EDIBILITY: edible but not highly regarded, sometimes bitter.
  COMMENTS: The bright orange to bright orange-red cap colors that fade to dull orange, cinnamon, and finally tan in age are most distinctive.

Tylopilus conicus var. **conicus** (Ravenel) Beardslee Illus. p. 56
  COMMON NAME: none.
  CAP: 1–3¾" (2.5–9.5 cm) wide, bluntly conical when young, becoming convex in age; surface dry, shaggy or scaly when young and often developing a network of ridges and small depressions in age, pinkish tan to golden yellow, yellow-brown or salmon-tinged, with white flesh showing between the scales or beneath the ridges; margin with thin flaps of sterile tissue, at least when young.
  FLESH: white, not blueing when cut; odor fruity or not distinctive; taste not distinctive.
  PORE SURFACE: white when young, becoming dingy pinkish or pinkish tan in age, unchanging when bruised; pores circular to angular, 1–2 per mm; tubes 8–14 mm deep.
  STALK: 1⅝–2¾" (4–7 cm) long, ¼–¾" (6–18 mm) thick, nearly equal, often curved, slender, dry, solid, white or yellow, sometimes with pinkish tones especially toward the midportion, smooth to minutely wrinkled.
  SPORE PRINT COLOR: pinkish brown to reddish brown.
  MACROCHEMICAL TESTS: cap surface stains brown with the application of KOH or

NH_4OH; flesh stains yellow and finally brown with the application of KOH; pore surface stains gray and finally steel gray with the application of $FeSO_4$.
- MICROSCOPIC FEATURES: spores 14–21 × 4–6 μm, fusoid, smooth, hyaline to honey yellow.
- FRUITING: scattered on the ground under pine and in mixed woods, often in bottomlands on sandy soil along streams; July–September; North Carolina south to Florida, western distribution limits yet to be established; occasional to fairly common.
- EDIBILITY: edible.
- COMMENTS: Also known as *Mucilopilus conicus*. *Tylopilus conicus* var. *reticulatus* = *Mucilopilus conicus* var. *reticulatus* (not illustrated), edibility unknown, described from Florida, is similar but has prominent reticulation on the stalk.

Tylopilus plumbeoviolaceus (Snell and Dick) Singer Illus. p. 56
- COMMON NAME: Violet-gray Bolete.
- CAP: 1½–5⅞" (4–15 cm) wide, convex, becoming broadly convex to nearly plane in age; surface velvety-subtomentose when young, dry, smooth, violet or purple when young, fading to purple-brown, purple-gray, brown, dull cinnamon, or tan, occasionally overlaid with a whitish bloom; margin even.
- FLESH: white, not staining blue when cut or bruised; odor not distinctive; taste very bitter.
- PORE SURFACE: white when young, becoming pinkish or dull pinkish tan in age, unchanging when bruised; pores nearly circular, 1–2 per mm; tubes 4–18 mm deep.
- STALK: 3⅛–4¾" (8–12 cm) long, ⅜–¾" (1–2 cm) thick, equal or enlarging slightly downward, solid, predominantly violet or purple, sometimes marbled with white, fading to dull purple, purple-gray, purple-brown, or brown, occasionally developing olive or olive brown stains when bruised, sometimes slightly reticulate at the apex.
- SPORE PRINT COLOR: pinkish brown.
- MACROCHEMICAL TESTS: cap surface stains orange to dull amber with the application of KOH and pale brownish amber with NH_4OH or $FeSO_4$; flesh stains yellowish to pinkish with the application of $FeSO_4$ and is negative with KOH or NH_4OH; stipitipellis stains dull yellow with the application of KOH, stains pale yellow with NH_4OH, and is negative with $FeSO_4$.
- MICROSCOPIC FEATURES: spores 10–13 × 3–4 μm, elliptic, smooth, pale brown.
- FRUITING: scattered or in groups on the ground under broadleaf trees, especially oak; June–September; widely distributed in the Southeast; occasional to frequent.
- EDIBILITY: inedible, bitter.
- COMMENTS: *Tylopilus violatinctus* (p. 243), edibility unknown, is similar, but it has a paler and more lilac cap and has smaller spores, 7–10 × 3–4 μm. It also stains rusty violet to dark violet when bruised. *Tylopilus rubrobrunneus*, inedible because of its bitter-tasting flesh, has a purple cap when young that becomes purple-brown to dull brown in age and a white to brown stalk that develops olive or olive brown stains from the base upward as specimens mature or are handled (see *MNE*, p. 372).

Tylopilus rhoadsiae (Murrill) Murrill Illus. p. 56
- COMMON NAME: none.
- CAP: 2⅜–3½" (6–9 cm) wide, convex, becoming broadly convex and finally plane in age; surface dry, subtomentose to glabrous, sometimes shiny, white to whitish, often with buff, grayish buff, pinkish, or pinkish tan tinges; margin even.

FLESH: white, not staining blue or brownish when bruised; odor not distinctive, taste bitter.

PORE SURFACE: white when young, becoming dull pinkish in age, unchanging when bruised, depressed around the stalk in age; pores somewhat irregular, 1–2 per mm; tubes 9–16 mm deep.

STALK: 2–4" (5–10 cm) long, 5/8–1 1/8" (1.6–2.8 cm) thick, nearly equal or enlarged above a pinched base, dry, solid, white or colored like the cap, prominently reticulate over at least the upper half.

SPORE PRINT COLOR: vinaceous pink to brown.

MACROCHEMICAL TESTS: cap surface stains yellow with the application of KOH or NH_4OH.

MICROSCOPIC FEATURES: spores 11–13.5 × 3.5–4.5 μm, oblong-elliptic, smooth, hyaline to pale yellow.

FRUITING: scattered or in groups on the ground especially in sandy soil, near or under pine and oak; July–November; widely distributed in the Southeast; occasional to fairly common.

EDIBILITY: inedible, bitter.

COMMENTS: *Tylopilus rhodoconius* (not illustrated), edibility unknown, has a pale ochraceous to brownish orange cap when young that becomes darker brown at maturity, white to creamy white flesh with hyaline marbling, and a whitish pore surface that stains dark brown when bruised. *Tylopilus felleus*, distinctly bitter and inedible, is very similar but has a darker cap (see *MNE*, p. 371).

Tylopilus tabacinus var. **tabacinus** (Peck) Singer   Illus. p. 57

COMMON NAME: none.

CAP: 1 3/4–7" (4.5–17.5 cm) wide, obtuse when young, becoming broadly convex and finally plane in age; surface dry, smooth, subvelutinous, becoming rimose-areolate in age, yellow-brown to orange-brown or tobacco brown; margin even and wavy.

FLESH: white, usually staining slowly purplish buff or pinkish buff when exposed, often brown at maturity; odor variously described as not distinctive, fruity, fishy, or pungent; taste slightly bitter or not distinctive.

PORE SURFACE: whitish or sometimes brown at first, becoming brown to yellow-brown, with darker brown patches and stains, not blueing, usually depressed about the stalk in age; pores circular to angular, 1–2 per mm; tubes 1–1.4 cm deep.

STALK: 1 1/2–6 1/2" (4–16.5 cm) long, 1–2 3/8" (2.5–6 cm) thick, bulbous when young, becoming nearly equal in age, dry, solid, colored like the cap, subvelutinous, prominently reticulate over the upper portion or sometimes overall, usually smooth on the lower portion.

SPORE PRINT COLOR: pinkish brown to reddish brown.

MACROCHEMICAL TESTS: cap surface stains rusty brown with the application of KOH and produces a dull vinaceous flash that becomes amber orange on the dot with NH_4OH.

MICROSCOPIC FEATURES: spores 10–17 × 3.5–4.5 μm, fusoid to elliptic, smooth, hyaline to pale honey yellow.

FRUITING: scattered or in groups in sandy soil in woods and at their edges or around trees in lawns, usually with oak and pine; June–August; widely distributed in the Southeast; occasional to fairly common.

EDIBILITY: unknown.

COMMENTS: *Tylopilus tabacinus* var. *amarus* (see photo, p. 57), inedible, is very similar but has bitter-tasting flesh.

Tylopilus variobrunneus Roody, A. R. Bessette, and A. E. Bessette   Illus. p. 57
COMMON NAME: none.
CAP: 1½–4¾" (4–12 cm) wide, convex, becoming broadly convex to nearly plane in age; surface dry, velvety-tomentose and dark green to brownish green when young, becoming subtomentose to nearly glabrous and dull medium brown to chestnut brown at maturity, remaining darker over the disc, fading to pale chestnut brown in age; margin incurved at first, with a narrow band of sterile tissue when young, becoming decurved and even at maturity.
FLESH: dull white, staining dull pinkish rose to pale brownish pink when exposed; odor not distinctive; taste somewhat bitter or not distinctive.
PORE SURFACE: dull white to creamy white at first, becoming brownish pink at maturity, depressed near the stalk in age, staining brownish rose to pinkish cinnamon when bruised; pores circular to angular, 2–3 per mm; tubes 5–13 mm deep, dull white, staining dull pinkish rose to pale brownish pink when cut.
STALK: 1½–4" (4–10 cm) long, ½–1⅛" (1.2–3 cm) thick, clavate to subclavate or nearly equal, usually with a ventricose base, dry, solid, dull white on the upper portion, becoming pale brown below and dark brown toward the base, with conspicuous reticulation over the upper two-thirds or at least the upper portion; reticulation white near the apex, brown below; basal mycelium white; flesh dull white, staining dull pinkish rose to pale brownish pink when exposed.
SPORE PRINT COLOR: dull pinkish brown to cocoa powder brown.
MACROCHEMICAL TESTS: cap surface stains grayish green with the application of $FeSO_4$ and reddish brown to orange-brown with KOH or NH_4OH; flesh stains grayish green with the application of $FeSO_4$ and is negative with KOH or NH_4OH.
MICROSCOPIC FEATURES: spores 9–13 × 3–4.5 μm, subfusiform to subelliptic, smooth, pale ochraceous, lacking an apical pore.
FRUITING: solitary, scattered, or in groups on sandy soil under oaks or in mixed oak-pine woods; July–September; North Carolina north to Connecticut and New York, distribution limits yet to be established; occasional to fairly common.
EDIBILITY: unknown.
COMMENTS: The name *variobrunneus* means "various shades of brown," a reference to the color of the cap, stalk, and bruising reactions of the pore surface and flesh. The compact stature of this bolete and the prominent reticulation on the stalk are distinctive. *Tylopilus indecisus*, edible, is similar, but it has an ochraceous brown to pale brown cap when young that becomes dull cinnamon in age; it also has white, mild-tasting flesh that slowly stains brownish to pinkish when exposed (see *MNE*, p. 371). *Tylopilus ferrugineus* (not illustrated), edible, is also similar, but it has a reddish brown cap and stalk and mild-tasting flesh. *Tylopilus felleus*, inedible, is also similar, but it has bitter-tasting flesh (see *MNE*, p. 371).

Tylopilus violatinctus Baroni and Both   Illus. p. 57
COMMON NAME: none.
CAP: 3–5½" (7.5–14 cm) wide, hemispheric, becoming broadly convex to nearly plane in age; surface dry, opaque, velvety-tomentose when young, becoming subglabrous and somewhat shiny in age, grayish to bluish violet at first, becoming pale purplish to grayish purplish or at times purplish pink, paler as the cap expands, staining rusty violet to dark violet when bruised; margin strongly incurved at first, becoming decurved, with a narrow band of sterile tissue, becoming pale purplish drab then mostly pale brownish with grayish violet tones on the margin in age.

FLESH: white, unchanging or slowly staining very pale slate color when exposed; odor not distinctive; taste extremely bitter.

PORE SURFACE: white at first and sometimes remaining so for some time, becoming pinkish and finally deep cocoa brown at maturity, deeply depressed with a short-decurrent tooth in age, unchanging or sometimes staining brownish olive when bruised; pores nearly round, 1–2 per mm; tubes up to 15 mm deep.

STALK: 3⅛–6" (8–15 cm) long, ⅜–1½" (1–3.8 cm) thick, nearly equal or often strongly clavate or clavate-bulbous to abruptly bulbous, sometimes curved and constricted at the midportion, subglabrous, dry, solid, colored like the cap or paler, typically pale lilac to lilac or rusty lilac, becoming brownish in age, apex and base usually white, typically staining dull yellow to yellow-brown when bruised; basal mycelium white; lacking reticulation or reticulate only at the apex.

SPORE PRINT COLOR: reddish brown.

MACROCHEMICAL TESTS: cap surface stains yellowish to yellowish brown with the application of NH_4OH and dingy pale yellowish to amber with KOH; flesh does not stain with the application of NH_4OH or KOH.

MICROSCOPIC FEATURES: spores 7–10 × 3–4 μm, subfusiform, smooth, pale yellow.

FRUITING: solitary, scattered, or in groups on the ground under mixed wood with spruce and beech as well as under mixed stands of red oak, beech, and spruce, sometimes with maples and hemlock present; July–November; widely distributed in the Southeast; fairly common.

EDIBILITY: unknown.

COMMENTS: The name *violatinctus* means "tinged violet." *Tylopilus plumbeoviolaceus* (p. 241), inedible and bitter, is similar, but its young cap is darker, and it has longer spores that measure 10–13 × 3–4 μm. It also stains rusty to dark violet when bruised.

Xanthoconium affine var. maculosus Singer   Illus. p. 57

COMMON NAME: Spotted Bolete.

CAP: 1⅜–4" (3.5–10 cm) wide, convex, becoming broadly convex to nearly plane; surface dry, velvety-subtomentose, smooth to finely wrinkled, sometimes pitted and rimose in age, red-brown varying to dark brown, brown, yellow-brown, or orange-brown, mottled with white to pale yellow spots; margin even.

FLESH: white, often pale yellow around larval tunnels, not blueing when cut or bruised; odor and taste not distinctive.

PORE SURFACE: white to pinkish white when young, becoming yellow in age, bruising olive ochre; pores circular, 1–2 per mm; tubes 8–20 mm deep.

STALK: 1⅛–4" (3–10 cm) long, ⅜–¾" (1–2 cm) thick, nearly equal or often enlarged at the base, dry, solid, whitish, tinged yellowish or pinkish tan or brown when bruised or in age, glabrous or sometimes reticulate at the apex.

SPORE PRINT COLOR: bright yellow-brown.

MACROCHEMICAL TESTS: mature pore surface stains orange to cinnabar with the application of KOH.

MICROSCOPIC FEATURES: spores 9–16 × 3–5 μm, nearly oblong, smooth, pale yellow.

FRUITING: scattered or in groups on the ground in mixed woodlands, often with beech and oak; June–October; widely distributed in the Southeast; fairly common.

EDIBILITY: edible.

COMMENTS: Also known as *Boletus affinis* var. *maculosus*. *Xanthoconium affine* var. *affine* (not illustrated), edible, is nearly identical but lacks the numerous white to pale yellow spots on its cap.

Xanthoconium separans (Peck) Halling and Both Illus. p. 57
  COMMON NAME: none.
  CAP: 2⅜–8" (6–20 cm) wide, convex, becoming broadly convex; surface dry, wrinkled-pitted to somewhat corrugated at maturity, color extremely variable but nearly always with lilac tones, possibly dark purplish brown to reddish brown, vinaceous purple to vinaceous red, brownish lilac, or pale pinkish brown, sometimes pale pink, becoming yellow-brown in age; margin with a very narrow band of sterile tissue or even.
  FLESH: white, unchanging when exposed or bruised; odor not distinctive; taste sweet and nutty or not distinctive.
  PORE SURFACE: white when young, becoming yellowish or yellowish ochre and finally ochre-brown in age, unchanging when bruised; pores circular, 1–2 per mm; tubes 1–3 cm deep.
  STALK: 2⅜–6" (6–15.5 cm) long, ⅜–1⅛" (1–3 cm) thick, equal or enlarging slightly downward, solid, colored like the cap but often paler or darker, with a pinkish, lilac, or wine-colored tinge especially on the midportion that persists well into maturity; apex and base typically whitish, becoming pale pinkish brown in age; reticulate at least over the upper half, unchanging when cut or bruised; basal mycelium white.
  SPORE PRINT COLOR: brownish ochraceous to pale reddish brown.
  MACROCHEMICAL TESTS: lilac areas of the cap surface and stipitipellis stain aquamarine to deep blue with the application of NH_4OH and aquamarine, dingy slate greenish, bright green, or dark green with KOH; brown and yellowish areas of the cap surface stain brick red to blood red with the application of NH_4OH and dark rusty brown with KOH.
  MICROSCOPIC FEATURES: spores 12–16 × 3.5–5 μm, narrowly subfusiform, smooth, pale brown.
  FRUITING: solitary, scattered, or in groups on the ground under broadleaf trees, especially oak, occasionally under pine or mixed conifers; June–October; widely distributed in the Southeast; fairly common.
  EDIBILITY: edible, choice.
  COMMENTS: Also known as *Boletus separans*. Compare with *Boletus edulis* (p. 213), which typically has a reddish brown cap and brown spores and lacks lilac tones on the cap and stalk.

Xanthoconium stramineum (Murrill) Singer Illus. p. 57
  COMMON NAME: none.
  CAP: 1½–3½" (4–9 cm) wide, convex, becoming broadly convex to nearly plane in age; surface dry, glabrous or nearly so when young, sometimes rimose-areolate in age, white at first, becoming whitish to pale straw-colored or tinged brownish; margin incurved at first, with a narrow band of sterile tissue.
  FLESH: white, unchanging when exposed; odor slightly fruity or not distinctive; taste not distinctive.
  PORE SURFACE: white to whitish at first, becoming buff to yellowish buff or pale yellow-brown in age, unchanging when bruised; pores circular to angular, 2–4 per mm, often elongated near the stalk; tubes 3–10 mm deep.
  STALK: 1⅛–2¾" (3–7 cm) long, ⅜–1⅜" (1–3.5 cm) thick, enlarged downward or sometimes nearly equal, with a ventricose base, dry, solid, glabrous, white to whitish, not staining when bruised; lacking reticulation.
  SPORE PRINT COLOR: brownish yellow to yellowish rusty brown.
  MACROCHEMICAL TESTS: cap surface stains bluish gray with the application of $FeSO_4$.

MICROSCOPIC FEATURES: spores 10–15 × 2.5–4 μm, cylindric, smooth, yellow to hyaline.
FRUITING: solitary, scattered, or in groups in sandy soil or among grasses, with oak and pine; June–November; coastal plain of North Carolina south to Florida, west to Texas; fairly common.
EDIBILITY: edible.
COMMENTS: Also known as *Boletus stramineus*. *Boletus albisulphureus* (p. 209), edible, has a reticulated stalk and a pore surface that is white to buff at first and becomes yellow to olive yellow in age. *Gyroporus subalbellus* (p. 232), edible, has a brittle stalk that is hollow at maturity.

Polypores

Members of this very large group of fungi form fruiting bodies with small cylindric tubes on the underside of the cap in which spores are produced. Spores are discharged through a tiny mouth-like opening, called a *pore*, at the bottom of each tube. The fruit body forms many tubes, each with a pore, which accounts for the name *polypore*. Several other common names have been used to describe various species in this diverse group, including *bracket fungi, shelf fungi,* and *conks*. Some polypores have a central to eccentric or lateral stalk, but others are stalkless. Most species grow on wood, but a few, which are attached to buried wood, appear to grow on soil or humus. Several wood-inhabiting polypores are destructive pathogens, whereas others live on decaying or dead material and play an important role in the recycling of nutrients.

Many polypores are hard and woody or corky to leathery, but some are fleshy to fibrous. A polypore's tube layer usually does not separate cleanly and easily from the supporting cap tissue. Several species are excellent edibles if collected when they are young and tender. Although no polypores are known to have caused fatalities, most are inedible. Some polypores resemble boletes, which also produce their spores in tubes. However, most boletes grow on the ground, are soft and fleshy, and have tube layers that are usually cleanly and easily separated from the supporting cap tissue.

Albatrellus cristatus (Schaeff.) Kotl. and Pouzar   Illus. p. 58
   COMMON NAME: Crested Polypore.
   FRUIT BODY: annual; single or often compound with multiple caps fused together.
   CAP: 2–8" (5–20 cm) wide (compound fruit bodies are often wider), convex to flat, more or less circular in outline to irregularly lobed; surface dry, glabrous to slightly velvety, often separating into coarse scales or cracks, especially at the center; brown to yellowish green or olivaceous.
   FLESH: tough, white to buff, staining yellowish green when cut; odor not distinctive; taste bitter or mild.
   PORE SURFACE: often decurrent, white with yellowish or greenish yellow stains, becoming pinkish on drying; pores small, 2–4 per mm.
   STALK: 1–3" (2.5–7.5 cm) long, ⅜–1" (1–2.5 cm) thick, central or eccentric, solid, irregular in shape, sometimes branched, smooth, whitish to greenish or colored like the cap.
   SPORE PRINT COLOR: white.
   MICROSCOPIC FEATURES: spores 5–7 × 4–5 μm, broadly elliptic to subglobose, smooth, hyaline, weakly amyloid in mass.
   FRUITING: solitary or more often in fused clusters on the ground in broadleaf and mixed

woods, especially with oak, also reported to occur infrequently in conifer woods; June–December; widely distributed in the Southeast; fairly common.

EDIBILITY: inedible.

COMMENTS: *Phaeolus schweinitzii*, inedible, is somewhat similar and typically grows in large rosettes at the base of conifer trees and stumps (see p. 265).

Albatrellus pes-caprae (Pers.) Pouzar
Illus. p. 58

COMMON NAME: Goat's Foot.

FRUIT BODY: annual; simple or compound with one or more caps developing from a single stalk.

CAP: 1–8" (2.5–20 cm) wide, irregular or kidney-shaped to more or less circular, convex at first, then expanding; surface dry, wooly to scaly, reddish brown, olive brown, dark yellow brown, or blackish brown, somewhat yellowish between the scales; margin lobed.

FLESH: soft when fresh, creamy white, staining pinkish on exposure; odor usually pleasant, like hazelnuts, or not distinctive; taste mild.

PORE SURFACE: short decurrent, white to yellowish or pinkish, bruising grayish yellow or pale greenish yellow; pores large, 1–2 mm wide, angular.

STALK: 1–3" (2.5–7.5 cm) long, ⅜–2⅜" (1–6 cm) thick, sometimes branched, eccentric or lateral, rarely central, cylindrical, often with an enlarged base, smooth or somewhat roughened with minute scales, white to yellowish or orangish.

SPORE PRINT COLOR: white.

MICROSCOPIC FEATURES: spores 8–11 × 6–8 μm, broadly elliptic with a distinct apiculus, smooth, hyaline, inamyloid.

FRUITING: solitary or in groups on the ground arising from buried wood beneath conifers and in mixed conifer-broadleaf woods; June–December; widely distributed in the Southeast; uncommon.

EDIBILITY: edible.

COMMENTS: The Goat's Foot is superficially similar to a bolete; however, it differs in that the pore layer is not separable from the cap flesh.

Albatrellus subrubescens (Murr.) Pouz.
Illus. p. 58

COMMON NAME: none.

FRUIT BODY: annual; a single cap with a central to eccentric stalk, or sometimes two to several fused caps on a stalk.

CAP: 1⅛–4¾" (3–12 cm) wide, circular to kidney-shaped; surface dry, glabrous to minutely tomentose, smooth but becoming wrinkled or cracked and scaly in age, whitish to pale buff at first, becoming light yellow to greenish yellow at maturity.

FLESH: brittle to fibrous-tough, cream-colored, drying yellowish buff; odor and taste not distinctive.

PORE SURFACE: strongly decurrent, whitish at first, becoming light yellow to pale greenish yellow at maturity, sometimes developing pale reddish orange tints in some areas; pores angular, 2–4 per mm.

STALK: ¾–2" (2–5 cm) long, ¼–1⅛" (6–30 mm) thick, central to eccentric, dry, solid, vinaceous to buff with pale reddish orange or brownish orange areas that develop on drying.

SPORE PRINT COLOR: white.

MICROSCOPIC FEATURES: spores 3.5–4.5 × 2.5–3.5 μm, subglobose to ovoid, usually flattened on one side, smooth, hyaline, distinctly amyloid.

FRUITING: solitary, scattered, or in groups on the ground in broadleaf woods or mixed woods; September–January; occurs along the Gulf Coast region from Florida to Texas; occasional to fairly common.

EDIBILITY: inedible.

COMMENTS: *Albatrellus ovinus* (not illustrated), edible when immature, is nearly identical macroscopically, but it has inamyloid spores, and it is a more northern species reported from Tennessee and New England north to Canada. *Albatrellus confluens,* edible but sometimes bitter, forms pinkish buff to pale orange caps that typically occur in fused clusters under conifers, and has creamy white flesh that dries pinkish buff, a cabbage-like or bitter taste, and weakly amyloid spores (see *MNE,* p. 395).

Antrodia albida (Fr.) Donk　　　　　　　　　　　　　　　　　　　　　　Illus. p. 58

COMMON NAME: none.

FRUIT BODY: annual; up to 4" (10 cm) or more long, resupinate, often effused-reflexed or forming numerous narrow overlapping caps on a decurrent pore surface.

CAP: up to 1⅛" (3 cm) wide, fan-shaped to semicircular, shelf-like, fibrous-tough to somewhat woody; upper surface white to cream, at first azonate, matted, and somewhat velvety, becoming glabrous and distinctly zonate in age, sometimes sulcate; margin acute.

FLESH: up to ⅛" (3 mm) thick, white, fibrous-tough; odor and taste not distinctive.

PORE SURFACE: white to cream; pores variable, angular to sinuous and elongated or sometimes somewhat gill-like, 2–3 per mm.

STALK: absent.

SPORE PRINT COLOR: white.

MICROSCOPIC FEATURES: spores 10–14 × 3.5–5 μm, cylindrical to somewhat variable, smooth, hyaline.

FRUITING: resupinate or forming one or more overlapping caps on numerous species of broadleaf trees and, rarely, conifers; June–December; often overwintering; widely distributed in the Southeast; fairly common.

EDIBILITY: inedible.

COMMENTS: *Antrodia juniperina* (not illustrated), inedible, is similar but has much larger daedaleoid to labyrinthine pores, 1–3 mm wide, and smaller spores, 6.5–9 × 2.5–3.5 μm; and it grows only on living Juniper species. *Antrodia serialis* (not illustrated), inedible, is also similar but has somewhat smaller spores, 7–10 × 2.5–4 μm, and it grows on decaying conifers in very long sheets, often several feet long.

Bjerkandera adusta (Willd.: Fr.) Karst.　　　　　　　　　　　　　　　　Illus. p. 58

COMMON NAME: Smoky Polypore.

FRUIT BODY: annual; fibrous-tough, spreading, effused-reflexed, often forming overlapping stalkless caps with a grayish pore surface, or sometimes resupinate.

CAP: ¾–3" (2–7.5 cm) wide, convex to nearly flat, shelf-like or sometimes forming rosettes, fibrous-tough to somewhat woody; upper surface whitish to tan, dull ochre, or grayish brown to pale smoky gray, tomentose when young, becoming nearly glabrous in age, azonate to faintly zonate; margin whitish when young, darkening when bruised, acute.

FLESH: up to ¼" (6 mm) thick, pale buff, azonate, lacking a dark layer at the base of the tubes, fibrous-tough; odor not distinctive; taste somewhat sour.

PORE SURFACE: pale gray at first, becoming dark smoky gray to blackish at maturity; pores angular to irregularly rounded, 4–7 per mm.

STALK: absent.

SPORE PRINT COLOR: white.

MICROSCOPIC FEATURES: spores 5–6 × 2.5–3.5 μm, short-cylindric, smooth, hyaline.

FRUITING: overlapping caps or rosettes on decaying broadleaf trees and sometimes conifers, sometimes resupinate on the undersurface of decaying logs; June–December, often overwintering; widely distributed in the Southeast; common.

EDIBILITY: inedible.

COMMENTS: *Bjerkandera fumosa* (not illustrated), inedible, is similar but forms larger fruit bodies with thicker flesh, up to 5⁄8" (1.5 cm) thick with a dark brown line near the tube layer, and it has a paler pore surface and often an anise-like odor when fresh.

Bondarzewia berkeleyi (Fries) Bondarzew et Singer   Illus. p. 58

COMMON NAME: Berkeley's Polypore.

FRUIT BODY: annual; a large overlapping cluster of flattened fused caps or sometimes a solitary cap, attached to a solid central stalk arising from an underground sclerotium.

CAP: 2 3⁄8–10" (6–25.5 cm) wide, fan-shaped, laterally fused and typically forming a rosette, sometimes lobed, fleshy-tough; surface densely matted and wooly or nearly smooth, dry, radially wrinkled and pitted, obscurely to conspicuously zoned, variously colored whitish to grayish, pale yellow to yellow-brown; margin blunt, wavy.

FLESH: up to 1 1⁄8" (3 cm) thick, corky to fibrous-tough, white; odor not distinctive when fresh; taste mild, bitter in age.

PORE SURFACE: white to creamy white; pores angular, frequently torn and irregular, 0.5–2 per mm.

STALK: above ground portion 1 3⁄4–4 3⁄4" (4.5–12 cm) long, 1 1⁄8–2" (3–5 cm) thick, central, roughened, dingy yellow to yellowish brown.

SPORE PRINT COLOR: white.

MICROSCOPIC FEATURES: spores 7–9 × 6–8 μm, round, with prominent amyloid ridges, hyaline.

FRUITING: solitary or scattered on the ground at the base of broadleaf trees and stumps, especially oak; July–October; widely distributed in the Southeast; occasional.

EDIBILITY: edible when young and easily sectioned, becoming fibrous-tough and bitter in age.

COMMENTS: Mature specimens may be more than 40" (102 cm) wide.

Cerrena unicolor (Bulliard : Fries) Murrill   Illus. p. 58

COMMON NAME: Mossy Maze Polypore.

FRUIT BODY: 1 1⁄2–4" (4–10 cm) wide, annual, semicircular to fan-shaped or somewhat irregular in outline, nearly plane, stalkless, sometimes laterally fused and forming extensive rows; upper surface dry, sometimes distinctly lobed, usually with prominent grooves, covered with a dense layer of short, stiff hairs arranged in conspicuous zones, often covered by green algae, whitish, grayish, pale brown, or green; margin fairly sharp, wavy, and often lobed.

FLESH: up to 1⁄8" (3 mm) thick, corky to fibrous-tough, whitish to pale brown with a thin dark zone separating the cap surface from the flesh.

PORE SURFACE: whitish to pale buff when young, becoming smoky gray to grayish brown at maturity; pores maze-like, 1–4 per mm, often splitting and becoming tooth-like in age.

SPORE PRINT COLOR: white.

MICROSCOPIC FEATURES: spores 5–7 × 2.5–4 μm, cylindric-ellipsoid, smooth, hyaline.

FRUITING: solitary, scattered, or typically in overlapping clusters on broadleaf trees and more rarely conifers; year-round; widely distributed in the Southeast; fairly common.

EDIBILITY: inedible.

COMMENTS: Previously known as *Daedalea unicolor*. The hairy cap surface, maze-like pore surface, and thin dark zone in the flesh are the distinctive features. *Gloeophyllum sepiarium* (p. 255), inedible, has a yellowish red to reddish brown cap with concentric zones and furrows, fibrous-tough yellow-brown to rusty brown flesh, and a golden brown to rusty brown gill-like to maze-like pore surface, and it grows solitary, in groups, or in rosette-like clusters on decaying wood, usually conifer.

Coltricia cinnamomea (Persoon) Murrill   Illus. p. 59
COMMON NAME: Shiny Cinnamon Polypore.
FRUIT BODY: annual; a small to medium polypore with a circular, thin cap and a central stalk that grows on the ground.
CAP: ½–2" (1.2–5 cm) wide, circular to irregular, plane to depressed, sometimes laterally fused with other caps, fibrous-tough; surface concentrically zoned, silky, shiny, bright reddish cinnamon to amber brown and dark rusty brown; margin faintly striate, often torn, thin, and sharp.
FLESH: up to ¹⁄₃₂" (1 mm) thick, fibrous-tough, rusty brown, black in KOH.
PORE SURFACE: yellowish brown to reddish brown, not decurrent; pores angular, 2–4 per mm.
STALK: ⅜–1½" (1–4 cm) long, ¹⁄₁₆–¼" (1.5–6 mm) thick, central, nearly equal or tapering downward, velvety, dark reddish brown.
SPORE PRINT COLOR: yellowish brown.
MICROSCOPIC FEATURES: spores 6–10 × 4.5–7 μm, elliptic, smooth, hyaline.
FRUITING: solitary, in groups, or fused together on the ground usually along roadsides and on paths in woods; June–November; widely distributed in the Southeast; fairly common.
EDIBILITY: inedible.
COMMENTS: *Coltricia perennis*, inedible, is larger, has a brownish orange to pale cinnamon brown dull cap and a decurrent pore surface (see *MNE*, p. 395).

Daedaleopsis confragosa (Bolton : Fries) Schroeter   Illus. p. 59
COMMON NAME: Thin-maze Flat Polypore.
FRUIT BODY: annual; a semicircular to kidney-shaped, stalkless polypore with a creamy white to grayish or pale brown concentrically zoned cap, growing on decaying wood.
CAP: 1⅛–6" (3–15.5 cm) wide, semicircular to kidney-shaped, flat to slightly convex, stalkless, fibrous-tough; surface coarsely wrinkled, finely velvety to smooth, with distinct concentric zones; variously colored, creamy white, tan, grayish, or pale brown; margin sharp, thin.
FLESH: up to ½" (1.2 cm) thick, fibrous-tough, whitish to pale brown.
PORE SURFACE: whitish to pale brown, bruising pinkish brown, usually labyrinthine, often with gill-like and elongated pores; fibrous-tough.
SPORE PRINT COLOR: white.
MICROSCOPIC FEATURES: spores 7–11 × 2–3 μm, cylindric to sausage-shaped, smooth, hyaline.
FRUITING: solitary, scattered, or in groups on decaying wood, especially broadleaf trees; year-round; widely distributed in the Southeast; common.
EDIBILITY: inedible.
COMMENTS: Also known as *Daedalia confragosa*. Compare with *Daedalia quercina*, inedible, which forms a larger and thicker fruit body with a conspicuously labyrinthine pore

surface, grows on decaying broadleaf trees, especially oak, and has much smaller spores that measure 5–6 × 2–3.5 μm (see *MNE*, p. 395).

Fistulina hepatica Schaeffer : Fries

Illus. p. 59

COMMON NAME: Beefsteak Polypore.
FRUIT BODY: annual; a fan-shaped, gelatinous, reddish orange to dark red cap with a short, lateral to eccentric stalk, growing on oak trees.
CAP: 2¾–10" (7–25.5 cm) wide, fan- to spoon-shaped; surface smooth to velvety, gelatinous, often sticky to slimy, reddish orange to pinkish red or dark red to purplish brown; often exuding a red juice when squeezed; margin rounded or sharp, often wavy or lobed.
FLESH: ¾–2" (2–5 cm) thick, fleshy and juicy when fresh, becoming fibrous in age, dingy white to pinkish or reddish, zoned with darker and paler areas, slowly darkening when exposed; taste sour to acidic.
PORE SURFACE: whitish to pinkish yellow, becoming reddish brown in age or when bruised; pores circular, 1–3 per mm; tubes crowded but distinctly independent when viewed with a hand lens.
SPORE PRINT COLOR: pinkish salmon.
STALK: up to 3⅛" (8 cm) long, lateral to eccentric or sometimes absent, colored like the cap.
MICROSCOPIC FEATURES: spores 4–6 × 2.5–4 μm, oval to tear-shaped, smooth, hyaline.
FRUITING: solitary or in groups on oak trunks and stumps; July–October; widely distributed in the Southeast; infrequent.
EDIBILITY: edible.
COMMENTS: The marbled flesh (like beef) and acidic taste are distinctive. *Pseudofistulina radicata* (p. 268), inedible, has a smaller, up to 3" (7.5 cm) wide, pale yellowish brown cap; a whitish pore surface; and a rooting stalk up to 4" (10 cm) long.

Fomes fasciatus (Swartz : Fries) M. C. Cooke

Illus. p. 59

COMMON NAME: none.
FRUIT BODY: 2¾–7" (7–18 cm) wide, hoof- to fan-shaped or semicircular, convex, stalkless; surface finely tomentose and slightly roughened when young, becoming hard and nearly smooth at maturity, concentrically sulcate, typically zonate, grayish with concentric zones of reddish brown and grayish brown, often darker brown to blackish brown in age; margin somewhat sharp, curved.
FLESH: up to 1½" (4 cm) thick at the base, hard and crusty near the upper surface, fibrous to granular and corky below, golden brown.
PORE SURFACE: pale brown at first, becoming dark grayish brown at maturity; pores circular, 4–5 per mm.
SPORE PRINT COLOR: white.
MICROSCOPIC FEATURES: spores 10–14 × 4–5 μm, cylindric, smooth, hyaline.
FRUITING: solitary or in groups or overlapping clusters on decaying broadleaf trees; year-round, perennial; North Carolina south to Florida, west to Texas; fairly common.
EDIBILITY: inedible.
COMMENTS: *Fomes fomentarius*, inedible, is very similar, but it has a more hoof-shaped fruit body and larger spores that measure 12–20 × 4–7 μm (see *MNE*, p. 396).

Fomitopsis nivosa (Berk.) Gilbn. and Ryv. Illus. p. 59
> COMMON NAME: none.
> FRUIT BODY: annual to biennial; a white stalkless cap that becomes sordid brown and produces a resinous dark cuticular layer that spreads from the base outward in age; grows on decaying broadleaf trees.
> CAP: 1⅛–6" (3–15.5 cm) wide, shelf-like, semicircular to fan-shaped, fibrous-tough when fresh, woody when dry; upper surface white when young, becoming pale brown and eventually developing a resinous dark layer that spreads from the base outward in older specimens, glabrous, smooth to slightly roughened, azonate; margin uneven, acute.
> FLESH: up to ¾" (2 cm) thick, zonate, whitish to pale brown, fibrous-tough to woody; odor and taste not distinctive.
> PORE SURFACE: whitish at first, becoming light brownish gray to grayish brown, usually glancing; pores circular to angular, 6–8 per mm.
> STALK: absent or a rudimentary extension of the cap.
> SPORE PRINT COLOR: white.
> MICROSCOPIC FEATURES: spores 6–9 × 2–3 μm, cylindrical, smooth, hyaline.
> FRUITING: solitary, scattered, or in overlapping clusters on decaying broadleaf trees; June–December or year-round; reported from the Gulf Coast region from Florida to Texas; occasional.
> EDIBILITY: inedible.
> COMMENTS: The formation of a dark resinous cuticular layer in older specimens is diagnostic. *Fomitopsis palustris* (not illustrated), inedible, is similar but lacks the resinous layer and has larger pores, 1–4 per mm; fresh specimens also have an unpleasant odor. *Fomitopsis meliae* (not illustrated), inedible, is also similar but lacks the resinous layer, and fresh specimens also lack an unpleasant odor.

Ganoderma applanatum (Persoon) Patouillard Illus. p. 59
> COMMON NAME: Artist's Conk.
> FRUIT BODY: perennial; a shelf-like, stalkless conk with concentric furrows and a white pore surface that stains brown when bruised.
> CAP: 2–26" (5–65 cm) wide, shelf-like to somewhat hoof-shaped, stalkless, woody; surface hard, thick, crusty, concentrically furrowed, thickened at the central point of attachment, finely cracked and roughened, gray to grayish black or brown, dull; margin thin, often white.
> FLESH: up to 2⅜" (6 cm) thick, corky to woody, brown; odor and taste not distinctive.
> PORE SURFACE: white, staining brown when bruised; pores circular, 4–6 per mm.
> SPORE PRINT COLOR: brown.
> MICROSCOPIC FEATURES: spores 7–11 × 5–7.5 μm, broadly elliptic, truncate, with a thick double wall, pale brown.
> FRUITING: solitary, scattered, or in overlapping clusters on decaying wood, especially of broadleaf trees; year-round; widely distributed in the Southeast; common.
> EDIBILITY: inedible.
> COMMENTS: This species is often used by artists, who score pictures on the pore surface with knives or other sharp objects. *Ganoderma zonatum* (see photo, p. 60), inedible, is similar, but its distinctly zonate, somewhat shiny upper surface is wood brown to dark purplish brown or mahogany; its pore surface is cream-colored when fresh and bruises or dries pale purplish brown; it has dark purplish brown flesh; and it grows on palm trees. *Heterobasidion annosum* (not illustrated), inedible, is smaller, up to 10" (25.5 cm) wide; it has a grayish white cap that becomes gray-brown to blackish brown in

age, a very rough and furrowed surface, and white to pale cream flesh; and it grows on conifers.

Ganoderma lucidum (W. Curt.: Fr.) Karst. Illus. p. 59
COMMON NAME: Ling Chih.
FRUIT BODY: annual; a single cap with a central or lateral stalk, or an overlapping cluster of caps, coated with a thin varnished crust.
CAP: 2–6" (5–15.5 cm) wide, circular to kidney-shaped, fibrous-tough; surface dry, shiny, covered with a thin varnished crust, usually concentrically zoned, whitish at first, becoming yellow to ochre, especially when arising from a stalk, and finally yellowish brown to dark reddish brown; margin fairly acute.
FLESH: up to 1⅛" (3 cm) thick, creamy white at first, becoming dark purple-brown in age, corky to fibrous-tough; odor and taste not distinctive.
PORE SURFACE: creamy white when young, becoming pale brown, bruising dark brown to dark purple-brown; pores circular to angular, 4–5 per mm.
STALK: 1½–4" (4–10 cm) long, ¼–1⅝" (6–40 mm) thick, nearly equal or tapered in either direction, central or lateral, colored like the cap; covered with a thin, shiny, varnished crust; sometimes absent.
SPORE PRINT COLOR: brown.
MICROSCOPIC FEATURES: spores 9–12 × 5.5–8 μm, ellipsoid, truncate at the apex, smooth, pale brown.
FRUITING: solitary, scattered, or in overlapping clusters on and around decaying broadleaf trees and stumps, especially oak; April–December, sometimes overwintering; widely distributed in the Southeast; common.
EDIBILITY: the fruit bodies are used to make tea or a powdered herb that is reported to be a tonic.
COMMENTS. Also known as *Ganoderma curtisii*. *Ganoderma meredithae* (p. 254), inedible, is very similar but grows at the base of pines or on pine stumps. *Ganoderma zonatum* (see photo, p. 60), inedible, reported only from Florida, has a conspicuously zonate dark brown to dark reddish brown cap at maturity and dark purplish brown flesh, and it grows on palms. *Ganoderma colossum* (not illustrated), inedible, also reported only from Florida, has a yellowish cap surface and creamy white to pale buff flesh; grows on broadleaf trees and palms; and has larger spores that measure 14.5–17.5 × 9.5–11.5 μm.

Ganoderma meredithae Adaskaveg and Gilbertson Not Illustrated
COMMON NAME: Pine Varnish Conk.
FRUIT BODY: annual; a circular to kidney-shaped cap coated with a shiny, thin, varnished crust; centrally or laterally stalked or sometimes stalkless.
CAP: 2–6" (5–15.5 cm) wide, circular to kidney-shaped, fibrous-tough; surface dry, shiny, covered with a thin varnished crust, cream-colored to yellowish buff at first, becoming reddish brown at maturity, concentrically zoned and shallowly sulcate; margin somewhat blunt.
FLESH: up to ¾" (2 cm) thick, light buff near the upper surface, gradually becoming pale purplish brown toward the tubes, corky to fibrous-tough; odor and taste not distinctive.
PORE SURFACE: creamy white at first, becoming pale vinaceous cream at maturity; pores circular to angular, 4–6 per mm.
STALK: 1⅛–4" (3–10 cm) long, ⅜–1⅛" (1–3 cm) thick, nearly equal or tapered in either

direction, central or lateral, colored like the cap; coated with a thin, shiny, varnished crust; sometimes rudimentary or absent.

SPORE PRINT COLOR: brown.

MICROSCOPIC FEATURES: spores 9.5–11.5 × 5.5–7 μm, ellipsoid, apex truncate with a germ pore, smooth, pale brown.

FRUITING: solitary, scattered, or in groups at the base of living pines and on dead pines and stumps; July–December, sometimes overwintering; widely distributed in the Gulf Coast region; occasional to fairly common.

EDIBILITY: inedible.

COMMENTS: This species was named in honor of Dr. Meredith Blackwell of Louisiana State University. All other similar species grow on broadleaf trees. *Ganoderma tsugae* (p. 255), inedible, grows primarily on hemlock and has a more northern distribution.

Ganoderma tsugae Murrill Illus. p. 59

COMMON NAME: Hemlock Varnish Shelf.

FRUIT BODY: annual; a fan- to kidney-shaped reddish cap coated with a thin, shiny, varnished crust, with a lateral stalk, growing on hemlock.

CAP: 2⅜–12" (6–31 cm) wide, fan- to kidney-shaped; soft and corky when fresh, covered with a thin crust; surface smooth to wrinkled, shiny, appearing varnished or dull and powdery when covered with spores, concentrically zoned and shallowly furrowed; brownish red to mahogany near the center or overall, brownish orange to reddish orange outward, and bright whitish on the margin or rarely blue to bluish green.

FLESH: up to 1⅛" (3 cm) thick, soft and corky to fibrous-tough, whitish; odor and taste not distinctive.

PORE SURFACE: white to creamy white, becoming brown in age or when bruised; pores circular to angular, 4–6 per mm.

STALK: 1⅛–6" (3–15 cm) long, ⅜–1½" (1–4 cm) thick, often lateral but sometimes eccentric to central or absent, shiny, appearing varnished, brownish red to mahogany or blackish brown.

SPORE PRINT COLOR: brown.

MICROSCOPIC FEATURES: spores 13–15 × 7.5–8.5 μm, elliptic, truncate, with a thick double wall, appearing rough, pale brown.

FRUITING: solitary or in groups on decaying conifer wood, especially hemlock; May–December, sometimes overwintering; North Carolina west to Tennessee and northward; fairly common.

EDIBILITY: inedible.

COMMENTS: *Ganoderma meredithae* (p. 254), inedible, grows at the base of pines and has a more southern distribution in the Gulf Coast region. *Ganoderma lucidum* (p. 254), edible when prepared as a tea, has a dark reddish brown cap with a creamy white margin and grows on decaying broadleaf trees, especially oak.

Gloeophyllum sepiarium (Fries) Karsten Illus. p. 60

COMMON NAME: Yellow-red Gill Polypore.

FRUIT BODY: annual; solitary or several fused and rosette-shaped, or overlapping semicircular to kidney-shaped; bright yellowish red to reddish brown stalkless caps and lamellate pore surface, usually growing on decaying conifer wood.

CAP: 1–4" (2.5–10 cm) wide, semicircular to kidney-shaped, flat or slightly convex, stalkless, fibrous-tough; surface covered with short and stiff hairs, becoming matted and felty or nearly smooth in age, with distinct concentric zones and furrows, bright yellowish red to reddish brown; margin whitish to orange-yellow or brownish yellow, uneven, with tufts of tiny hairs.

FLESH: up to ¼" (6 mm) thick, fibrous-tough, yellow-brown to rusty brown, black in KOH; odor and taste not distinctive.

PORE SURFACE: golden brown to rusty brown, thick and gill-like to labyrinthine, often with both and sometimes with elongated pores; pores 1–2 per mm, fibrous-tough.

SPORE PRINT COLOR: white.

MICROSCOPIC FEATURES: spores 9–13 × 3–5 μm, cylindric, smooth, hyaline.

FRUITING: solitary or in groups or rosette-like clusters on decaying wood, usually conifer; year-round; widely distributed in the Southeast; common.

EDIBILITY: inedible.

COMMENTS: *Gloeophyllum striatum* (p. 256), inedible, has a pale yellowish brown to umber brown cap, gill-like pores that sometimes fork and anastomose, and smaller spores, 6–10 × 2.5–3.5 μm; and it grows on decaying broadleaf trees. *Lenzites betulina* (p. 263), inedible, has white flesh and usually grows on decaying broadleaf wood.

Gloeophyllum striatum (Swartz : Fr.) Murr. Illus. p. 60

COMMON NAME: none.

FRUIT BODY: annual; solitary or several fused and rosette-shaped or overlapping semicircular to fan-shaped brown caps, with a rudimentary stalk-like base and lamellate pore surface, growing on broadleaf trees.

CAP: ⅜–3⅛" (1–8 cm) wide, fused and rosette-shaped or semicircular to fan-shaped, flexible, fibrous-tough; upper surface finely velvety at first, becoming glabrous in age, roughened and uneven, conspicuously zonate when fresh, pale yellowish brown to umber brown when moist, becoming paler brown to grayish brown and inconspicuously zonate to azonate when dry, often slightly sulcate; margin thin, acute, often lobed or wavy.

FLESH: up to ⅛" (1.5 mm) thick at the point of attachment and papery thin at the margin, fibrous-tough, yellowish brown to dark rusty brown; odor and taste not distinctive.

PORE SURFACE: brown to dark brown, becoming grayish brown in age; pores conspicuously lamellate and sometimes forking and anastomosing, sometimes splitting to form flattened teeth in older specimens.

STALK: either a broad rudimentary extension of the cap at the point of attachment or absent.

SPORE PRINT COLOR: whitish.

MICROSCOPIC FEATURES: spores 6–10 × 2.5–3.5 μm, oblong-ellipsoid to cylindric, smooth, hyaline.

FRUITING: solitary, scattered, or in overlapping or fused and rosette-shaped clusters on decaying broadleaf trees; September–January, sometimes overwintering; occurs along the Gulf Coast from Florida to Texas; occasional.

EDIBILITY: inedible.

COMMENTS: *Gloeophyllum trabeum* (not illustrated), has a weakly zonate to azonate smooth cap, and its pore surface has a combination of small pores, 2–4 per mm, and gills. *Gloeophyllum sepiarium* (p. 255), inedible, has a bright yellowish brown then darker reddish brown cap that becomes grayish to black in age, dense gills on the undersurface, and larger spores, 9–13 × 3–5 μm; and it grows on conifers.

Gloeoporus dichrous (Fries) Bresadola   Illus. p. 60
  COMMON NAME: none.
  FRUIT BODY: annual; composed of fused patches in rows or shelves; individual caps up to 1½" (4 cm) wide and up to 4" (10 cm) long; variable from resupinate to effused-reflexed or forming bracket-like caps.
  CAP: upper surface finely wooly to nearly smooth, somewhat zonate, whitish to ochraceous or grayish ochre; margin thin, wavy.
  FLESH: white, thicker than the tubes, soft and fibrous or cottony when fresh, resinous hard when dry.
  PORE SURFACE: gelatinous when fresh, separable from the flesh, drying resinous-horny, grayish pink to purplish or reddish brown with age; margin whitish when young; pores small, round to angular, 4–6 per mm.
  STALK: absent.
  SPORE PRINT COLOR: white.
  MICROSCOPIC FEATURES: spores 3.5–5.5 × 0.5–1.5 μm, allantoid to cylindrical, smooth, hyaline, inamyloid.
  FRUITING: firmly attached to the substrate, typically in overlapping, fused groups on decaying wood of broadleaf trees, also occasionally on conifer wood and on old, woody polypores.
  EDIBILITY: inedible.
  COMMENTS: Some species of *Stereum* are superficially similar, but they lack pores on the underside.

Hapalopilus croceus (Fries) Donk   Illus. p. 60
  COMMON NAME: none.
  FRUIT BODY: annual; stalkless, broadly attached, fan-shaped to semicircular, bright orange to brownish orange cap; bright reddish orange to brownish orange pore surface; growing on decaying broadleaf wood.
  CAP: up to 8" (20.5 cm) wide, fan-shaped to semicircular, convex, stalkless, soft and watery when fresh, becoming corky to brittle when dry; surface finely velvety to pubescent, bright orange when young, becoming brownish orange and somewhat smooth in age; margin sharp, curved.
  FLESH: up to 1⅛" (3 cm) thick at the base, spongy, watery and bright orange when fresh, becoming hard and darker orange to brownish when dry; odor not distinctive; taste slightly bitter.
  PORE SURFACE: bright reddish orange when fresh, becoming brownish when dry; pores angular, 2–3 per mm.
  SPORE PRINT COLOR: white.
  MICROSCOPIC FEATURES: spores 4–7 × 3–4.5 μm, broadly ellipsoid, smooth, hyaline.
  FRUITING: solitary or in groups or overlapping clusters on decaying broadleaf wood; June–November; along the Gulf Coast from Florida west to Texas; occasional.
  EDIBILITY: inedible.
  COMMENTS: All parts of this mushroom instantly stain red with KOH. Formerly known as *Aurantioporus croceus*. Compare with *Hapalopilus nidulans* (p. 258), inedible, which is smaller, up to 4¾" (12 cm) wide; has an ochraceous to cinnamon brown pore surface; and stains bright violet with KOH. *Hapalopilus salmonicolor* (not illustrated), inedible, has a resupinate light orange to pink fruit body and grows on conifer wood. *Hapalopilus albo-citrinus* (not illustrated), inedible, reported from the Gulf Coast, has a resupinate pale yellowish fruit body and grows on broadleaf wood.

Hapalopilus nidulans (Fries) Karsten Illus. p. 60
- COMMON NAME: Tender Nesting Polypore.
- FRUIT BODY: annual; stalkless, broadly attached, fan-shaped to semicircular, dull brownish orange to cinnamon cap; ochraceous to cinnamon brown pore surface; growing on decaying broadleaf wood.
- CAP: 1–4¾" (2.5–12 cm) wide, fan-shaped to semicircular, convex, stalkless, soft and watery when fresh, becoming corky to brittle when dry; surface coated with tiny matted hairs, becoming smooth in age, often with one or more shallow and concentric furrows, dull brownish orange to cinnamon; margin sharp, curved.
- FLESH: up to 1⅛" (3 cm) thick at the base, soft and watery when fresh, pale cinnamon.
- PORE SURFACE: ochraceous to cinnamon brown; pores angular, 2–4 per mm.
- SPORE PRINT COLOR: white.
- MICROSCOPIC FEATURES: spores 3.5–5 × 2–3 μm, elliptic to cylindric, smooth, hyaline; setae absent.
- FRUITING: solitary or in groups or overlapping clusters on decaying broadleaf wood; June–November, sometimes overwintering; widely distributed in the Southeast; occasional.
- EDIBILITY: inedible.
- COMMENTS: Also known as *Hapalopilus rutilans*. All parts instantly stain bright violet with KOH (see photo). This mushroom is very popular with those who dye wool. *Hapalopilus croceus* (p. 257), inedible, is larger, up to 8" (20.5 cm) wide; has a bright reddish orange pore surface when fresh; and stains red with the addition of KOH. *Phellinus gilvus* (p. 266), inedible, has an ochre to bright rusty yellow and fibrous-tough cap; yellowish brown flesh that stains black in KOH; a grayish brown to dark brown pore surface; and setae.

Hexagonia hydnoides (Fries : Sw.) M. Fidalg. Illus. p. 61
- COMMON NAME: none.
- FRUIT BODY: annual; solitary or overlapping fan-shaped caps with conspicuous dark brown to blackish erect hairs; growing on decaying broadleaf trees.
- CAP: 1⅛–8" (3–20 cm) wide, shelf-like, convex or flat, fan-shaped to semicircular, sometimes fused, flexible and leathery when fresh, becoming rigid when dry; upper surface dark brown to blackish, at first covered with conspicuous erect, stiff hairs that fall off in age; margin thin, acute.
- FLESH: up to ½" (1.3 cm) thick but typically much thinner, cinnamon brown to dark brown; odor and taste not distinctive.
- PORE SURFACE: fulvous to dark brown with a distinct grayish tint; pores round to somewhat irregular, 3–5 per mm.
- STALK: absent.
- SPORE PRINT COLOR: white.
- MICROSCOPIC FEATURES: spores 11–14.5 × 3.5–5 μm, cylindrical, smooth, hyaline.
- FRUITING: solitary, scattered, or in overlapping clusters on decaying broadleaf logs, stumps, and standing trees; July–December, sometimes overwintering; Florida west to Texas, also reported from Kansas; fairly common.
- EDIBILITY: inedible.
- COMMENTS: Formerly known as *Polyporus hydnoides*. The conspicuous stiff, erect, dark brown to blackish hairs are most distinctive.

Hydnopolyporus fimbriatus (Fries) Reid   Illus. p. 61
 COMMON NAME: none.
 FRUIT BODY: annual, consisting of numerous fan-shaped to spatula-like caps, with rudimentary stalks; growing in rosette-like clusters up to 4¾" (12 cm) in diameter.
 CAP: individual lobes up to 1" (2.5 cm) wide, 1/16–1/8" (1.5–3 mm) thick, flexible, fibrous to tough, erect; surface somewhat velvety, becoming glabrous in age, azonate to concentrically zoned, smooth to radially striate, white to pale tan, becoming darker when dried; margin wavy, fringed to incised.
 FLESH: thin, fibrous, white; odor and taste not distinctive.
 PORE SURFACE: variable, from nearly smooth, warted, and reticulate to toothed with flat teeth or distinctly poroid; white; pores angular to sinuous, 2–5 per mm.
 STALK: rudimentary on each cluster.
 SPORE PRINT COLOR: white.
 MICROSCOPIC FEATURES: spores 3.5–5 × 2.5–3.5 μm, broadly ellipsoid to subglobose, smooth, hyaline.
 FRUITING: in groups or clusters or sometimes solitary on the ground arising from buried wood and on stumps of broadleaf trees; June–December; along the Gulf Coast from Florida west to Texas; occasional.
 EDIBILITY: reported to be edible, but somewhat chewy.
 COMMENTS: Sometimes mistakenly called *Hydnopolyporus palmatus*. *Sparassis crispa* (p. 307), edible, is somewhat similar but typically larger, has a distinct rooting stalk, and is completely smooth on the underside of its lobes.

Inonotus dryadeus (Pers.:Fr.) Murrill   Illus. p. 61
 COMMON NAME: none.
 FRUIT BODY: annual; shelf-like, a semicircular to fan-shaped conk with a smooth whitish to dark brown upper surface that becomes cracked in age.
 CAP: 5–14" (13–36 cm) wide, convex, semicircular to fan-shaped, sessile; upper surface very finely tomentose or glabrous, azonate, becoming cracked in age, buff to dark brown; margin concolorous or sometimes ivory.
 FLESH: up to 10" (25.5 cm) thick, soft to fibrous, zonate, bright yellowish brown at first, becoming reddish brown in older specimens, appearing distinctly mottled when cut; odor and taste not distinctive.
 PORE SURFACE: buff, often with exuding droplets of amber liquid in fresh specimens, becoming dark brown and cracking in age; pores circular or angular, 4–6 per mm.
 STALK: absent.
 SPORE PRINT COLOR: white.
 MICROSCOPIC FEATURES: spores 6–8 × 5–7 μm, subglobose, smooth, hyaline, thick-walled, dextrinoid; setae usually frequent, rare in some specimens, ventricose, usually hooked, 25–40 × 9–16 μm.
 FRUITING: solitary or in overlapping clusters on oaks; June–December, often overwintering; widely distributed in the Southeast; occasional to fairly common.
 EDIBILITY: inedible.
 COMMENTS: The combination of a large fruit body, subglobose hyaline spores, and strongly ventricose hooked setae is diagnostic of this species. Fruit bodies typically develop at the ground line at the base of infected trees or from roots at some distance from the base.

Inonotus hispidus (Bull.: Fr.) Karst. Illus. p. 61
  COMMON NAME: none.
  FRUIT BODY: annual; shelf-like, a semicircular to fan-shaped conk with conspicuous, coarse, bright reddish orange hairs that darken in age.
  CAP: 4–12" (10–30 cm) wide, shelf-like, convex, semicircular to fan-shaped, soft to fibrous-tough; upper surface azonate, broadly attached to the substrate, covered with coarse, bright reddish orange hairs that become reddish brown to blackish in age; margin bright sulfur yellow when young, rounded and blunt.
  FLESH: up to 3" (7.5 cm) thick, soft to fibrous, somewhat zonate, yellow to ochre with darker zones, immediately staining brownish when cut; odor somewhat acidic and pleasant; taste not distinctive.
  PORE SURFACE: yellow at first, becoming yellow-ochre then yellowish brown or blackish in age, uneven; pores angular to irregularly rounded, 1–3 per mm.
  STALK: absent.
  SPORE PRINT COLOR: brown.
  MICROSCOPIC FEATURES: spores 8–11 × 6–8 μm, subglobose to ovoid, thick-walled, smooth, brown; setae present or absent, up to 30 μm long, apex pointed, dark brown.
  FRUITING: usually solitary or sometimes in groups on living broadleaf trees, especially oak and walnut, often emanating from tree wounds; June–December, often overwintering; widely distributed in the Southeast; fairly common.
  EDIBILITY: inedible.
  COMMENTS: The coarse, bright reddish orange hairs on the upper surface of growing specimens are distinctive. Compare with *Inonotus quercustris* (p. 260), inedible, which has a golden yellow cap that becomes rusty brown in age. *Inonotus texanus* (not illustrated), reported only from Texas and Arizona, has a light brown to buff cap that becomes cracked radially and concentrically into angular scales, and it grows on living mesquite and acacia. *Inonotus cuticularis* (not illustrated), has a yellowish brown tomentose cap that becomes glabrous and finally blackened and rimose in age, pointed setae that are often curved, and abundant pointed and branching setal hyphae on the cap surface; it grows on many broadleaf species.

Inonotus quercustris M. Blackwell and Gilbertson Illus. p. 61
  COMMON NAME: none.
  FRUIT BODY: annual; a shelf-like, stalkless conk with overlapping caps growing on living water oaks.
  CAP: 4¾–8" (12–20 cm) wide, convex, shelf-like to hoof-shaped, soft when young, fibrous-tough at maturity; surface moist or dry, matted-tomentose or with short and erect hairs, azonate, golden yellow at first, becoming rusty brown, usually with darker splotches.
  FLESH: moist, soft and spongy at first, becoming dry, firm, and fibrous-tough at maturity; dark reddish brown, with faint concentric zones; odor and taste not distinctive.
  PORE SURFACE: yellow to golden yellow, with a bright golden luster when viewed obliquely, becoming brownish in age; pores angular, 3–5 per mm.
  STALK: absent.
  SPORE PRINT COLOR: yellowish brown.
  MICROSCOPIC FEATURES: spores 7.5–10 × 5.5–8 μm, ellipsoid, thick-walled, smooth, pale golden yellow; setal hyphae thick-walled, tapering to a point, dark reddish brown, up to 200 μm or slightly longer; hymenial setae absent.

FRUITING: solitary or in overlapping clusters on living water oak, *Quercus nigra*; August–January, sometimes year-round; reported only from Louisiana and Mississippi; rare.

EDIBILITY: inedible.

COMMENTS: *Inonotus rickii* (not illustrated), inedible, is similar, but it has pointed, thick-walled, dark brown hymenial setae, conspicuous setal hyphae up to 250 µm long, and smaller spores that measure 6–8.5 × 4.5–5.5 µm. Also compare with *Inonotus hispidus* (p. 260), inedible, which has a cap with conspicuous coarse, bright reddish orange hairs that darken in age.

Ischnoderma resinosum (Fries) Karsten — Illus. p. 61

COMMON NAME: Resinous Polypore.

FRUIT BODY: annual; a shelf-like, semicircular to fan-shaped stalkless conk with brownish, overlapping velvety caps that become crusty in age.

CAP: 3–10" (7.5–25.5 cm) wide, semicircular to fan-shaped, flattened to convex, stalkless, fleshy-soft when young and fresh, becoming fibrous-tough to brittle in age; surface concentrically and radially furrowed, faintly to distinctly zoned, velvety when young, later covered with a thin, glossy, resinous crust; dull brownish orange to dark brown; margin thick, rounded, whitish to ochre, frequently exuding drops of water when fresh.

FLESH: up to ¾" (2 cm) thick, soft, becoming fibrous in age, whitish to pale yellow.

PORE SURFACE: white bruising brown, becoming pale brown in age; pores angular to circular, 4–6 per mm.

SPORE PRINT COLOR: whitish.

MICROSCOPIC FEATURES: spores 4.5–7 × 1.5–2.5 µm, cylindric to sausage-shaped, smooth, hyaline.

FRUITING: solitary or in overlapping clusters on decaying wood; August–December, often overwintering, sometimes producing fresh fruit bodies in spring; widely distributed in the Southeast; fairly common.

EDIBILITY: inedible.

COMMENTS: Some authors consider *Ischnoderma resinosum* to be a species that grows only on hardwoods and recognize *Ischnoderma benzoinum*, inedible, as a similar species that grows only on conifers. Other authors consider these two species to be synonymous.

Laetiporus cincinnatus (Morgan) Burdsall, Banik, and Volk — Illus. p. 62

COMMON NAME: White-pored Sulphur Shelf.

FRUIT BODY: annual; a large rosette of flattened to convex, laterally fused, and lobed caps attached to a solid central branching stalk.

CAP: 1⅛–10" (3–25.5 cm) wide, petal- to fan-shaped, soft, fleshy when young, fibrous-tough in age; surface velvety to densely matted and woolly, dry, radially wrinkled, pinkish orange to pinkish brown; margin pale pinkish cream to brownish, blunt, wavy, sometimes lobed.

FLESH: up to ¾" (2 cm) thick, soft, fleshy-fibrous, whitish to pale pinkish yellow; odor nutty or meaty; taste not distinctive.

PORE SURFACE: white to pinkish cream, bruising brownish; pores circular, 3–4 per mm.

STALK: 1½–3½" (4–9 cm) long, up to 2" (5 cm) thick, whitish to pinkish yellow.

SPORE PRINT COLOR: white.

MICROSCOPIC FEATURES: spores 6.5–8 × 4–5 µm, oval to elliptic, smooth, hyaline.

FRUITING: solitary or scattered on the ground attached to roots at the base of oaks and (rarely) pines or clustered around decaying stumps; July–March; widely distributed in the Southeast; occasional.

EDIBILITY: edible and choice.

COMMENTS: Formerly known as *Laetiporus sulphureus* var. *semialbinus*. Compare with *Laetiporus sulphureus* (p. 262), edible, which has bright to dull orange caps and a bright sulfur yellow pore surface when fresh. *Laetiporus persicinus* (p. 262), edibility unknown, is similar, but it has a conspicuous stalk that is often branched, one or more caps with a pinkish brown upper surface and a darker brown band around the margin, and a pinkish cream pore surface, and it grows at the base of oaks and pines.

Laetiporus persicinus (Berk. and Curt.) Gilbertson   Illus. p. 62

COMMON NAME: none.

FRUIT BODY: annual; solitary or several caps attached to a solid, central, branching stalk.

CAP: 4–10" (10–25.5 cm) wide, circular to fan-shaped, soft and fleshy when young, becoming fibrous-tough in age; surface dry, tomentose, azonate to faintly zonate, pinkish brown with a darker brown marginal band; margin blunt, wavy, sometimes lobed.

FLESH: up to ⅜" (8 mm) thick, soft; odor of fresh specimens like ham or bacon; taste unpleasant or not distinctive.

PORE SURFACE: pinkish cream when fresh, becoming dull brown in age, staining brown when bruised; pores circular, 3–4 per mm.

STALK: up to 2¾" (7 cm) long, up to 2" (5 cm) thick, simple or branched at the base, dry, solid, pinkish brown.

SPORE PRINT COLOR: white.

MICROSCOPIC FEATURES: spores $6.5–8 \times 4–5$ μm, ovoid to ellipsoid, smooth, hyaline.

FRUITING: solitary or scattered, attached to roots at the base of living oak trees and sometimes pine trees; July–December, sometimes overwintering; widely distributed in the Southeast; occasional.

EDIBILITY: unknown.

COMMENTS: Compare with *Laetiporus sulphureus* (p. 262), edible, which has an orange cap and a bright sulfur yellow pore surface, and *Laetiporus cincinnatus* (p. 261), edible, which has a pinkish orange cap and a white pore surface.

Laetiporus sulphureus (Bulliard : Fries) Murrill   Illus. p. 62

COMMON NAME: Chicken Mushroom, Sulphur Shelf.

FRUIT BODY: a large overlapping cluster of flattened, laterally fused, and lobed caps, sometimes forming rosettes or a solitary cap; stalkless or with a rudimentary stalk.

CAP: 2–12" (5–31 cm) wide, fan- to petal-shaped, soft, fleshy when young, fibrous-tough in age; surface velvety to densely matted and wooly, dry, radially wrinkled and roughened, bright to dull orange, fading to orange-yellow then whitish in age; margin pale orange, blunt, wavy, often lobed.

FLESH: up to ¾" (2 cm) thick, fleshy-fibrous, white; odor nutty or not distinctive; taste not distinctive.

PORE SURFACE: bright sulfur yellow; pores angular, 3–4 per mm.

SPORE PRINT COLOR: white.

MICROSCOPIC FEATURES: spores $5–8 \times 3.5–5$ μm, oval to elliptic, smooth, hyaline.

FRUITING: solitary or more often in overlapping clusters or rosettes on broadleaf trees, especially oak and cherry, occasionally on conifers; May–December; widely distributed in the Southeast; occasional to fairly common.

EDIBILITY: edible and choice for most people when collected on hardwoods; reported to cause gastrointestinal upset when gathered from conifer wood or when alcohol is consumed at the same meal.

COMMENTS: The flesh of this mushroom has the consistency and flavor somewhat reminiscent of white chicken meat. Compare with *Laetiporus cincinnatus* (p. 261), edible and choice, which has a pinkish orange cap and a white pore surface and which forms rosettes.

Lenzites betulina (Fries) Fries   Illus. p. 62

COMMON NAME: Multicolor Gill Polypore.

FRUIT BODY: annual; a semicircular to kidney-shaped stalkless cap with multicolored concentric zones; growing on decaying wood.

CAP: 1⅛–4" (3–10 cm) wide, semicircular to kidney-shaped, nearly plane, stalkless, fibrous-tough; surface velvety to hairy, with distinct multicolored concentric zones; colors variable, often white, pink, gray, yellow, orange, or brown, sometimes green when covered with algae.

FLESH: up to 1/16 (1.5 mm) thick, fibrous-tough, white.

PORE SURFACE: white to creamy white, conspicuously gill-like, sometimes forking, occasionally with elongated pores near the margin, fibrous-tough.

SPORE PRINT COLOR: white.

MICROSCOPIC FEATURES: spores 4–6 × 2–3 μm, cylindric to sausage-shaped, smooth, hyaline.

FRUITING: solitary or in groups on decaying wood, especially broadleaf trees; July–December, sometimes year-round; widely distributed in the Southeast; common.

EDIBILITY: inedible.

COMMENTS: Compare with *Gloeophyllum sepiarium* (p. 255), inedible, which has yellow-brown to rusty brown flesh and usually grows on conifer wood.

Meripilus sumstinei (Murrill) Larsen in Lombard   Illus. p. 62

COMMON NAME: Black-staining Polypore.

FRUIT BODY: a large dense cluster of overlapping shelf-like caps attached to a short, thick, common stalk.

CAP: 2⅜–8" (6–20.5 cm) wide, fan- to spoon-shaped, fleshy-fibrous; upper surface yellowish tan to grayish yellow or sometimes yellow-orange to yellow when young, becoming pale brownish yellow to grayish yellow and then dark ochraceous to grayish brown at maturity; margin thin, sharp, wavy, often lobed, blackening when bruised or in age.

FLESH: up to ⅝" (1.6 cm) thick, fleshy-fibrous to fibrous-tough, white.

PORE SURFACE: white to creamy white; pores angular, 3–6 per mm.

STALK: ⅜–1⅜" (1–3 cm) long, up to 4¼" (11.5 cm) thick, ochre to reddish brown.

SPORE PRINT COLOR: white.

MICROSCOPIC FEATURES: spores 6–7 × 4.5–6 μm, oval to nearly round, smooth, hyaline.

FRUITING: solitary or in groups on the ground at the base of trees and stumps of broadleaf trees, especially oak; June–December; widely distributed in the Southeast; occasional.

EDIBILITY: edible when young and tender.

COMMENTS: Fruit bodies may attain a diameter of 16" (41 cm) or more. This species is commonly and incorrectly labeled in many older field guides as *Meripilus giganteus*, which is a European species. *Grifola frondosa*, edible and choice, has many gray to brownish gray caps with thinner flesh and a white pore surface that does not stain black (see *MNE*, p. 397).

Microporellus dealbatus (Berk. and Curtis) Murr.   Illus. p. 62

COMMON NAME: none.

FRUIT BODY: annual; kidney-shaped or circular to fan-shaped cap supported by a central to lateral stalk.

CAP: ¾–4" (2–10 cm) wide, kidney-shaped or circular to fan-shaped, fibrous-tough when fresh, becoming hard and woody when dry; upper surface buff to pale brown, light grayish brown or pale smoky gray, distinctly zonate, dry, somewhat velvety to tomentose at first, becoming glabrous in age; margin thin, acute.

FLESH: up to ⅛" (3 mm) thick, white, fibrous-tough; odor and taste not distinctive.

PORE SURFACE: white at first, becoming cream and finally ochraceous at maturity, often wrinkled when dry; pores angular to irregularly rounded, minute, 8–10 per mm.

STALK: 1⅛–4" (3–10 cm) long, ⅛–½" (3–12 mm) thick, nearly equal or tapered in either direction, often twisted, colored like the cap, dry and rigid.

SPORE PRINT COLOR: white.

MICROSCOPIC FEATURES: spores 4.5–6 × 3.5–4.5 μm, ellipsoid to drop-shaped, smooth, hyaline, slightly dextrinoid; tramal skeletal hyphae dextrinoid.

FRUITING: solitary, scattered, or in groups attached to buried wood in broadleaf and mixed woodlands; June–December, often overwintering; widely distributed in the Southeast; fairly common.

EDIBILITY: inedible.

COMMENTS: Compare with the smaller *Microporellus obovatus* (p. 264), inedible, which has a lateral stalk, smaller subglobose to ellipsoid spores, and tramal skeletal hyphae that are negative in Melzer's reagent, and which grows above ground on decaying wood.

Microporellus obovatus (Jungh.) Ryv.   Illus. p. 62

COMMON NAME: none.

FRUIT BODY: annual; circular, kidney-shaped, spoon-shaped, or fan-shaped cap supported by a lateral or sometimes central to eccentric stalk.

CAP: ½–2¾" (1.2–7 cm) wide, circular, kidney-shaped, spoon-shaped, or fan-shaped, sometimes pendant, fibrous-tough when fresh, hard and brittle when dry; upper surface finely velvety to tomentose at first, becoming glabrous in age, white when young, becoming cream to ochraceous, zonate with darker grayish to umber zones; margin thin, acute.

FLESH: up to ⅛" (3 mm) thick at the base, white, fibrous-tough; odor and taste not distinctive.

PORE SURFACE: white at first, becoming cream to pale straw-colored at maturity; pores angular, 6–8 per mm.

STALK: up to 2¾" (7 cm) long, ¹⁄₁₆–¼" (2–6 mm) thick, typically colored like the cap, becoming somewhat wrinkled and irregular in age, sometimes absent.

SPORE PRINT COLOR: white.

MICROSCOPIC FEATURES: spores 3.5–5 × 2–4.5 μm, subglobose to ellipsoid, smooth, hyaline; tramal skeletal hyphae negative in Melzer's reagent.

FRUITING: solitary or in small groups or clusters attached to buried wood in broadleaf woodlands, rarely on conifers; June–December; widely distributed in the Southeast; fairly common.

EDIBILITY: inedible.

COMMENTS: Compare with the larger *Microporellus dealbatus* (p. 264), inedible, which has a central to lateral stalk, larger ellipsoid to drop-shaped spores, and dextrinoid tramal skeletal hyphae, and which grows on the ground attached to buried wood.

Nigroporus vinosus (Berk.) Murr. Illus. p. 63
COMMON NAME: none.
FRUIT BODY: annual; semicircular to fan-shaped violaceous to vinaceous brown caps on decaying wood.
CAP: ¾–4¾" (2–12 cm) wide, broadly attached, leathery to fibrous-tough when fresh, brittle when dry; upper surface somewhat velvety when young, becoming glabrous in age, zonate or sometimes azonate, violaceous to vinaceous brown at first, becoming purplish brown to dark violet at maturity; margin thin and acute.
FLESH: up to ¼" (6 mm) thick at the base and much thinner toward the margin, umber to vinaceous brown or paler; odor and taste not distinctive.
PORE SURFACE: purplish brown to dark violet or smoky black; pores irregularly rounded, 7–8 per mm.
STALK: absent.
SPORE PRINT COLOR: white.
MICROSCOPIC FEATURES: spores 3.5–4.5 × 1–1.5 μm, allantoid to cylindrical, smooth, hyaline.
FRUITING: solitary or in groups or overlapping clusters on decaying broadleaf trees and sometimes conifers, especially pine; June–December; widely distributed in the Southeast; fairly common.
EDIBILITY: inedible.
COMMENTS: This is the only species of *Nigroporus*, a common genus in tropical areas, known to occur in North America. *Nigrofomes melanoporus* (not illustrated), inedible, reported only from tropical regions and Florida, forms a larger, up to 8" (20 cm) wide, and very hard dark brown to purplish black cap that often has sulcate zones or cracks in age; a dark brown to blackish pore surface with tiny pores arranged like a honeycomb; and broadly ellipsoid spores, 4–5 × 3–3.5 μm.

Phaeolus schweinitzii (Fries) Patouillard Illus. p. 63
COMMON NAME: Dye Polypore.
FRUIT BODY: annual; a large overlapping cluster of flattened fused caps or sometimes a solitary cap attached to a solid central stalk.
CAP: 1½–10" (4–25.5 cm) wide, fan- to petal-shaped or circular, fibrous-tough; surface densely matted and wooly or hairy, faintly to distinctly zoned, dull orange to ochre when young, rusty brown to dark brown in age; margin yellow-orange to brownish orange, sharp, wavy, sometimes lobed.
FLESH: up to 1⅛" (3 cm) thick, fibrous-tough, yellowish to reddish brown; odor and taste not distinctive.
PORE SURFACE: yellow to greenish yellow or orange when young, bruising brown and becoming yellowish brown to dark rusty brown in age; pores angular, 0.5–3 per mm.
STALK: ¾–2¾" (2–7 cm) long, up to 2" (5 cm) thick, branched or unbranched, enlarging upward, pale to dark brown.
SPORE PRINT COLOR: whitish.
MICROSCOPIC FEATURES: spores 5–9 × 3–5 μm, elliptic, smooth, hyaline.
FRUITING: solitary or overlapping clusters or rosettes on roots at the base of trees and on

decaying conifer wood; June–November, sometimes overwintering; widely distributed in the Southeast; fairly common.

EDIBILITY: inedible.

COMMENTS: This mushroom is frequently used to dye wool. *Inonotus tomentosus,* inedible, has a smaller, up to 6½" (16.5 cm) wide cap, thinner flesh, and a brown pore surface (see *MNE,* p. 397).

Phellinus chrysoloma (Fries) Donk   Illus. p. 63

COMMON NAME: Golden Spreading Polypore.

FRUIT BODY: perennial; a reddish brown spreading crust with projecting and overlapping laterally fused caps growing on living or fallen trunks of conifers.

CAP: ⅜–3" (1–7.5 cm) wide, thin, flattened, dry, tawny to reddish brown, zonate, with tomentose ridges; margin wavy, sharp, yellowish tawny when young, becoming dark brown to blackish in age.

FLESH: up to ⅛" (3 mm) thick, tawny to ochre; odor and taste not distinctive.

PORE SURFACE: bright tawny ochre, darkening in age or when bruised; pores 2–5 per mm, round to angular or irregular.

SPORE PRINT COLOR: light brown.

MICROSCOPIC FEATURES: spores $4.5–6 \times 4–5$ μm, subglobose, smooth, dull yellowish; setae abundant.

FRUITING: a spreading crust with projecting caps on living and dead conifer trunks; year-round; widely distributed in the Southeast.

EDIBILITY: inedible.

COMMENTS: Formerly known as *Phellinus pini* var. *abietus.*

Phellinus gilvus (Schweinitz) Patouillard   Illus. p. 63

COMMON NAME: Mustard-yellow Polypore.

FRUIT BODY: 1⅛–4¾" (3–12 cm) wide, perennial, fan- to shell-shaped, somewhat flattened, stalkless to slightly effused-reflexed; upper surface slightly tomentose to glabrous, often wrinkled, zonate or azonate, ochre to bright rusty yellow at first, becoming dark yellowish brown to reddish brown at maturity; margin sharp.

FLESH: up to ¾" (2 cm) thick, fibrous-tough, bright yellowish brown, zonate; odor and taste not distinctive.

PORE SURFACE: dark purplish brown to dull yellowish brown; pores circular, 6–8 per mm.

SPORE PRINT COLOR: white.

MICROSCOPIC FEATURES: spores $4–5 \times 3–3.5$ μm, ellipsoid to ovoid, smooth, hyaline.

FRUITING: solitary or in groups or overlapping clusters on broadleaf trees and stumps, especially oak, and sometimes on conifers; year-round; widely distributed in the Southeast; fairly common.

EDIBILITY: inedible.

COMMENTS: Previously known as *Polyporus gilvus.*

Polyporus tenuiculus (Beauv.) Fr.   Illus. p. 63

COMMON NAME: none.

FRUIT BODY: annual; a cluster of white to pale ochraceous overlapping caps with concolorous hexagonal to elongated pores, growing on decaying broadleaf trees.

CAP: ¾–4" (2–10 cm) wide, fan-shaped to spoon-shaped or semicircular, soft and flexible when fresh, brittle when dry; upper surface white at first, becoming cream to pale ochraceous in age, glabrous or sometimes roughened, azonate; margin thin, acute.

FLESH: up to ⅛" (3 mm) thick, white to very pale ochraceous; odor and taste not distinctive.
PORE SURFACE: white at first, becoming cream to pale ochraceous in age, strongly decurrent; pores hexagonal to radially elongated, often finely incised, 1–3 per mm or sometimes up to 2 mm wide.
STALK: rudimentary or sometimes short and stout, central to lateral.
SPORE PRINT COLOR: white.
MICROSCOPIC FEATURES: spores 9–12 × 2–3.5 μm, cylindrical to subnavicular with tapering ends, smooth, hyaline.
FRUITING: solitary, scattered, or in overlapping clusters on decaying broadleaf logs and stumps; August–January, sometimes year-round; Florida west to Texas; common.
EDIBILITY: unknown.
COMMENTS: Formerly known as *Favolus brasiliensis*, this species is common throughout the tropics and extends northward to the Gulf states.

Polyporus varius Fries Illus. p. 63
COMMON NAME: none.
FRUIT BODY: annual; a centrally to laterally stalked cap attached to decaying wood.
CAP: 1–3⅛" (2.5–8 cm) wide, convex to broadly convex, circular to fan-shaped, sometimes depressed and funnel-shaped; surface glabrous, azonate, pale buff with radially aligned darker striations; margin acute.
FLESH: up to ⅛" (3 mm) thick, corky to woody, buff; odor and taste not distinctive.
PORE SURFACE: whitish to pale buff, becoming darker in age, decurrent; pores circular to angular, 7–9 per mm.
STALK: up to ¾" (2 cm) long, ⅛–¼" (3–7 mm) thick, central to lateral, dark brown to black with a paler apex.
SPORE PRINT COLOR: white.
MICROSCOPIC FEATURES: spores 9–12 × 2.5–3 μm, cylindric to slightly curved, smooth, hyaline.
FRUITING: solitary or several from a branched base, attached to decaying broadleaf and conifer wood; June–December, sometimes overwintering; North and South Carolina west to Tennessee and northward, the southern distribution limits yet to be determined; occasional to fairly common.
EDIBILITY: inedible.
COMMENTS: *Polyporus elegans* (see photo, p. 63), inedible, is very similar but has a tan to chestnut brown cap that lacks a striate upper surface.

Pseudofavolus cucullatus (Mont.) Pat. Illus. p. 64
COMMON NAME: none.
FRUIT BODY: annual; a stalkless, convex, fan- to kidney-shaped cap attached to wood by a small whitish disc.
CAP: ¾–3⅛" (2–8 cm) wide, convex, becoming nearly plane, fan- to kidney-shaped; surface glabrous, smooth, whitish to ochraceous when young, soon becoming ochre to dull ochre, often with a dark reddish tint along the margin, darkening when dry; attached to wood by a small whitish disc.
FLESH: up to 1/16" (2 mm) thick, fibrous-tough, straw-colored to pale ochraceous; odor and taste not distinctive.
PORE SURFACE: dark ochraceous to dull reddish or reddish brown; pores angular to hexagonal, 1–3 per mm.

STALK: absent.

SPORE PRINT COLOR: white.

MICROSCOPIC FEATURES: spores 11.5–16 × 4–6 μm, cylindrical, smooth, hyaline.

FRUITING: solitary, scattered, or in groups on decaying broadleaf trees; May–December; widely distributed along the Gulf Coast region; occasional to fairly common.

EDIBILITY: inedible.

COMMENTS: The only member of this tropical genus reported from North America. Formerly known as *Favolus cucullatus* and *Hexagonia taxodii*. *Polyporus mori* = *Polyporus alveolaris* (not illustrated), inedible, is similar but lacks the conspicuous whitish disc.

Pseudofistulina radicata (Schw.) Wright and Burdsall   Illus. p. 64

COMMON NAME: none.

FRUIT BODY: annual; a rounded to kidney-shaped cap, usually with a long radicating lateral stalk, typically attached to buried wood.

CAP: 1⅛–2¾" (3–7.5 cm) wide, rounded to kidney-shaped or irregularly lobed; surface densely tomentose, azonate, yellowish brown, darkening in age; margin acute.

FLESH: up to ¼" (7 mm) thick, fibrous-tough, white when fresh, becoming pale buff when dry; odor and taste not distinctive.

PORE SURFACE: white to cream-colored at first, becoming pinkish buff to ochraceous; pores circular, 5–7 per mm.

STALK: 1½–4" (4–10 cm) long, up to ⅜" (1 cm) thick, usually tapered downward, distinctly lateral, typically radicating, colored like the cap or darker, tomentose on the upper portion, sometimes whitish toward the base.

SPORE PRINT COLOR: white.

MICROSCOPIC FEATURES: spores 3–4 × 2–3 μm, ovoid, smooth, hyaline.

FRUITING: solitary or scattered on the ground, typically attached to buried wood; June–December; widely distributed in the Southeast; occasional.

EDIBILITY: inedible.

COMMENTS: Formerly called *Fistulina radicata*. Beefsteak Polypore, *Fistulina hepatica* (p. 252), edible, has a much larger reddish to reddish brown cap, up to 8" (20 cm); a short lateral stalk; and reddish flesh with a red sap.

Pycnoporus sanguineus (Linnaeus) Murrill   Illus. p. 64

COMMON NAME: none.

FRUIT BODY: annual; a stalkless, thin, fibrous-tough cap attached to decaying broadleaf trees.

CAP: 1⅛–3⅛" (3–8 cm) wide, semicircular to fan-shaped, broadly convex to nearly plane, thin, fibrous-tough when fresh, becoming woody when dry; surface dry, finely tomentose near the margin, subglabrous toward the disc, azonate, orange-red color usually persisting or sometimes fading to salmon buff in age; margin acute.

FLESH: up to ⅛" (3 mm) thick, fibrous-tough, concentrically zoned with pale orange and pale buff in some specimens, azonate and orange-buff in others.

PORE SURFACE: dark red; pores circular, 5–6 per mm.

STALK: absent.

SPORE PRINT COLOR: white.

MICROSCOPIC FEATURES: spores 5–6 × 2–2.5 μm, cylindric, slightly curved, smooth, hyaline.

FRUITING: solitary or in groups or overlapping clusters on decaying broadleaf trees; year-round; widely distributed in the Southeast; fairly common.

EDIBILITY: inedible.

COMMENTS: *Pycnoporus cinnabarinus*, inedible, is very similar but has thicker flesh (up to 1.5 cm), less intense and persistent orange-red pigmentation, and larger spores that measure 6–8 × 2.5–3 μm (see *MNE*, p. 399).

Schizopora paradoxa (Fries) Donk   Illus. p. 64

COMMON NAME: Split-pore Polypore, Creamy Maze Crust.

FRUIT BODY: a spreading crust with small nodules that is flat when growing on horizontal surfaces and somewhat projecting and bracket-like when growing on vertical substrates; upper surface dry, white to cream-colored when fresh, darkening in age, with fine hairs at the margin; fertile surface maze-like with irregular and angular to elongated pores that may separate and form tooth-like projections, creamy white when fresh, darkening to ochraceous yellow in age; pores 1–3 per mm.

FLESH: thin, leathery, white to cream; odor and taste not distinctive.

SPORE PRINT COLOR: white.

MICROSCOPIC FEATURES: spores 5–6.5 × 3.5–4 μm, ellipsoid, smooth, hyaline.

FRUITING: tightly attached to decaying branches of broadleaf trees, especially beech, oak, and American hornbeam *(Carpinus caroliniana)*, rarely on conifer wood; year-round; widely distributed in the Southeast; common.

EDIBILITY: inedible.

COMMENTS: Also known as *Poria versipora*. This common wood-decaying fungus is soft when fresh and moist, but becomes hard when dry. Some tooth-like forms might be confused with the Milk-white Toothed Polypore, *Irpex lacteus*, inedible, which often forms rows of bracket-like caps with a white to buff, densely tomentose to hairy upper surface and a whitish, tooth-like fertile surface (see *MNE*, p. 397).

Spongipellis pachyodon (Pers.) Kotl. and Pouz.   Illus. p. 64

COMMON NAME: none.

FRUIT BODY: annual; leathery to fibrous-tough, spreading, effused-reflexed, forming white to creamy white overlapping shelf-like caps with white to creamy white flattened teeth that darken at maturity on the undersurface.

CAP: ¾–2" (2–5 cm) wide, convex, shelf-like, leathery to fibrous-tough; upper surface white to creamy white at first, becoming ochraceous to brownish in age, finely tomentose to glabrous, azonate; margin acute and somewhat incurved.

FLESH: ⅛–5⁄16" (3–8 mm) thick, white to pale cream, leathery to fibrous-tough; odor and taste not distinctive.

PORE SURFACE: white to creamy white at first, darkening in age to ochraceous or brownish; pores gill-like to labyrinthine or porose near the margin, breaking up and forming conspicuous flattened teeth in age; teeth up to ½" (1.2 cm) long.

STALK: absent.

SPORE PRINT COLOR: white.

MICROSCOPIC FEATURES: spores 5–7 × 5–6.5 μm, subglobose, thick-walled, smooth, hyaline; clamp connections present in the flesh.

FRUITING: solitary or in overlapping groups on living broadleaf trees, especially oak; June–December, sometimes overwintering; widely distributed in the Southeast; occasional.

EDIBILITY: inedible.

COMMENTS: The combination of the conspicuous teeth, small thick-walled spores, and presence of clamp connections is distinctive. *Irpex lacteus*, inedible, also has a white to creamy white cap and conspicuous teeth on the undersurface, but it has cylindric spores, and its flesh lacks clamp connections (see *MNE,* p. 397). *Spongipellis unicolor* (not illustrated), inedible, forms whitish to pale buff caps that become brownish or ochraceous tawny in age, large circular to angular pores that become somewhat daedaleoid at maturity but do not form teeth, and larger ovoid to ellipsoid spores that measure 7–9 × 6–7 μm.

Trametes elegans (Spreng.: Fr.) Fr. Illus. p. 64
  COMMON NAME: none.
  FRUIT BODY: annual or perennial; a whitish to tan or grayish semicircular to circular or fan-shaped cap growing on decaying broadleaf trees.
  CAP: 1½–14" (4–36 cm) wide, shelf-like, semicircular to circular or fan-shaped, corky and flexible when fresh, more rigid when dry; upper surface white to cream or buff ochraceous, often developing gray tints or becoming blackish from the base of older specimens, finely tomentose at first, becoming glabrous in age, concentrically sulcate or warted to somewhat uneven, sometimes smooth; margin thin, acute, even or lobed.
  FLESH: up to ⅝" (1.5 cm) thick, white to pale cream, fibrous-tough, woody when dry.
  PORE SURFACE: highly variable, partly poroid, partly sinuous-daedaleoid, and sometimes partly lamellate; pores round to angular, 1–2 per mm.
  STALK: absent or short and stubby, up to 1⅛" (3 cm) long, up to ⅝" (1.5 cm) wide, colored like the cap.
  SPORE PRINT COLOR: white.
  MICROSCOPIC FEATURES: spores 5–7 × 2–3 μm, cylindric to oblong-ellipsoid, smooth, hyaline.
  FRUITING: solitary, scattered, or in groups on decaying broadleaf trees; May–December, sometimes overwintering; widely distributed in the Southeast; fairly common.
  EDIBILITY: inedible.
  COMMENTS: *Trametes cubensis* (not illustrated), inedible, reported along the Gulf Coast from Florida to Texas, is similar, but its cap lacks the gray coloration; it becomes reddish to bay from the base outward in age; and it has very small pores, 5–7 per mm, and larger spores that measure 7–9.5 × 3–3.5 μm.

Trametes menziesii (Berk.) Ryv. Illus. p. 65
  COMMON NAME: none.
  FRUIT BODY: annual; semicircular to fan-shaped or sometimes circular caps growing on decaying broadleaf logs and stumps.
  CAP: 2–12" (5–30 cm) wide, shelf-like, broadly attached, fibrous-tough and flexible when young and fresh, becoming woody in age; upper surface whitish to pale brown or grayish, obscurely to distinctly zonate, especially along the margin, dry, glabrous or sometimes roughened; margin rounded.
  FLESH: up to ½" (1.2 cm) thick, white, fibrous-tough; odor and taste not distinctive.
  PORE SURFACE: white to cream or pale brown; pores irregularly rounded to angular, 3–4 per mm.
  STALK: absent.
  SPORE PRINT COLOR: white.
  MICROSCOPIC FEATURES: spores 6–8.5 × 2.5–3 μm, elliptic to slightly allantoid, smooth, hyaline.
  FRUITING: solitary, scattered, or in overlapping clusters on logs and stumps of decaying broadleaf trees and occasionally on conifers; March–November, sometimes over-

wintering; reported from Florida west to Texas, the exact range not yet determined; occasional.
EDIBILITY: inedible.
COMMENTS: This little known species has not previously been illustrated or described in any books that feature the polypores of North America. It has also been reported from Africa.

Trametes versicolor (Linnaeus : Fries) Pilát   Illus. p. 65
COMMON NAME: Turkey-tail.
FRUIT BODY: annual to perennial; fan- to kidney-shaped stalkless cap typically growing in clusters on decaying wood.
CAP: ¾–4" (2–10 cm) wide, fan- to kidney-shaped, flattened, sometimes laterally fused and forming extensive rows, fibrous-tough, thin; surface velvety to silky, with conspicuous concentric zones; zones contrasting and variously colored, often with shades of brown, blue, gray, orange, and green; margin thin, sharp, wavy, sometimes folded or lobed.
FLESH: up to ⅛" (3 mm) thick, fibrous-tough, white to creamy white.
PORE SURFACE: white to grayish; pores angular to circular, 3–5 per mm.
STALK: absent.
SPORE PRINT COLOR: white.
MICROSCOPIC FEATURES: spores 5–6 × 1.5–2 μm, cylindric to sausage-shaped, smooth, hyaline.
FRUITING: solitary or in overlapping clusters, rows, or rosettes on decaying wood; year-round; widely distributed in the Southeast; very common.
EDIBILITY: inedible.
COMMENTS: Compare with *Stereum* species, which lack pores on their lower surfaces. *Trametes hirsuta* (not illustrated), inedible, has a grayish to yellowish or brownish zoned cap, usually with a brown margin. *Trametes pubescens* (not illustrated), inedible, has a finely hairy to smooth, creamy white to yellowish buff, azonate or faintly zoned cap.

Trichaptum biforme (Fries) Ryvarden   Illus. p. 65
COMMON NAME: Violet Toothed Polypore.
FRUIT BODY: annual; shell- to petal-shaped or semicircular zoned cap attached to decaying broadleaf trees.
CAP: ⅜–2⅜" (1–6 cm) wide, shell- to petal-shaped or semicircular, convex to nearly plane, leathery to stiff, sometimes smooth in age; surface distinctly zoned and variously colored, white to grayish, reddish brown, green if covered with algae; margin often violet and wavy.
FLESH: up to 1/16" (1.5 mm) thick, fibrous-tough, white to ochre.
PORE SURFACE: violet to purple-brown or sometimes fading to buff; pores angular, 2–5 per mm; tubes and pores splitting and becoming tooth-like or jagged in age.
STALK: rudimentary or absent.
SPORE PRINT COLOR: white.
MICROSCOPIC FEATURES: spores 5–8 × 2–2.5 μm, cylindric, smooth, hyaline.
FRUITING: in overlapping clusters on decaying broadleaf trees; year-round; widely distributed in the Southeast; very common.
EDIBILITY: inedible.
COMMENTS: Sometimes incorrectly spelled *Trichaptum biformis*. The species *Trichaptum*

abietinum (not illustrated), inedible, is smaller, has stiff white hairs on the cap, and grows on conifer wood. *Irpex lacteus,* inedible, has a white to creamy white cap and a whitish tooth-like pore surface, and it grows on decaying broadleaf wood (see *MNE*, p. 397). *Spongipellis pachyodon* (p. 269), inedible, is much larger, has a white to ochraceous cap and tooth-like pore surface, and grows on broadleaf trees.

Stinkhorns

Stinkhorns are members of the Gasteromycetes, commonly known as the stomach fungi. These mushrooms are unable to discharge their spores forcibly and have evolved alternative strategies for spore dispersal, such as rain and vectors including insects and mammals. The immature stage of a stinkhorn is an egg-like structure that resembles a small puffball. As the stinkhorn matures, a hollow stalk with a head or arms arises from the egg. A slimy, foul-smelling spore mass is formed within the egg. At maturity, this mass is exposed on an elevated stalk or on the inner portion of the stinkhorn's arms. Mushroom hunters often smell the foul odor, which attracts many insects, before they ever see the fungus.

One member of this group, *Phallogaster saccatus* (p. 277), does not form a stalk. It retains its pear-shaped to nearly round appearance until it becomes perforated and releases its spores.

Aseroe rubra Labillard   Illus. p. 65
    COMMON NAME: Fungus Flower, Starfish Stinkhorn.
    FRUIT BODY: at first globose to obovate with basal rhizomorphs, ¾–1½" (2–4 cm) wide, off-white or cream or pale grayish brown, at times with grayish spots; interior gelatinous; outer peridium soon rupturing at the top, giving rise to a cylindrical stalk and a central disk with radiating tapered arms and spore mass at the stalk apex, and leaving a sac-like volva at the base.
    STALK: 1½–2½" (4–6.5 cm) long, ⅝–1¼" (1.5–3 cm) thick, sponge-like, hollow, forming a flattened disk with 5–11 radiating chambered arms that are often forked near the midportion and curled at the tip; central stalk pinkish to reddish, usually paler toward the base; arms up to 1¾" (4.5 cm) long, bright red.
    SPORE MASS: dark olive brown, covering the disk and interior base of the arms, mucilaginous; odor fetid.
    MICROSCOPIC FEATURES: spores 4.5–7 × 1.7–2.5 μm, elliptic-cylindric with a truncate base, smooth, hyaline or tinted pale brown.
    FRUITING: solitary or in groups on rich soil, leaf litter, and wood mulch; May–September; widely distributed in the Southeast, infrequent but may be locally common.
    EDIBILITY: of no culinary interest.
    COMMENTS: Although variable in size and color, this distinctive stinkhorn is easily recognized. Originally described from Australia (Tasmania), it has spread to many parts of the world, and one form or another has now been recorded from Asia, South Africa, Europe, Hawaii, and the Americas.

Blumenavia angolensis (Welwitsch and Currey) Dring Illus. p. 66
  COMMON NAME: none.
  FRUIT BODY: at first egg-like, resembling a puffball, subglobose, up to 2¾" (7 cm) wide, dark gray to blackish, cracking into large scales and soon developing longitudinal furrows; opening by a number of large, irregular apical lobes as the fruit body expands; internally gelatinous; giving rise to the stalk and terminal columns.
  STALK: 2–4" (5–10 cm) long, up to 2¾" (7 cm) wide, long-ovoid, white, terminating in 3–5 unbranched columns that are united at the apex to form a lattice.
  VOLVA: whitish, but often darker when remnants of the egg remain attached, wrinkled and tough.
  SPORE MASS: grayish green to grayish olive at first, darkening in age, slimy, foul smelling, deposited on the underside of the columns.
  MICROSCOPIC FEATURES: spores 2–3.5 × 1.5 μm, ellipsoid, smooth, hyaline.
  FRUITING: solitary or in groups on mulch or fertile soil; June–December; to date only reported from the Aline McAshan Arboretum at Memorial Park in Houston, Texas; believed to have been transported to the arboretum in mulch or soil accompanying exotic plants.
  EDIBILITY: of no culinary interest.
  COMMENTS: Formerly known as *Laternea angolensis* and *Clathrus angolensis*. This fungus was originally described from Angola and has been reported from other locations in Africa, including Tanzania and South Africa, as well as several sites in South America. *Blumenavia* is a reference to Blumenau, an area in Brazil.

Dictyophora duplicata (Bosc) Fischer Illus. p. 66
  COMMON NAME: Netted Stinkhorn.
  FRUIT BODY: at first egg-like, resembling a puffball, nearly round to somewhat fattened, 1¾–2¾" (4.5–7 cm) high and wide; whitish to pale flesh color or pinkish brown; often grooved on the lower portion; with one or more thick, whitish to pinkish, often branched rhizomorphs; internally gelatinous; giving rise to a distinct head and stalk.
  HEAD: 1⅜–2" (3.5–5 cm) wide, 2–2¾" (5–7 cm) high, oval to conic or bell-shaped, pendant, deeply pitted, with a white-rimmed opening at the apex; covered with a slimy spore mass.
  STALK: 3½–7¼" (9–18.5 cm) long, 1⅜–2⅜" (3.5–6 cm) wide, nearly equal, roughened, spongy, hollow, white; surrounded at the apex by a white, net-like, flaring veil that emerges from beneath the head.
  VOLVA: whitish to pale flesh color or pinkish brown, wrinkled to folded, tough.
  SPORE MASS: greenish brown to brownish olive, slimy, foul smelling.
  MICROSCOPIC FEATURES: spores 3.5–4.5 × 1–2 μm, elliptic, smooth, hyaline.
  FRUITING: solitary or in groups on the ground in broadleaf or mixed woods; June–October; widely distributed in the Southeast; occasional.
  EDIBILITY: of no culinary interest.
  COMMENTS: This species is the largest stinkhorn in the Southeast. Compare with *Phallus ravenelii* (p. 277) and *Phallus hadriani* (p. 277).

Linderia columnata (Bosc) G. H. Cunningham Illus. p. 66
  COMMON NAME: Columned Stinkhorn.
  FRUIT BODY: at first egg-like, nearly round, resembling a puffball; 1–2¾" (2.5–7 cm) wide, 2–6" (5–15.5 cm) high; composed of 2–5 spongy, erect, curved, delicate, orange to red-

dish orange or rosy red columns that are fused at their tips and extend from a sac-like volva.

STALK: absent.

VOLVA: whitish, attached at the base by one or more cord-like strands, becoming wrinkled and tough.

SPORE MASS: olive brown, slimy, foul smelling, deposited on the underside of the columns.

MICROSCOPIC FEATURES: spores 3.5–5 × 1.5–2.5 μm, smooth, elliptic, yellowish brown.

FRUITING: solitary, scattered, or in groups on the ground in woods, woodland margins, and especially grassy areas; October–March; widely distributed in the Southeast; common.

EDIBILITY: unknown; hogs have reportedly been poisoned after consuming this species.

COMMENTS: Also known as *Clathrus columnatus*.

Lysurus gardneri Berkeley   Illus. p. 66

COMMON NAME: Lizard's Claw Stinkhorn.

FRUIT BODY: at first egg-like, white, globose, resembling a puffball; up to 1¼" (3 cm) wide, arising from a white mycelium, splitting open and giving rise to the stalk and terminal columns.

STALK: 2–6" (5–15.5 cm) high, ⅜–¾" (1–2 cm) wide, hollow, white to pale cream, terminating in 4–6 short columns that are fused at their tips, but may become free in older specimens.

VOLVA: white, membranous, egg-shaped.

SPORE MASS: dark grayish brown to olive black, slimy, foul smelling, deposited on the outer portions of the columns.

MICROSCOPIC FEATURES: spores 4–5 × 1.5 μm, elliptical, smooth, yellowish brown.

FRUITING: often solitary or sometimes several on loose, manured, mulched soil; June–December; reported from Texas, exact distribution unknown; occasional.

EDIBILITY: of no culinary interest.

COMMENTS: Also known as *Lysurus borealis*, this fungus has also been reported from Sri Lanka, India, Indonesia, and Africa.

Lysurus periphragmoides (Klotzsch) Dring   Illus. p. 67

COMMON NAME: none.

FRUIT BODY: at first egg-like, globose to elongated, white to buff, resembling a puffball, arising from whitish mycelial strands, up to 1¾" (4.5 cm) wide, splitting open and giving rise to a head and stalk.

HEAD: a conspicuous swollen and rounded terminal network of anastomosing arms that form a lattice; arms of the lattice sharply angled and distinctly corrugated; pinkish to pinkish orange, yellowish or sometimes whitish.

STALK: 2–6" (5–15.5 cm) high, ⅜–¾" (1–2 cm) wide, hollow, pinkish to pinkish orange, yellowish or sometimes whitish.

VOLVA: white to buff, membranous, egg-shaped.

SPORE MASS: dark olive green, slimy; odor sweet when fresh and first exposed, soon becoming foul-smelling; filling the entire interior of the head and extending outward between the arms of the lattice.

MICROSCOPIC FEATURES: spores 4–4.5 × 1.5–2 μm, elliptical, smooth, pale brown.

FRUITING: solitary to numerous on fertile ground and on mulch; October–June; reported from Texas, exact distribution unknown; occasional.

EDIBILITY: of no culinary interest.

COMMENTS: Formerly known as *Simblum texense* and *Simblum sphaerocephalum*, this stinkhorn has been reported from many locations, including Mauritius, Africa, India, Asia, Indonesia, and Australia.

Mutinus elegans Montagne — Illus. p. 67

COMMON NAME: Elegant Stinkhorn.

FRUIT BODY: at first egg-like, resembling a small puffball, nearly round to oval then elongated, ½–1" (1.2–2.5 cm) wide and high, white, with a strong white basal cord; internally gelatinous; giving rise to a stalk; head absent.

STALK: 4–7" (10–18 cm) high, ½–1" (1.2–2.5 cm) wide, roughened, tapered from the middle in both directions, with a narrow opening at the apex, hollow, spongy, orange to pinkish orange or pinkish red.

VOLVA: whitish, tough, wrinkled, sac-like.

SPORE MASS: olive green to dull greenish, slimy, foul-smelling, covering the upper third or half of the stalk.

MICROSCOPIC FEATURES: spores 4–7 × 2–3 μm, elliptic, smooth, pale yellowish green.

FRUITING: scattered or in groups, sometimes in dense clusters on wood chips, leaf litter, and soil; July–November; widely distributed in the Southeast; fairly common.

EDIBILITY: reported to be edible in the egg stage.

COMMENTS: Formerly known as *Mutinus curtisii* and *Mutinus bovinus*. *Mutinus ravenelii* (p. 276), edible, is similar but has a distinct head at the apex of its conspicuously pitted, deep carmine to rosy red stalk.

Mutinus ravenelii (Berkeley and Curtis) E. Fischer — Illus. p. 68

COMMON NAME: Little Red Stinkhorn, Ravenel's Mutinus.

FRUIT BODY: At first egg-like, ovoid, up to 1¼" (3.5 cm) long, ⅜–¾" (1–2 cm) wide; whitish with rhizomorphs at the base; gelatinous within; eventually rupturing at the apex as the stalk expands upward, leaving a saccate volva at the base.

HEAD: clearly differentiated, a somewhat swollen spore mass zone on the upper portion.

STALK: 2–3½" (5–9 cm) high, ⅜–⅝" (1–1.5 cm) thick, spongy, hollow, conspicuously pitted overall, more or less equal or tapering slightly downward below the somewhat enlarged fertile head, which is bluntly pointed at the apex, deep carmine or rosy red below and beneath the slime layer, becoming paler toward the base.

VOLVA: whitish, membranous, tough.

SPORE MASS: olive green to olive brown, slimy, foul-smelling, covering the upper quarter or less of the stalk.

MICROSCOPIC FEATURES: spores 1.55–2.2 × 3.5–5 μm, elliptic, smooth, pale yellowish green.

FRUITING: solitary or in groups or clusters on soil, well-decayed wood, and mulch, often around decaying tree stumps; July–November; widely distributed in the Southeast; fairly common.

EDIBILITY: generally considered to be of no culinary interest, but reported to be edible in the egg stage.

COMMENTS: *Mutinus elegans* (p. 276), edible, is similar, but it has an orange to pinkish orange or pinkish red and roughened stalk that is tapered from the middle in both directions; its spore mass covers the upper third or half of the stalk; and it lacks a clearly differentiated swollen fertile head. The Dog Stinkhorn, *Mutinus caninus* (not illustrated), of no culinary interest, is very similar, but its stalk is pale yellowish buff to orange with a darker orange to orange-red head. This species is commonly found in Europe and has also been reported from Great Britain and Switzerland. Its occurrence, if any, in the United States is likely owing to import on wood chips and mulch. It has often been misidentified and is incorrectly illustrated in numerous field guides published in the United States. Excellent photographs of *Mutinus caninus* may be found on page 256 of *Mushrooms and Other Fungi of Great Britain and Europe* (Phillips 1981) and on page 400 of *Fungi of Switzerland*, vol. 2 (Breitenbach and Kranzlin 1986).

Phallogaster saccatus Morgan Illus. p. 68
  COMMON NAME: Club-shaped Stinkhorn.
  FRUIT BODY: at first egg-like, resembling a small puffball, 1–2" (2.5–5 cm) high, ⅜–1⅜" (1–3.5 cm) wide, typically pear-shaped and narrowed toward the base or nearly round; white to pink or pinkish lilac on the upper portion, white toward the base; surface smooth then forming irregular depressions that eventually perforate to expose the spore mass; base attached by numerous intertwined and whitish to pinkish rhizomorphs; internally gelatinous.
  SPORE MASS: dark green, slimy, foul-smelling.
  MICROSCOPIC FEATURES: spores 4–5.5 × 1.5–2 μm, subcylindric, smooth, green-tinted.
  FRUITING: solitary or in groups on decaying wood, often wood chip mulch; May–September; widely distributed in the Southeast; occasional to fairly common.
  EDIBILITY: of no culinary interest.

Phallus hadriani Venturi : Persoon Illus. p. 68
  COMMON NAME: Witch's Egg, Dune Stinkhorn.
  FRUIT BODY: globose to egg-shaped at first, typically with rhizomorphs at the base, 1¼–2¾" (3–7 cm) wide, white when embedded in the soil but changing to violaceus pink when exposed; rupturing as it matures and giving rise to an erect, hollow stalk and head; gelatinous within.
  HEAD: 1¼–2" (3–5 cm) wide, more or less bell-shaped, surface pitted or chambered, at first covered with a foul-smelling olive green spore-bearing slime, apex with a white rim that is often covered with a flap of membranous volval material.
  STALK: 4–10" (10–25.5 cm) long, ¾–1¼" (2–3 cm) thick, nearly equal or tapered toward the apex, spongy, fragile, hollow, whitish to tinged pinkish near the base.
  VOLVA: membranous, whitish with pinkish to pinkish purple tints.
  SPORE MASS: olive green.
  MICROSCOPIC FEATURES: spores 3–6 × 2–3.5 μm, elongated elliptic, smooth, yellowish.
  FRUITING: solitary to gregarious or clustered in sandy soil, dunes, and cultivated areas; August–December; widely distributed in the Southeast; occasional.
  EDIBILITY: of no culinary interest.
  COMMENTS: *Phallus impudicus* (not illustrated), of no culinary interest, is very similar but has a white egg and volva. *Phallus ravenelii* (p. 277), edible, is also similar, but the surface of its head is granular to wrinkled and lacks pits. *Dictyophora duplicata* (p. 274), of no culinary interest, is similar but has a net-like skirt below the head.

Phallus ravenelii Berkeley and Curtis Illus. p. 68
  COMMON NAME: Ravenel's Stinkhorn.
  FRUIT BODY: at first egg-like, resembling a small puffball, oval to pear-shaped, 1⅜–2⅜" (3.5–6 cm) high, 1⅛–1¾" (3–4.5 cm) wide; whitish to pinkish lilac, with pinkish lilac

and often branched rhizomorphs; internally gelatinous; giving rise to a distinct head and stalk.

HEAD: 1⅛–1¾" (3–4.5 cm) long, ⅝–1½" (1.5–4 cm) wide, conic, with a granular to wrinkled surface, lacking distinct pits, with a white-rimmed opening at the apex; covered with a slimy spore mass.

STALK: 4–6¼" (10–16 cm) long, ⅝–1⅛" (1.5–3 cm) wide; nearly equal; roughened, hollow, spongy, whitish.

VOLVA: whitish to pinkish lilac, wrinkled, tough.

SPORE MASS: greenish brown to olive brown, slimy, foul-smelling.

MICROSCOPIC FEATURES: spores 3–4 × 1–1.5 μm, cylindric, smooth, hyaline.

FRUITING: solitary to scattered or clustered on or near decaying logs and stumps, woody debris and wood chips; August–April; widely distributed; fairly common.

EDIBILITY: reported to be edible in the egg stage.

COMMENTS: *Phallus impudicus* (not illustrated), of no culinary interest, is very similar, but the surface of its head is deeply pitted or chambered.

Morels, False Morels, and Allies

Morels are hollow-stalked, single-chambered mushrooms with sponge-like, conic to bell-shaped caps that have distinct pits and ridges. False morels, also known as lorchels, include mushrooms with a brain-like, saddle-shaped, trilobate, miter-shaped, shield-shaped, or irregularly lobed cap. The stalks of some species are small and terete to compressed or ribbed, whereas others are massive and multichambered. Many species in this group are considered to be choice edibles, but some are poisonous, and one has caused fatalities.

Gyromitra caroliniana (Bosc : Fries) Fries Illus. p. 69
  COMMON NAME: Carolina False Morel.
  CAP: 2–7" (5–18 cm) wide and tall, convoluted and brain-like, typically with several prominent seam-like vertical ridges, usually not saddle-shaped; moist and lubricous; fertile surface reddish brown; sterile surface white to whitish; margin appressed against the stalk, wavy and irregular.
  STALK: 1½–6" (4–15.5 cm) long, 1–3⅛" (2.5–8 cm) thick, enlarged at the base, distinctly ribbed, white; interior multichambered.
  MICROSCOPIC FEATURES: spores 22–35 × 10–16 μm, elliptic, reticulate, with one or more short projections at each end, hyaline.
  FRUITING: solitary or in groups on the ground in broadleaf and mixed woods; March–June; widely distributed in the Southeast; occasional.
  EDIBILITY: questionable and not recommended; although eaten by some individuals, suspected to contain toxins.

Gyromitra esculenta (Persoon : Fries) Fries Illus. p. 69
  COMMON NAME: Brain Mushroom.
  CAP: 1⅜–4" (3.5–10 cm) wide, 1½–4" (4–10 cm) tall, brain-like to irregularly lobed, deeply wrinkled to convoluted, moist to dry; fertile surface pinkish tan to dark reddish brown or orange-brown, lubricous when fresh; sterile surface pale pinkish tan to yellowish tan; margin undulating to contorted, often curved toward the stalk.
  STALK: ¾–2¾" (2–7 cm) long, ¾–1⅛" (2–3 cm) thick, enlarging downward or nearly equal, hollow or stuffed with cottony hyphae, sometimes chambered; surface smooth and waxy to slightly granular, dingy white to pinkish tan or tan, often ribbed near the base.
  MICROSCOPIC FEATURES: spores 18–28 × 9–13 μm, elliptic, smooth, with two oil drops.
  FRUITING: solitary, scattered, or in groups on the ground under conifers, especially white pine; April–June; widely distributed in the Southeast; occasional.
  EDIBILITY: poisonous; contains hydrazines that can cause serious illness or death.

COMMENTS: The interior of the cap is chambered, and its flesh is very brittle. Many common names have been assigned to this mushroom, including Conifer False Morel, Beefsteak Morel, and Lorchel.

Gyromitra infula (Schaeffer : Fries) Quélet   Illus. p. 69
COMMON NAME: Saddle-shaped False Morel.
CAP: 1–4" (2.5–10 cm) wide, ¾–4" (2–10 cm) tall, usually saddle-shaped or sometimes trilobate; fertile surface wrinkled to convoluted or sometimes nearly smooth, moist when fresh, reddish brown to dark brown, lacking distinct violet to lavender tints; interior hollow or chambered; flesh brittle; margin incurved.
STALK: ¾–2⅜" (2–6 cm) long, ¾–1" (2–2.5 cm) thick, dry, hollow, finely granular, whitish to pinkish buff.
MICROSCOPIC FEATURES: spores 18–23 × 7–10 μm, elliptic, smooth, hyaline, with two large oil drops when mounted in water.
FRUITING: solitary, scattered, or in groups on decaying wood or humus; July–October; widely distributed in the Southeast; occasional.
EDIBILITY: poisonous.
COMMENTS: This false morel is unusual for the genus *Gyromitra* in that it typically fruits in the fall rather than in the spring.

Helvella albella Quélet   Illus. p. 69
COMMON NAME: none.
CAP: ⅜–1" (1–2.5 cm) wide, weakly to moderately saddle-shaped when young, becoming strongly saddle-shaped to trilobate at maturity; fertile (upper) surface gray-brown to brown, smooth to slightly wrinkled; sterile (lower) surface creamy white to white, pubescent when young, becoming nearly glabrous in age; margin curving over the fertile surface when young, expanding and flaring in age, entire or lacerate.
STALK: ¾–2" (2–5 cm) long, ⅛–¼" (2–6 mm) thick, rounded, somewhat enlarged downward, glabrous to finely pruinose; whitish.
MICROSCOPIC FEATURES: spores 20–24 × 12–15 μm, oblong to ellipsoid, with one large central oil droplet and up to four small droplets at each end, smooth, hyaline.
FRUITING: scattered or in groups on the ground in woodlands; August–November; widely distributed in the Southeast; occasional.
EDIBILITY: unknown.
COMMENTS: *Helvella elastica* (p. 281), edibility unknown, is very similar, but its cap is typically paler, and its sterile surface is smooth when young. *Helvella macropus* (p. 281), edibility unknown, has a gray, cup-shaped cap with a downy outer surface; a gray stalk covered with dense tufts of downy gray hairs; and larger spores that measure 20–30 × 10–12 μm.

Helvella crispa Fries   Illus. p. 70
COMMON NAME: Fluted White Helvella.
CAP: ¾–2⅜" (2–6 cm) wide, ⅜–1⅝" (1–4cm) tall, saddle-shaped to irregularly lobed, dry; fertile surface pale cream to pale buff, smooth to somewhat wrinkled; sterile surface pale cream to buff, with tiny short hairs; margin rolled inward when young then gradually unrolling and expanding at maturity, entire or somewhat lacerated.
STALK: 1–3½" (2.5–9 cm) long, ⅜–1⅛" (1–3 cm) thick, typically enlarged in the middle and tapering toward the base and apex, sometimes nearly equal; surface whitish to pale buff

or pinkish buff, deeply pitted and ribbed; ribs branching, anastomosing, and extending onto the sterile surface of the cap.

MICROSCOPIC FEATURES: spores 14–23 × 10–14 μm, elliptic, smooth, with one to five oil drops.

FRUITING: solitary, scattered, or in groups on the ground in hardwood and conifer forests; July–December; widely distributed in the Southeast; fairly common.

EDIBILITY: not recommended; although consumed in Europe, its edibility in North America has not yet been established.

COMMENTS: The pale cream to buff colors of the cap and stalk are diagnostic field characters. This species sometimes fruits on moss-covered decaying wood. *Helvella lacunosa*, edibility unknown, is similar, but its cap is pale to dark gray or nearly black (see *MNE*, p. 479).

Helvella elastica Fries   Illus. p. 70

COMMON NAME: Smooth-stalked Helvella.

CAP: ¾–2" (2–4 cm) wide, depressed saddle-shaped to irregularly lobed with lobes typically bent downward toward the stalk; fertile (upper) surface tan to pale grayish tan or dark grayish brown, smooth to somewhat wrinkled or sometimes pitted, dry or moist; sterile (lower) surface white to pale tan, smooth; margin entire or somewhat lacerated.

STALK: 1⅛–4" (3–10 cm) long, ⅛–⅜" (3–10 mm) thick, rounded, somewhat enlarged downward, whitish to buff, smooth overall or slightly roughened near the apex.

MICROSCOPIC FEATURES: spores 18–24 × 11–14 μm, elliptic to oblong, smooth or rarely slightly roughened, containing one large central oil drop and zero to five smaller droplets at each end, hyaline.

FRUITING: solitary, scattered, or in groups on the ground and on extremely rotted wood in conifer and broadleaf forests; July–November; widely distributed in the Southeast; occasional.

EDIBILITY: unknown.

COMMENTS: *Helvella albella* (p. 280), edibility unknown, is very similar, but its cap is typically darker, and its sterile surface is pubescent when young. *Helvella macropus* (p. 281), edibility unknown, has a gray, cup-shaped cap with a downy outer surface; a gray stalk covered with dense tufts of downy gray hairs; and larger spores that measure 20–30 × 10–12 μm.

Helvella macropus (Fries) Karsten   Illus. p. 70

COMMON NAME: none.

CAP: ⅝–1½" (1.5–4 cm) wide, cup- or saucer-shaped; fertile (upper) surface gray to gray-brown, smooth to slightly wrinkled, dry or moist; sterile (lower) surface gray to gray-brown, villose especially near the margin; margin curving over the fertile surface when young.

STALK: ¾–2" (2–5 cm) long, 1/16–¼" (1.5–5 mm) thick, rounded, somewhat enlarged downward, dry, solid, gray to gray-brown with a whitish base, pubescent to villose.

MICROSCOPIC FEATURES: spores 19–28 × 10–13 μm, fusiform-elliptic to subfusiform, with one large central oil droplet and two smaller droplets, one at each end, usually finely punctate, hyaline.

FRUITING: solitary, scattered, or in groups on the ground, among mosses, and on decaying wood in broadleaf woods and sometimes in conifer forests; July–November; widely distributed in the Southeast; occasional.

EDIBILITY: unknown.

COMMENTS: Also known as *Macroscyphus macropus*. Both *Helvella albella* (p. 280) and *Helvella elastica* (p. 281) have saddle-shaped to trilobate brown caps.

Morchella elata Fries complex Illus. p. 70
COMMON NAME: Black Morel.
CAP: ¾–2⅜" (2–6 cm) wide, 1–3½" (2.5–9 cm) tall, conic to oval, sponge-like, hollow, divided into pits and ridges, continuous with the stalk below; pits elongated and irregular, yellow-brown to gray-brown; ridges anastomosing, dark brown to brownish black.
STALK: 2–4" (5–10 cm) long, ⅜–1½" (1–4 cm) thick, enlarged near the base, hollow; surface white to dingy yellow, granular, sometimes ribbed.
MICROSCOPIC FEATURES: spores 18–25 × 11–15 μm, elliptic, smooth.
FRUITING: solitary, scattered, or in groups on soil in a variety of habitats: in mixed woods associated with cherry, poplars, and pines, in mixed broadleaf woodlands, and in burned areas; March–May; widely distributed in the Southeast; fairly common.
EDIBILITY: edible and choice, but must be thoroughly cooked.
COMMENTS: This species is usually the first morel to appear in spring. Although choice, it has caused gastrointestinal distress in some individuals, especially when consumed with alcohol. The Black Morel is a complex of several varieties or species that are nearly indistinguishable. Other commonly applied names include *Morchella angusticeps* and *Morchella conica*. Much work remains to be done before the taxonomic problems of this complex are resolved.

Morchella esculenta Fries complex Illus. p. 70
COMMON NAME: Common Morel.
CAP: ¾–2¾" (2–7 cm) wide, ¾–7" (2–17.5 cm) tall, oval to conic or somewhat cylindric, sponge-like, hollow, divided into pits and ridges of variable color, continuous with the stalk below; pits round to elongated and irregular, gray to brown or yellowish; ridges anastomosing, whitish to grayish, yellow or yellow-brown.
STALK: 1–4½" (2.5–11.5 cm) long, ¾–2¾ (2–7 cm) thick, nearly equal or enlarged (sometimes massively) toward the base, hollow; surface whitish, granular, often ribbed.
MICROSCOPIC FEATURES: spores 20–26 × 12–16 μm, elliptic, smooth.
FRUITING: solitary, scattered, or in groups on soil in a variety of habitats: near dead elms, in old apple orchards, in burned areas, in mixed broadleaf woodlands, and sometimes under conifers; March–June; widely distributed in the Southeast; fairly common.
EDIBILITY: edible and choice, but must be thoroughly cooked; one of the most highly prized and sought after mushrooms for the table.
COMMENTS: This morel has many additional common names, including Yellow Morel, Sponge Mushroom, Land Fish, Pine Cone Mushroom, and Honeycomb. Some authors recognize various color forms and statures as distinct species. In her book *A Morel Hunter's Companion* (1988), Nancy Weber notes that more than one hundred names exist in the literature for species in the genus *Morchella*. Extensive work must be done to resolve the taxonomic problems of this genus.

Morchella semilibera De Chambre : Fries Illus. p. 70
COMMON NAME: Half-free Morel.
CAP: ⅝–1½" (1.5–4 cm) wide, ⅜–1½" (1–4 cm) tall, broadly conic with a round to blunt apex, sponge-like, hollow, divided into pits and ridges, attached to the stalk about midway and flaring below; pits elongated and irregular, grayish tan to yellow-brown;

ridges anastomosing, brownish, darkening to grayish brown or brownish black in age; underside of flaring cap whitish and granular.

STALK: 2–6" (5–15 cm) long, ⅜–1" (1–2.5 cm) thick, enlarging downward or nearly equal overall, hollow; surface whitish, slightly granular, commonly ribbed.

MICROSCOPIC FEATURES: spores 22–30 × 12–17 μm, elliptic, smooth, hyaline.

FRUITING: solitary or scattered on the ground in broadleaf forests, in old apple orchards, and sometimes associated with a variety of other trees, including poplars and oaks; March–May; widely distributed in the Southeast; fairly common.

EDIBILITY: edible.

COMMENTS: This species typically fruits before the Common Morel and is easily recognized because of its half-free cap and hollow stalk. The Wrinkled Thimble-cap, *Ptychoverpa bohemica* (not illustrated), is similar but has a yellow-brown wrinkled cap that lacks true pits and is attached only at the top of the stalk. It can also be differentiated from the Half-free Morel by the smooth undersurface of the flaring skirt-like cap, a stalk stuffed with cottony hyphae, and extremely large, elliptic, smooth spores that measure 55–85 × 15–18 μm. *Ptychoverpa bohemica* is edible for some, but poisonous to many others, causing variable reactions including severe gastrointestinal upset and temporary loss of coordination. The Smooth Thimble-cap, *Verpa conica,* poisonous for some individuals, has a smooth, thimble- to bell-shaped cap that is rounded to flattened at the center and attached to the stalk only at the top of the cap (see *MNE,* p. 479).

Fiber Fans and Vases

Fiber fans and vases, members of the genus *Thelephora*, are leathery, fibrous-tough fungi with split or torn margins and brown spores that are usually warted or spiny. Their fruit bodies are typically a shade of brown at maturity. Some members have distinct caps and stalks; some are stalkless; and others produce coral-like tufts of forking or spoon-shaped to flattened branches.

Thelephora palmata Scopoli : Fries Illus. p. 71
 COMMON NAME: none.
 FRUIT BODY: 1½–4" (4–10 cm) high, 1½–2¾" (4–7 cm) wide, coral-like, consisting of numerous branches arising from a short stalk; stalk short, up to ⅝" (1.5 cm) long and wide, colored like the branches, divided on the upper portion; branches smooth, flattened, palm-shaped, richly divided, whitish when very young, soon becoming grayish brown to lilac-brown or dark brown, often with purple to violet tints; branch tips usually whitish, conspicuously flattened and spatula-shaped, often finely fringed.
 FLESH: corky to fibrous-tough, brown; odor strongly fetid and disagreeable, especially after drying, rarely not distinctive; taste not determined.
 SPORE PRINT COLOR: purplish brown.
 MACROCHEMICAL TESTS: tramal tissue stains deep blue with the application of KOH.
 MICROSCOPIC FEATURES: spores 8–12 × 6–9 μm, angular-lobate, strongly echinulate, brown.
 FRUITING: solitary, scattered, or in groups on the ground in conifer and mixed woods; July–October; widely distributed in the Southeast; occasional.
 EDIBILITY: inedible.
 COMMENTS: *Thelephora anthocephala* var. *anthocephala* (not illustrated), inedible, is similar, but its flesh lacks the fetid odor, and its tramal tissue does not stain deep blue in KOH. *Thelephora anthocephala* var. *americana* (see photo, p. 71), inedible, is also similar, and it, too, has a fetid odor, but its tramal tissue does not stain deep blue in KOH.

Thelephora terrestris f. **concrescens** Lundell Illus. p. 71
 COMMON NAME: none.
 FRUIT BODY: partially erect and enveloping the stems and branches of host plants, ½–2¾" (1.3–7 cm) wide, composed of circular to fan-shaped or funnel-shaped, stalkless caps in overlapping clusters, sometimes laterally fused and forming patches up to 10" (25 cm) or more in diameter; upper surface covered with short, stiff hairs, often matted and wooly or scaly, somewhat concentrically zoned, rusty brown to dark brown or grayish brown, becoming blackish brown in age; lower surface somewhat wrinkled and finely warted to nearly smooth, grayish to pinkish brown when young, becoming dull brown in age; margin wavy or sometimes lobed, white to grayish when fresh.

FLESH: thin, leathery, brown; odor somewhat earthy or absent; taste not distinctive.
SPORE PRINT COLOR: purplish brown.
MICROSCOPIC FEATURES: spores 8–12 × 6–9 μm, angularly oval to elliptic, warted and spiny, brown.
FRUITING: in overlapping clusters that envelope and clasp the stems and branches of woody plants, especially oak; July–December; Georgia and Florida west to Texas; occasional.
EDIBILITY: inedible.
COMMENTS: *Thelephora terrestris* f. *terrestris*, inedible, grows on the ground attached to roots, branches, seedlings, and mosses in conifer woods, but does not envelop the substrate (see *MNE*, p. 14).

Thelephora vialis Schweinitz   Illus. p. 71
COMMON NAME: Vase Thelephore.
FRUIT BODY: erect, large, 1–4" (2.5–10 cm) tall, highly variable, typically composed of ascending lobes or caps clustered and arising from a common central stalk.
CAP: funnel- to spoon-shaped or fused and somewhat vase-shaped, arising from a common central base; inner (upper) surface striate or minutely scaly, whitish to yellowish; outer (lower) surface wrinkled, dingy yellow then grayish brown.
FLESH: thick, leathery, whitish to grayish; odor and taste not distinctive when fresh, odor becoming somewhat disagreeable on drying.
STALK: ⅜–2" (1–5 cm) long, ¼–1½" (5–40 mm) thick, erect, enlarged downward, dry, solid, whitish to grayish, minutely pubescent.
SPORE PRINT COLOR: brown.
MICROSCOPIC FEATURES: spores 4.5–8 × 4.5–6.5 μm, angular and warted, minutely spiny, olive buff.
FRUITING: solitary, scattered, or in groups on the ground in broadleaf woods, especially oak; July–December; widely distributed in the Southeast; fairly common.
EDIBILITY: inedible.

Branched and Clustered Corals

This group includes species with two different kinds of fruit bodies. The first type has spindle-shaped to worm-like, erect, and unbranched or infrequently branched stalks that may or may not be fused at their bases and that typically grow in colonies or clusters. The second type has erect, repeatedly branched, coral-like stalks and grows solitary or in groups. Both types have brittle flesh and are easily broken, but some are fibrous to tough and flexible. Spores are produced on portions of the outer surface of the stalks and branches. Most species grow on the ground, but a few occur on decaying wood and on the bark of standing trees.

Several species are edible; some are poisonous; and the others are inedible or of unknown edibility. A few cases of severe gastrointestinal upset have been caused by some members of this group. However, no fatalities have been reported. Similar species that are erect and unbranched are included in the Earth Tongues, Earth Clubs, and Allies group (beginning on p. 291).

Clavaria vermicularis Fries   Illus. p. 72
    COMMON NAME: White Worm Coral.
    FRUIT BODY: 1⅛–4¾" (3–12 cm) high, ⅛–¼" (2–5 mm) thick, erect, tapering, spindle-shaped to worm-like cylinders, typically unbranched, white, sometimes yellowish at the tips.
    FLESH: thin, brittle, white; odor and taste not distinctive.
    SPORE PRINT COLOR: white.
    MICROSCOPIC FEATURES: spores 4–7 × 3–5 μm, elliptic, smooth, hyaline.
    FRUITING: small to large clusters or troops on the ground in woodlands and grassy areas; July–December; widely distributed in the Southeast; occasional to locally common.
    EDIBILITY: reported to be edible.
    COMMENTS: The fruit bodies of this species often become flattened and curved with age and nearly translucent when wet.

Clavaria zollingeri Léveille   Illus. p. 72
    COMMON NAME: Magenta Coral.
    FRUIT BODY: up to 3½" (9 cm) high, erect and coral-like, with repeatedly forked branches arising from a short stalk; branches sparingly branched on the upper one-third; surface smooth, reddish purple, darkest toward the tips; branch tips rounded to blunt or tapering.
    FLESH: reddish purple, very brittle; odor not distinctive; taste somewhat radish-like.
    SPORE PRINT COLOR: white.

MICROSCOPIC FEATURES: spores 5–7.5 × 3–4.5 μm, elliptic to oval, smooth, hyaline, inamyloid.
FRUITING: solitary, scattered, or in groups on the ground in mixed woods; July–October; widely distributed in the Southeast; occasional.
EDIBILITY: edible.
COMMENTS: The Violet-branched Coral, *Clavulina amethystina* (not illustrated), edible, has lilac-purple branches when young; brittle, lilac-purple flesh that lacks a distinctive odor and taste; and larger spores, 7–12 × 6–8 μm. The Purple Club Coral, *Clavaria purpurea* (not illustrated), edible, forms clusters of erect, typically unbranched, cylindrical to spindle-shaped, purple to grayish purple, brittle fruit bodies; and it has spores that measure 6–9 × 3–5 μm. The White Worm Coral, *Clavaria vermicularis* (p. 286), edible, has typically unbranched, erect, white, worm-like cylinders.

Clavicorona pyxidata (Fries) Doty Illus. p. 72
COMMON NAME: Crown-tipped Coral.
FRUIT BODY: up to 5⅛" (13 cm) high, erect and coral-like, with numerous repeatedly forked branches arising from a short stalk; branch tips distinctly crown-like; surface smooth, white to pale creamy white when young, becoming ochre-yellow or tan in age.
FLESH: tough to brittle, whitish; odor usually not distinctive; taste somewhat peppery.
SPORE PRINT COLOR: white.
MICROSCOPIC FEATURES: spores 4–5 × 2–3 μm, elliptic, smooth, hyaline, amyloid.
FRUITING: solitary, scattered, or in groups on decaying broadleaf trees; May–October; widely distributed in the Southeast; common.
EDIBILITY: edible.
COMMENTS: The crown-like tips are a most distinctive feature. *Ramaria concolor*, inedible, grows on decaying wood, but it has more erect, parallel branches that are pale yellowish tan to pale cinnamon, with creamy tan to beige tips (see *MNE*, p. 423).

Clavulina cristata (Fries) Schroeter Illus. p. 72
COMMON NAME: Crested Coral.
FRUIT BODY: up to 2" (5 cm) wide, 1⅛–3½" (3–9 cm) high, extremely variable in shape, typically sparingly branched except near the apex, which is crested with numerous tooth-like and jagged or sometimes more blunt-tipped points; branches smooth to wrinkled or longitudinally grooved, white to yellowish or grayish ochre, sometimes blackened from the base upward when attacked by a parasitic fungus.
FLESH: white, soft, and fragile; odor and taste not distinctive.
SPORE PRINT COLOR: white.
MICROSCOPIC FEATURES: spores 7–10 × 6–8 μm, subglobose, smooth, hyaline.
FRUITING: solitary, scattered, or in groups on the ground or among mosses in woodlands, especially in pine woods; June–November; widely distributed in the Southeast; fairly common.
EDIBILITY: edible.
COMMENTS: The Gray Coral, *Clavulina cinerea*, edible, is very similar, but its branch tips are pointed or blunt, not tooth-like and jagged (see *MNE*, p. 423). Both the Crested Coral and Gray Coral may be blackened from the base upward when attacked by *Spadicioides clavariae*, a parasitic fungus.

Clavulinopsis aurantio-cinnabarina (Schw.) Corner   Illus. p. 72
COMMON NAME: none.
FRUIT BODY: up to 6" (15.5 cm) high, 1/16–3/8" (1.5–10 mm) thick, spindle-shaped to worm-like, usually flattened, with a conspicuous longitudinal channel, hollow; apex usually pointed; surface smooth, reddish orange to pale orange, yellow to whitish near the base.
FLESH: thin, brittle, reddish orange; odor somewhat unpleasant or not distinctive; taste not distinctive.
SPORE PRINT COLOR: whitish to pale yellow.
MICROSCOPIC FEATURES: spores 4.5–6 × 5–6.5 μm, subglobose, smooth, hyaline.
FRUITING: solitary or more often in groups or clusters on the ground in woodlands and grassy areas; June–October; widely distributed in the Southeast; occasional.
EDIBILITY: reported to be edible.
COMMENTS: *Clavulinopsis laeticolor* = *Clavulinopsis pulchra* (not illustrated), edibility unknown, has an unbranched, deep golden yellow fruit body and spores that measure 6–7 × 4.5–5 μm.

Clavulinopsis fusiformis (Fries) Corner   Illus. p. 73
COMMON NAME: Spindle-shaped Yellow Coral.
FRUIT BODY: up to 5½" (14 cm) high, 1/16–3/8" (1.5–10 mm) thick, cylindric to worm-like or somewhat flattened, usually unbranched but sometimes branching near the apex; apex pointed to rounded; surface typically smooth, but sometimes wrinkled or grooved, bright to dull yellow.
FLESH: thin, brittle to fibrous, yellowish.
SPORE PRINT COLOR: white to pale yellow.
MICROSCOPIC FEATURES: 5–9 × 4–9 μm, broadly oval to globose, smooth, hyaline, inamyloid.
FRUITING: in dense clusters on soil and among grasses in woods and pastures; June–October; widely distributed in the Southeast; fairly common.
EDIBILITY: reported to be edible.
COMMENTS: *Clavulinopsis helveola* (not illustrated), edible, has a smaller, typically unbranched, and light buffy yellow fruit body and smaller spores that measure 7–8 × 2–3 μm.

Ramaria murrillii (Coker) Corner   Illus. p. 73
COMMON NAME: none.
FRUIT BODY: 1½–4¾" (4–12 cm) high, erect and coral-like with numerous branches arising from a distinct, rounded, central stalk; branches rounded, fibrous-tough; surface glabrous, dull brownish pink to pale rusty brown, darkening when bruised; branch tips somewhat pointed or blunt, whitish at first, becoming brown at maturity; stalk up to 2⅜" (6 cm) long, up to 3/8" (1 cm) thick, nearly equal or enlarged downward, sometimes divided, often twisted, pointed and rooting, fibrous-tough, covered over most of its lower portion with a whitish tomentum and whitish rhizomorphs that develop pinkish stains.
FLESH: moderately thick, whitish and sometimes staining pinkish, fibrous-tough; odor not distinctive; taste bitter.
SPORE PRINT COLOR: dull ochraceous tan.
MACROCHEMICAL TESTS: branches stain green with the application of $FeSO_4$.

MICROSCOPIC FEATURES: spores 6.5–9.5 × 3.5–5.5 μm, narrowly elliptic to bottle-shaped, with conspicuous sharp spines, rusty ochraceous.

FRUITING: solitary, scattered, or in groups on the ground in broadleaf and mixed broadleaf and conifer woodlands, usually with oak; June–October; widely distributed in the Southeast; occasional.

EDIBILITY: unknown.

COMMENTS: This mushroom was named after mycologist William Alphonso Murrill (1869–1957), who first found it in 1904.

Ramaria subbotrytis (Coker) Corner   Illus. p. 73

COMMON NAME: Rose Coral.

FRUIT BODY: 1⅛–4" (3–10 cm) high, erect and coral-like with numerous branches arising from a distinct, rounded, central stalk; branches rounded, fleshy to somewhat brittle; surface glabrous, rose coral to salmon pink at first, fading to creamy ochraceous at maturity; branch tips bluntly rounded, salmon orange to rose coral when young, becoming salmon pink to creamy ochraceous in age; stalk up to 1⅛" (3 cm) long, up to ¾" (2 cm) thick, nearly equal or somewhat enlarged downward, usually with a pointed base, colored like the branches or white.

FLESH: moderately thick, somewhat brittle, pinkish or white, often marbled with watery areas at the stalk base; odor and taste faintly reminiscent of sauerkraut or not distinctive.

SPORE PRINT COLOR: ochraceous.

MACROCHEMICAL TESTS: branches stain green with the application of $FeSO_4$.

MICROSCOPIC FEATURES: spores 8–11 × 3–4 μm, elliptic, minutely roughened, pale ochraceous.

FRUITING: solitary, scattered, or in groups on the ground in broadleaf and mixed broadleaf and conifer woodlands; June–October; widely distributed in the Southeast; occasional.

EDIBILITY: edible with caution, see Comments.

COMMENTS: The Yellow-tipped Coral, *Ramaria formosa* (not illustrated), poisonous, has pinkish orange to light red branches with yellowish tips that arise from a massive stalk-like base and has a golden-yellow spore print. *Ramaria conjunctipes* (not illustrated), edibility unknown, has pale salmon orange to yellow-orange branches with bright yellow tips, a stalk-like base composed of many fused branches coated with white rhizomorphs, and a yellowish spore deposit. *Ramaria fennica* (not illustrated), edibility unknown, forms a stout fruit body with lilac to violet tones on the upper part of the base and young branches that become dull grayish tan at maturity; the lilac to violet areas stain bright red when a drop of 20 percent KOH is applied.

Tremellodendropsis semivestitum (Berk. and Curt.) Petersen   Illus. p. 73

COMMON NAME: none.

FRUIT BODY: up to 2½" (6.5 cm) high, erect and coral-like with numerous branches arising from a distinct, rounded, central stalk; branches flattened, fibrous-tough; surface glabrous, whitish to straw-colored or pale tan; branch tips blunt at first, becoming somewhat pointed at maturity and darkening in age; stalk up to 2" (5 cm) tall, about ⅛" (3 mm) thick, sometimes branched, whitish to dull yellowish tan, fibrous-tough.

FLESH: thin, whitish, fibrous-tough.

SPORE PRINT COLOR: white.
MICROSCOPIC FEATURES: spores 12–18 × 6–7 μm, subelliptic to amygdaliform, smooth, hyaline.
FRUITING: scattered or in groups on the ground in woodlands, usually associated with broadleaf trees; July–September; widely distributed in the Southeast; fairly common.
EDIBILITY: inedible.
COMMENTS: Formerly known as *Lachnocladium semivestitum*.

Earth Tongues, Earth Clubs, and Allies

Members of this group form erect fruit bodies that resemble small tongues or clubs. Some coral fungi that are typically neither clustered nor branched are also included here. Most species have a distinct cap or a head-like fertile surface and a supporting stalk or stalk-like base. A few members lack a cap or head and consist of a club-shaped to cylindric or irregular stalk. *Cordyceps, Claviceps,* and Allies (beginning on p. 295), are similar, but their fertile surfaces are finely roughened like sandpaper.

Clavariadelphus pistillaris (Fries) Donk   Illus. p. 73
  COMMON NAME: Pestle-shaped Coral.
  FRUIT BODY: 2¾–8" (7–20.5 cm) high, ⅜–1¾" (1–4.5 cm) wide, cylindric when young, becoming club-shaped with age, unbranched or rarely forked, smooth to longitudinally wrinkled; apex usually rounded and inflated; yellowish to orange-yellow, becoming brownish orange to pale reddish brown at maturity, slowly staining brownish when bruised.
  FLESH: thick, firm or spongy, white, staining brownish when cut; odor not distinctive; taste mild to bitter; stalk poorly defined, white.
  SPORE PRINT COLOR: white to creamy white.
  MICROSCOPIC FEATURES: spores 9–16 × 5–9 µm, elliptic, smooth, hyaline with yellow oil drops.
  FRUITING: scattered or in groups on the ground, usually in broadleaf woodlands; July–November; widely distributed in the Southeast; occasional to frequent.
  EDIBILITY: edible, but often bitter or unpleasant.

Clavariadelphus truncatus (Quélet) Donk   Illus. p. 73
  COMMON NAME: Flat-topped Coral.
  FRUIT BODY: 2–6" (5–15.5 cm) high, 1–2¾" (2.5–7 cm) wide, club-shaped to top-shaped, narrowing downward; typically unbranched but sometimes forking, smooth near the base, becoming longitudinally wrinkled upward; apex flattened and often slightly depressed; golden yellow to orange-yellow or pale brownish orange, usually darkest toward the base.
  FLESH: thick, firm or spongy, white; odor not distinctive; taste sweet or bland or sometimes bitter; stalk poorly defined, white, often with a dense, white, basal mycelium imbedded in the substrate.
  SPORE PRINT COLOR: pale brownish yellow.
  MICROSCOPIC FEATURES: spores 9–12 × 5–8 µm, broadly elliptic, smooth, hyaline, with yellow oil drops.

FRUITING: scattered or in groups on the ground in conifer woods; August–November; widely distributed in the Southeast; occasional.

EDIBILITY: edible, but sometimes bitter.

Cudonia circinans (Pers.) Fries   Illus. p. 74

COMMON NAME: none.

CAP: ¼–¾" (5–20 mm) wide, irregularly rounded and flattened, smooth to folded or lobed with an inrolled margin, dry; flesh thin, cartilaginous, pale ochre to pale brown.

STALK: ¾–1¾" (2–4.5 cm) long, up to ¼" (6 mm) thick, more or less equal, cylindrical or compressed, colored like the cap toward the top, reddish brown toward the base.

MICROSCOPIC FEATURES: spores 30–45 × 1.8–2.5 μm, narrow, thread-like, sometimes curved, multiseptate, hyaline.

FRUITING: in groups or clusters on the ground and in needle litter, occasionally on well-decayed wood, under conifers; June–December; widely distributed in the Southeast; occasional.

EDIBILITY: unknown.

COMMENTS: *Cudonia lutea* (see photo, p. 74), edibility unknown, is nearly identical but yellow to pale brownish orange, and it grows in broadleaf woods. *Leotia lubrica* (p. 292), edibility unknown, is similar but has a slippery, gelatinous, ochre cap and stalk as well as spindle-shaped spores that measure 20–24 × 5–6 μm.

Leotia lubrica Persoon : Fries   Illus. p. 74

COMMON NAME: Ochre Jelly Club, Jelly Babies.

HEAD: ¼–1⅛" (6–30 mm) wide, irregularly rounded and flattened, smooth to distinctly furrowed or brain-like, with a strongly inrolled margin, moist, gelatinous, pale dull yellow to orange-yellow, sometimes with olive tints.

STALK: ¾–2" (2–5 cm) long, ¼–⅜" (6–10 mm) thick, enlarged downward, slippery, smooth or nearly so, pale dull yellow to orange-yellow.

MICROSCOPIC FEATURES: spores 18–25 × 4–6 μm, cylindric-oblong to spindle-shaped, often curved, multiseptate at maturity.

FRUITING: in groups or clusters on the ground and sometimes on well-decayed wood under conifer and broadleaf trees; July–October; widely distributed in the Southeast; fairly common.

EDIBILITY: unknown.

Leotia viscosa Fries   Illus. p. 74

COMMON NAME: Green-headed Jelly Club.

HEAD: ¼–¾" (6–20 mm) wide, irregularly rounded and flattened, smooth to distinctly furrowed or brain-like, with a distinctly inrolled margin, moist, gelatinous, medium to dark green or sometimes nearly black.

STALK: ¾–1½" (2–4 cm) long, ¼–⅜" (6–10 mm) thick, enlarged downward or nearly equal, slippery, smooth, pale greenish yellow to yellowish orange.

MICROSCOPIC FEATURES: spores 18–20 × 5–6 μm, narrowly elliptic to spindle-shaped with rounded ends, straight or curved, multiseptate at maturity, smooth, hyaline.

FRUITING: in groups or clusters on soil or among mosses in woods, also on sand dunes; July–February; widely distributed in the Southeast; occasional to frequent.

EDIBILITY: unknown.

COMMENTS: This species is common on old sand dunes along the Atlantic and Gulf Coasts during fall and winter. *Leotia atrovirens,* edibility unknown, is similar but has a pea green to bluish green cap and a pea green to bluish green or pale green to whitish stalk (see *MNE*, p. 21).

Microglossum rufum (Schweinitz) Underwood   Illus. p. 74
  COMMON NAME: Orange Earth Tongue.
  HEAD: ⅛–⅝" (3–16 mm) wide, ⅜–1⅜" (1–3.5 cm) high, spoon- to tongue-shaped or cylindric, compressed at the center or longitudinally furrowed, smooth, dull or shiny, yellow-orange to orange.
  STALK: ⅝–1¾" (1.5–4.5 cm) long, 1/16–3/16" (1.5–5 mm) thick, nearly equal, yellow to orange-yellow, scaly-granular.
  MICROSCOPIC FEATURES: spores 18–38 × 4–6 μm, sausage- to spindle-shaped, nonseptate when young, becoming 5–10 septate in age, smooth, hyaline.
  FRUITING: scattered or in groups or clusters on humus, on decaying wood, and among mosses; July–October; widely distributed in the Southeast; fairly common.
  EDIBILITY: unknown.
  COMMENTS: *Spathularia flavida* (p. 293), edibility unknown, has a pale yellow-ochre to deep yellow, fan-shaped to folded head, a whitish stalk, and longer spores.

Spathularia flavida Persoon : Fries   Illus. p. 74
  COMMON NAME: none.
  FRUIT BODY: up to 2" (5 cm) high, consisting of a fan-shaped head and a distinct stalk.
  HEAD: ⅜–1⅜" (1–3.5 cm) high, ⅜–1½" (1–4 cm) wide, flattened, fan-shaped, wavy to lobed or irregularly folded, dry, glabrous, pale yellow to yellow-ochre or deep yellow.
  STALK: ⅜–1" (1–2.5 cm) long, 3/16–¾" (4–20 mm) thick, tapered toward the base, dry, smooth, whitish.
  MICROSCOPIC FEATURES: spores 38–52 × 2–2.5 μm, spindle-shaped, multiseptate, smooth, hyaline; paraphyses slender, forked, with curved to spiral tips.
  FRUITING: scattered or in groups on the ground and among mosses in conifer woods; July–October; widely distributed in the Southeast; occasional.
  EDIBILITY: unknown.
  COMMENTS: Also known as *Spathulariopsis flavida*. *Spathularia velutipes* = *Spathulariopsis velutipes*, edibility unknown, is similar, but it has a yellowish to brownish yellow head, and its reddish brown stalk has a dense orange basal mycelium (see *MNE*, p. 21). *Microglossum rufum* (p. 293), edibility unknown, has a yellow-orange to orange, spoon- to tongue-shaped or cylindric head and smaller spores.

Trichoglossum walteri (Berkeley) Durand   Illus. p. 74
  COMMON NAME: none.
  FRUIT BODY: 1⅛–3⅛" (3–8 cm) high, brownish black to black, with a head and stalk.
  HEAD: ⅛–⅜" (3–10 mm) wide, ½–1⅛" (1.2–3 cm) high, spoon- to lance-shaped or tongue-shaped to cylindric, compressed in the center or irregularly folded, appearing minutely spiny when examined with a hand lens, brownish black.
  STALK: 1–2⅜" (2.5–6 cm) long, 1/16–3/16" (1.5–5 mm) thick, nearly equal or tapered in either direction, smooth when young, becoming velvety and often furrowed in age, appearing minutely spiny when examined with a hand lens.
  MICROSCOPIC FEATURES: spores 60–125 × 5–6 μm, needle-like, 7 or 8 septate, sometimes 4 or 0 septate when immature, smooth, grayish or brownish; setae 200–250 × 6–7.5 μm, tapering to a sharp point, dark brown to blackish.

FRUITING: scattered or in groups on the ground, on decaying wood, and among mosses; July–December; widely distributed in the Southeast; occasional.

EDIBILITY: unknown.

COMMENTS: Several similar species of *Trichoglossum* must be identified microscopically (see *MNE,* p. 506). Several species of *Geoglossum* are also similar, but they lack setae and must be identified microscopically (see *MNE,* p. 506).

Underwoodia columnaris Peck Illus. p. 75

COMMON NAME: Fluted-stalked Fungus.

FRUIT BODY: 1½–4" (4–10 cm) high, ⅜–1⅛" (1–3 cm) thick, cylindric to spindle-shaped, tapering upward to a rounded tip, with longitudinal grooves or wrinkles, lacking a distinct cap and stalk, cream to tan when young, becoming pale brown in age, interior chambered.

MICROSCOPIC FEATURES: spores 25–27 × 12–14 μm, elliptic, coarsely warted, hyaline.

FRUITING: in clusters with fused bases or sometimes scattered on the ground in broadleaf woods; June–August; widely distributed in the Southeast; rare.

EDIBILITY: unknown.

COMMENTS: Fruit bodies of this fungus are easily overlooked because their pale colors resemble leaves.

Xylocoremium flabelliforme (Schw.: Fr.) J. D. Rogers Illus. p. 75

COMMON NAME: Bubble Gum Fungus.

FRUIT BODY: erect, consisting of a loose, fertile, head-like mass supported by a more slender base, ¼–1⅛" (5–30 mm) high; fertile head ¹⁄₁₆–⅝" (1.5–15 mm) wide, dry, delicate, spongy to powdery, fan-shaped to leaf-like or convoluted and brain-like, whitish or yellowish or pinkish; supporting base a conical black bundle of closely adhering, long, shaggy hairs.

MICROSCOPIC FEATURES: conidia 3–7 × 1.5–4 μm, obovate to ellipsoid with a flattened basal scar, smooth, hyaline.

FRUITING: solitary or scattered on decaying wood; June–December; widely distributed in the Southeast; occasional.

EDIBILITY: inedible.

COMMENTS: Formerly known as *Isaria flabelliformis.* This fruit body is the asexual reproductive stage of *Xylaria cubensis* (not illustrated), inedible, a short, cylindric-clavate to clavate, typically unbranched, brown to brownish black fruit body that grows on decaying broadleaf wood and forms small, dark ascospores that appear to lack germ slits.

Cordyceps, Claviceps, and Allies

Members of this group belong to the class Ascomycetes and are commonly called the flask fungi. Their fertile surfaces are finely roughened like sandpaper owing to the protruding necks of numerous flask-shaped reproductive structures called *perithecia*, which contain ascospores. *Cordyceps* species are parasitic on other fungi, on the pupae and larvae of moths and butterflies, or on the larvae and adult stages of beetles. *Claviceps purpurea* is parasitic on the inflorescences of many species of grasses (see *MNE,* p. 22). One member of this group, *Podostroma alutaceum,* is not parasitic and grows on the ground or on decaying wood (see *MNE,* p. 519).

Cordyceps capitata (Holmskjold : Fries) Link Illus. p. 75
   COMMON NAME: Head-like Cordyceps.
   HEAD: ¼–¾" (6–20 mm) wide and tall, irregularly rounded, finely roughened like sandpaper, dark reddish brown to olive black.
   FLESH: white, thick, firm; odor and taste not distinctive.
   STALK: ¾–3⅛" (2–8 cm) long, ¼–⅝" (5–16 mm) thick, nearly equal overall, smooth to slightly ridged, fibrous, yellow to dull yellow, becoming olive brown in age.
   MICROSCOPIC FEATURES: spores 16–28 × 2.5–3.5 μm, cylindric to thread-like, smooth, hyaline, inamyloid.
   FRUITING: solitary or in groups in woods, attached to the buried, walnut-shaped, reddish brown fruit bodies of *Elaphomyces* species (false truffles); August–January; widely distributed in the Southeast; uncommon to locally common.
   EDIBILITY: unknown.
   COMMENTS: This fungus is parasitic and must be carefully dug up to retrieve the false truffle host. *Cordyceps longisegmentis* (not illustrated), edibility unknown, is macroscopically identical, but is easily identified by microscopic examination of its spores, which are 30–65 × 3–5 μm. *Cordyceps ophioglossoides,* edibility unknown, is very similar and also grows attached to fruit bodies of *Elaphomyces* species, but it has a spindle-shaped to oval or club-shaped head that is yellowish brown to dark reddish brown or olive black, and it is connected to the false truffle by thick strands of golden rhizomorphs (see *MNE,* p. 22).

Cordyceps melolonthae (Tulasne) Saccardo Illus. p. 75
   COMMON NAME: Beetle Cordyceps.
   HEAD: ⅜–⅝" (1–1.5 cm) wide, ¾–1⅛" (2–3 cm) high, oval, whitish to yellowish, finely roughened like sandpaper.
   FLESH: thin, whitish to pale yellow.

STALK: 2–2¾" (5–7 cm) long, ⅛–⅜" (3–10 mm) thick, nearly equal or club- to spindle-shaped, smooth, yellowish.

MICROSCOPIC FEATURES: spores 4–10 × 1–2.5 μm, elliptic, smooth, hyaline.

FRUITING: solitary or several attached to a buried beetle larva; May–October; widely distributed in the Southeast; occasional.

EDIBILITY: unknown.

COMMENTS: The Trooping Cordyceps, *Cordyceps militaris*, edibility unknown, has a cylindric to club-shaped or spindle-shaped, reddish orange, roughened fruit body that is attached to the pupae and larvae of shallowly buried moths and butterflies (see *MNE*, p. 22).

Cordyceps olivascens Mains   Illus. p. 75

COMMON NAME: none.

HEAD: ⅛–¼" (4–6 mm) wide, 1–1⅛" (2.5–3 cm) tall, cylindric, very light green to olive buff, roughened like sandpaper.

FLESH: thin, whitish, firm.

STALK: ¾–1⅛" (2–3 cm) long, 1/16–3/16" (1.5–4 mm) thick, enlarged downward or nearly equal, smooth, colored like the head or slightly darker olive, usually with a whitish base.

MICROSCOPIC FEATURES: spores filiform, multiseptate, soon breaking into oblong fragments, 3.5–6 × 1–1.5 μm, smooth, hyaline, inamyloid.

FRUITING: solitary or several attached to the remains of a buried or partially buried unidentified insect; June–November; reported from Alabama and Mississippi, distribution limits yet to be determined; rare.

EDIBILITY: unknown.

COMMENTS: *Cordyceps olivaceo-virescens* is a synonym. *Cordyceps militaris*, edibility unknown, is very similar, but it has a reddish orange fruit body (see *MNE*, p. 22).

Cordyceps sphecocephala (Berk.) Sacc.   Illus. p. 75

COMMON NAME: none.

HEAD: 1/16–3/16" (1.5–4 mm) wide, ⅛–5/16" (2–8 mm) tall, ovoid, obovoid, or subcylindric, finely roughened like sandpaper, light cream to brownish yellow at first, becoming yellowish brown at maturity, longitudinally irregularly wrinkled when dry.

FLESH: thin, firm, brown.

STALK: ¾–3½" (2–9 cm) long, 1/32–1/16" (0.3–1.5 mm) thick, dry, glabrous, colored like the head on the upper portion, often darker on the lower portion.

MICROSCOPIC FEATURES: spores filiform, multiseptate, breaking into one-celled, fusoid, hyaline segments, 8–14 × 1.5–2 μm.

FRUITING: solitary or in groups on wasps that are covered by leaves and pine straw; June–October; widely distributed in the Southeast; uncommon.

EDIBILITY: unknown.

COMMENTS: Specimens must be carefully excavated to recover the host organism.

Nomuraea atypicola (Yasuda) Samson   Illus. p. 76

COMMON NAME: none.

HEAD: ⅛–¼" (2.5–5 mm) wide, ⅝–1⅛" (1.5–3 cm) tall, cylindrical to subfusiform, dry, scurfy, grayish tan to light gray-purple.

FLESH: firm, whitish.
STALK: ¾–1⅛" (2–3 cm) long, ⅛–³⁄₁₆" (2.7–3.5 mm) thick.
MICROSCOPIC FEATURES: spores 4.5–6 × 1.5–2 μm, cylindric, usually with two oil drops, smooth, hyaline.
FRUITING: solitary or scattered on buried decaying spiders in woodlands; June–August; widely distributed in the Southeast; occasional.
EDIBILITY: unknown.
COMMENTS: Specimens must be carefully excavated to observe the dead host.

Bird's-nest Fungi

These very small fungi are related to the puffballs, with both groups placed in the Gasteromycetes (stomach fungi). They are essentially small, nest-like cups (the shape is generally cylindric to inverted-conic), each filled with numerous peridioles (eggs). The peridioles contain the reproductive spores and can be compared to miniaturized puffballs. When young, the cups are generally protected by some type of membrane-like lid. Bird's-nest mushrooms are decomposers and are especially common on wood-chip mulch, but are also found on dung, fallen branches, logs, leaves, and other kinds of organic debris.

Crucibulum laeve (Hudson) Kamby Illus. p. 76
   COMMON NAME: Common Bird's-nest.
   CUP: more or less cylindric, tapering somewhat toward the bottom, ¼–⅜" (6–10 mm) wide at the top, ¼–⅜" (6–10 mm) high; cup protected when immature by a white, membrane-like lid coated at first with yellowish orange fibers; interior whitish, smooth; exterior hairy to finely velvety, yellowish orange, becoming paler yellow.
   PERIDIOLES: white or pallid, lens-shaped, ¹⁄₁₆–³⁄₃₂" (1.5–2 mm) in diameter, each attached beneath by a tiny, coiled cord.
   MICROSCOPIC FEATURES: spores 8–10 × 4–5.5 μm, elliptic, smooth, hyaline.
   FRUITING: scattered to gregarious on wood chips, twigs, fallen branches, etc.; July–December, sometimes overwintering; widely distributed in the Southeast; common.
   EDIBILITY: inedible.
   COMMENTS: Formerly known as *Crucibulum vulgare*. The late mycologist Howard Bigelow reported finding this species on "old discarded leather and paper goods" (Bigelow 1974, 646).

Cyathus striatus Hudson : Persoon Illus. p. 76
   COMMON NAME: Splash Cups.
   CUP: inverted-conic, ¼–⁵⁄₁₆" (6–8 mm) wide, ¼–⅜" (6–10 mm) high; cup protected by a white, membranous lid and upper edge of the cap rolled inward when mushroom is immature; interior gray to grayish white, shiny, smooth, vertically lined; exterior reddish brown to chocolate brown or grayish brown, shaggy-hairy to wooly, sometimes faintly to distinctly fluted.
   PERIDIOLES: gray, flattened, ¹⁄₁₆–⅛" (1.5–3 mm) in diameter, often vaguely triangular, each attached to the cup interior by a tiny, coiled cord.
   MICROSCOPIC FEATURES: spores 15–20 × 8–12 μm, elliptic, smooth, hyaline.
   FRUITING: scattered to gregariously grouped on wood chips, twigs, bark, logs, etc.; July–December, sometimes overwintering; widely distributed in the Southeast; frequent to common.
   EDIBILITY: inedible.

Cup and Saucer Fungi

Members of this group produce fruit bodies that resemble small cups or saucers. These fungi are members of the class Ascomycetes, and their fertile (inner) surfaces, which are often brightly colored, are lined with asci and ascospores. The group includes several hundred species. Classification of the cup and saucer fungi is currently in transition. Considerable disagreement exists concerning the nomenclature and taxonomic position of many species in this group. A discussion of the majority of species included in the category is beyond the scope of this work, so we have listed only those species with distinctive features and those most commonly encountered.

The flesh of these fungi is often thin and brittle, and, with few exceptions, the edibility of members of this group is unknown.

Aleuria aurantia (Fries) Fuckel   Illus. p. 77
  common name: Orange Peel.
   fruit body: ⅜–4" (1–10 cm) wide, cup- to saucer-shaped, stalkless or with a rudimentary stalk; fertile surface bright orange to yellow-orange, smooth; outer surface pale yellowish orange to yellow, slightly scurfy; margin wavy and often torn at maturity.
   microscopic features: spores 17–24 × 9–11 μm, elliptic, covered by a coarse reticulum, usually with one or more projecting spines at each end, hyaline.
   fruiting: solitary or in groups or clusters in grassy areas on disturbed soil, in gardens, and along roadsides; June–November; widely distributed in the Southeast; fairly common.
   edibility: edible.
   comments: *Bisporella citrina*, edibility unknown, forms small, wide (up to ⅛" [3 mm]) wide saucer-shaped, bright lemon yellow to golden yellow fruit bodies that grow in groups or dense clusters on decaying wood (see *MNE*, p. 20).

Ascocoryne sarcoides (Jacquin : S. F. Gray) Groves and Wilson   Illus. p. 77
   common name: Purple Jelly Drops.
   fruit body: ¼–¾" (5–20 mm) wide, round when very young, becoming either turban- to cushion-shaped with a depressed center or saucer-shaped; stalkless or with a rudimentary stalk; fertile surface smooth to slightly wrinkled, reddish purple to violet-pink; flesh violet-pink, gelatinous; outer surface reddish purple to violet-pink, smooth; margin entire, even, usually darker than the disc.
   microscopic features: spores 11–18 × 3.5–5 μm, elliptical, smooth, with a single septum when mature, hyaline; paraphyses cylindric, branched, with few septa, tips occasionally thickened.
   fruiting: in dense clusters on decaying wood; August–December; widely distributed in the Southeast; common.

EDIBILITY: unknown.

COMMENTS: *Ascocoryne cylichnium*, edibility unknown, is very similar, but it often has a more wrinkled fertile surface when mature and larger spores, 20–25 × 5–6 μm, with several septa at maturity and unbranched paraphyses (see *MNE*, p. 498).

Chlorociboria aeruginascens (Nylander) Kanouse   Illus. p. 77

COMMON NAME: Blue Stain, Green Stain.

FRUIT BODY: ⅛–³⁄₁₆" (3–8 mm) wide, cup-shaped to nearly flat and saucer-shaped; fertile surface smooth, blue-green, sometimes tinted yellow; outer surface blue-green, finely roughened; stalk ⅛–¼" (3–6 mm) long, tapering downward, frequently off-center, blue-green.

MICROSCOPIC FEATURES: spores 6–10 × 1.5–2 μm, irregularly spindle-shaped with oil drops, smooth, hyaline.

FRUITING: in groups or clusters on decaying broadleaf wood; June–November; widely distributed in the Southeast; fairly common.

EDIBILITY: unknown.

COMMENTS: Also known as *Chlorosplenium aeruginascens*. The mycelium stains the substrate blue-green. *Chlorociboria aeruginosa* (not illustrated), edibility unknown, is nearly identical but has spores that measure 8–15 × 2–4 μm, and the decaying wood on which it is growing does not stain blue-green.

Galiella rufa (Schweinitz) Nannfeldt and Korf   Illus. p. 77

COMMON NAME: Hairy Rubber Cup.

FRUIT BODY: ½–1⅛" (1.2–3 cm) high, ⅜–1⅜" (1–3.5 cm) wide, cup-shaped; fertile surface pale orange to dull orange, reddish orange, reddish brown or tan, smooth; outer surface tough, brown near the margin, blackish brown below, covered with a dense layer of matted wooly hairs; margin finely toothed; flesh rubbery-gelatinous.

MICROSCOPIC FEATURES: spores 18–20 × 8–10 μm, elliptic with narrow ends, finely warted, hyaline.

FRUITING: in groups or clusters on decaying broadleaf wood; July–October; widely distributed in the Southeast; fairly common.

EDIBILITY: unknown.

COMMENTS: Also known as *Bulgaria rufa*. Compare with *Wolfina aurantiopsis* (p. 302).

Helvella acetabulum (Fries) Quélet   Illus. p. 77

COMMON NAME: Ribbed-stalked Cup, Elfin Cup.

FRUIT BODY: ¾–2¾" (2–7 cm) wide, cup-shaped; fertile surface pale to dark brown; outer surface pale to dark brown near the margin, becoming whitish near the base; stalk ⅜–2¾" (1–7 cm) long, ⅜–1" (1–2.5 cm) thick, conspicuously ribbed, white; ribs rounded, branched upward, extending over the outer surface.

MICROSCOPIC FEATURES: spores 16–19.5 × 11–14 μm, broadly elliptic, smooth, hyaline, with one oil drop.

FRUITING: scattered or in groups on the ground under broadleaf trees; March–June; widely distributed in the Southeast; occasional.

EDIBILITY: unknown.

COMMENTS: *Helvella griseoalba*, edibility unknown, is similar but has a pale to dark gray cup (see *MNE*, p. 499).

Humaria hemisphaerica (Wiggers : Fries) Fuckel Illus. p. 77
  COMMON NAME: Brown-haired White Cup.
  FRUIT BODY: ⅜–1⅛" (1–3 cm) wide, ⅜–¾" (1–2 cm) high, cup-shaped; fertile surface whitish to pale gray, smooth; outer surface brownish yellow, covered by a dense layer of brownish hairs that project from the margin over part of the fertile surface.
  MICROSCOPIC FEATURES: spores 25–27 × 12–15 μm, elliptic with or without tiny warts, hyaline, with two oil drops.
  FRUITING: scattered or in groups on soil, among mosses, and on decaying wood; June–November; widely distributed in the Southeast; fairly common.
  EDIBILITY: unknown.
  COMMENTS: The Bladder Cup, *Peziza vesiculosa*, edibility unknown, has a pale to dark yellowish brown, smooth, fertile surface and a pale yellowish brown, scurfy, outer surface; it grows on manure piles and manured soil (see *MNE*, p. 501).

Peziza repanda Persoon Illus. p. 77
  COMMON NAME: Recurved Cup.
  FRUIT BODY: 2– ½" (5–11.5 cm) wide, cup- to saucer-shaped, stalkless or with a rudimentary stalk; fertile surface pale to dark yellow-brown or dark brown, smooth; outer surface whitish; margin wavy, often split and turned downward.
  MICROSCOPIC FEATURES: spores 14–16 × 8–10 μm, elliptic, smooth, hyaline, lacking oil drops.
  FRUITING: solitary, scattered, or in groups on decaying broadleaf wood and adjacent soil; April–December; widely distributed in the Southeast; fairly common.
  EDIBILITY: unknown.
  COMMENTS: The Common Brown Cup, *Peziza phyllogena*, edible, has a purple-brown to reddish brown or olive brown, smooth, fertile surface and a pale reddish brown to purple-brown, scurfy, outer surface; it grows on soil or decaying wood (see *MNE*, p. 500). Several other similar species grow on the ground and must be differentiated microscopically.

Plicaria anthracina (Cooke) Boudier Illus. p. 78
  COMMON NAME: none.
  FRUIT BODY: ¼–1⅛" (5–30 mm) wide, cup-shaped at first, becoming flattened and saucer-shaped at maturity; fertile surface finely roughened to uneven and folded, dark brown to blackish, sometimes exuding droplets when injured; outer surface dark grayish brown, finely roughened; margin entire when young, becoming wavy to folded and sometimes ragged in age.
  MICROSCOPIC FEATURES: spores 12–16 μm (excluding the wart-like spines), round, with coarse, blunt, wart-like spines, pale brown; paraphyses cylindric, septate, with thickened, sometimes finely encrusted, rounded tips.
  FRUITING: scattered or in groups in grass and on burned ground; April–May; reported from Mississippi, distribution limits yet to be determined; uncommon.
  EDIBILITY: unknown.
  COMMENTS: The round spores with their coarse, blunt, wart-like spines are most distinctive.

Sphaerosporella brunnea (Albert. and Schw.) Svrcek and Kubicka Illus. p. 78
  COMMON NAME: none.
  FRUIT BODY: ⅛–¼" (2–6 mm) wide, saucer-shaped with a raised margin; fertile sur-

face smooth to slightly roughened, reddish brown at first, becoming dark brown at maturity; outer surface brownish to dark reddish brown, finely roughened with tufts of matted fine hairs; margin thick, raised, roughened with densely compacted fine hairs.

MICROSCOPIC FEATURES: spores 14–18 μm, round, smooth, hyaline, with one large oil drop; paraphyses cylindrical, septate, with clavate tips.

FRUITING: scattered or in groups or clusters on bare sandy soil near water oaks, on burned ground, and among mosses; March–April or August–September; reported from Alabama, Mississippi, and Massachusetts west to Iowa, distribution limits yet to be determined; rare to uncommon.

EDIBILITY: unknown.

COMMENTS: The round, smooth, hyaline spores are very unusual for species of cup fungi. *Peziza ammophila* (see photo, p. 77), edibility unknown, also grows in sandy habitats and forms dark brown cups with lobed edges. Before the mushroom is unearthed, only the uppermost portion of the cup rim is exposed, causing the fungus to resemble a brown hole in the sand.

Urnula craterium (Schweinitz) Fries   Illus. p. 78

COMMON NAME: Devil's Urn.

FRUIT BODY: 1¾–4⅛" (4.5–10.5 cm) high, ¾–2¾" (2–7 cm) wide, deeply cup-shaped with a distinct stalk; fertile surface blackish brown to black, smooth; outer surface black to brownish black, densely matted and wooly, becoming brownish gray, tough and leathery in age; margin toothed and irregularly torn; stalk ¾–1½" (2–4 cm) long, 3/16–⅜" (5–10 mm) thick, tapering downward, black to brownish black.

MICROSCOPIC FEATURES: spores 25–35 × 12–14 μm, broadly elliptic, smooth, hyaline.

FRUITING: in groups or clusters on decaying wood or on the ground and attached to buried wood; March–June; widely distributed in the Southeast; occasional to fairly common.

EDIBILITY: unknown.

COMMENTS: This species is one of the earliest mushrooms to appear in spring.

Wolfina aurantiopsis (Ellis) Seaver   Illus. p. 78

COMMON NAME: none.

FRUIT BODY: up to 2½" (7 cm) wide, stalkless and shallow cups; inner fertile surface pale yellow to ochraceous or somewhat orangish, smooth; exterior surface with projecting folds, minutely roughened, brownish to black; flesh tough, becoming corky on drying, whitish.

MICROSCOPIC FEATURES: spores 27–33 × 16–18 μm, broadly elliptic, granular within, smooth, hyaline to slightly yellowish.

FRUITING: solitary or in small groups or clusters on decaying wood in mixed woods, especially with pine and hemlock, also reported to grow on soil; July–December; widely distributed in the Southeast; uncommon.

EDIBILITY: inedible.

COMMENTS: Until its firm, corky texture is noted, this uncommon cup fungus is likely to be mistaken for a pale form of *Galiella rufa* (p. 300). *Aurantiopsis* means "appearing orange-colored."

Jelly Fungi

Members of this group typically produce fruit bodies with a gelatinous consistency. Some species are firm and rubbery-gelatinous, whereas others have a soft-gelatinous texture. In many cases, the fruit bodies are rubbery-gelatinous at first and become soft-gelatinous in age.

During dry periods, jelly fungi shrivel and become hard and horny or crust-like as they lose moisture. Considerable variation in shape and color can be seen within this group. Identification based on macroscopic features is usually possible. In some cases, however, examination of microscopic characters is needed for accurate identification.

Arrhytidia involuta (Schweinitz) Coker   Not Illustrated
   COMMON NAME: none.
   FRUIT BODY: ⅛–¼" (3–6 mm) wide, cushion-shaped to somewhat brain-shaped, waxy-gelatinous; surface smooth or more or less convoluted, bright orange-yellow to reddish orange with yellow tints when fresh, and forming irregular, rust-colored, flake-like reddish brown patches when dry.
   MICROSCOPIC FEATURES: spores 14–19 × 5–7 μm, allantoid, 1–3 septate, smooth, yellowish.
   FRUITING: in groups or clusters that fuse and form irregular, resupinate, continuous masses on decaying broadleaf and conifer wood; year-round; widely distributed in the Southeast; occasional to fairly common.
   EDIBILITY: unknown.
   COMMENTS: Formerly known as *Dacrymyces corticioides* and *Dacrymyces involutus*.

Auricularia auricula (Hooker) Underwood   Illus. p. 79
   COMMON NAME: Tree-Ear.
   FRUIT BODY: 1⅛–5⅞" (3–15 cm) wide, ear-shaped to irregularly cup-shaped, stalkless, rubbery-gelatinous; upper (outer) surface wrinkled, minutely hairy to velvety, reddish brown; lower (inner) surface smooth, yellowish brown to reddish brown or grayish brown to purplish brown.
   MICROSCOPIC FEATURES: spores 12–15 × 4–6 μm, sausage-shaped, smooth, hyaline.
   FRUITING: solitary or in groups or fused clusters on decaying conifer or broadleaf wood; May–December; widely distributed in the Southeast; fairly common.
   EDIBILITY: edible.
   COMMENTS: *Auricularia polytricha* (see photo, p. 79), edible, is similar, but its upper (outer) surface has a conspicuous dense layer of hairs.

Calocera cornea (Batsch : Fries) Fries   Illus. p. 79
  COMMON NAME: Club-like Tuning Fork.
  FRUIT BODY: up to ⅛" (3 mm) wide and ⅝" high, consisting of erect, pointed spikes that resemble antler tines, simple or sometimes forked; yellow to orange-yellow, rubbery-gelatinous.
  MICROSCOPIC FEATURES: spores 7–11 × 3–4.5 μm, cylindric to sausage-shaped, with a single septum at maturity, smooth, hyaline to yellowish.
  FRUITING: scattered or in groups or clusters on decaying wood, especially broadleaf trees; August–November; widely distributed in the Southeast; common.
  EDIBILITY: unknown.
  COMMENTS: *Calocera viscosa*, edibility unknown, forms larger, rubbery-gelatinous, golden yellow to orange-yellow, erect fruit bodies, 1⅜–4" (3.5–10 cm) high, that are repeatedly forked, antler-like tines (see *MNE*, p. 435).

Dacrymyces palmatus (Schweinitz) Bresadola   Illus. p. 79
  COMMON NAME: Orange Jelly, Orange Witches' Butter, Witches' Butter.
  FRUIT BODY: ⅜–2⅜" (1–6 cm) wide, up to 1" (2.5 cm) high, a brain-like to multilobed spreading mass; yellowish orange to orange or reddish orange, whitish near the point of attachment; rubbery-gelatinous at first, becoming soft-gelatinous in age.
  MICROSCOPIC FEATURES: spores 17–25 × 6–8 μm, cylindric to sausage-shaped, 7–9 septate, smooth, hyaline.
  FRUITING: a dense cluster on decaying conifer branches and logs; year-round; widely distributed in the Southeast; common.
  EDIBILITY: edible and rather bland; may be eaten raw or cooked.
  COMMENTS: *Tremella lutescens*, edible and rather bland, is similar but grows on decaying wood of broadleaf trees (see *MNE*, p. 436).

Dacryopinax elegans (Berk. and Curt.) Martin   Illus. p. 79
  COMMON NAME: none.
  FRUIT BODY: ¼–¾" (5–20 mm) wide, ½–1½" (1.2–4 cm) high, obliquely cup-shaped to spoon-shaped, or fan-shaped; dark amber brown to blackish brown, rubbery-gelatinous, smooth; undersurface smooth to slightly scurfy.
  STALK: round to somewhat flattened, dark amber brown to blackish brown, darkening in age.
  MICROSCOPIC FEATURES: spores 11–16 × 4.5–6.5 μm, sausage-shaped, with 1–3 septa at maturity, smooth, yellowish.
  FRUITING: scattered or in groups on decaying broadleaf trees; July–November; widely distributed in the Southeast; occasional.
  EDIBILITY: unknown.
  COMMENTS: *Dacryopinax spathularia* (p. 304), edibility unknown, forms spatula-shaped to shoehorn-shaped or fan-shaped yellow-orange fruit bodies with deeply cut lobes, and it has smaller spores, 8–12 × 3.5–5 μm.

Dacryopinax spathularia (Schweinitz) Martin   Illus. p. 79
  COMMON NAME: none.
  FRUIT BODY: 3⁄16–⅜" (5–10 mm) wide, ⅜–1" (1–2.5 cm) high, spatula-shaped to shoehorn-shaped with a wavy margin or fan-shaped with deeply cut lobes; yellow-orange to orange, rubbery-gelatinous, smooth; undersurface longitudinally ribbed.

STALK: round at the base, becoming flattened upward, yellow-orange to orange, darkening in age.
MICROSCOPIC FEATURES: spores 8–12 × 3.5–5 μm, sausage-shaped, with a single septum at maturity, smooth, yellowish.
FRUITING: in groups or clusters on decaying wood; July–October; widely distributed in the Southeast; infrequent.
EDIBILITY: unknown.
COMMENTS: Compare with *Dacryopinax elegans* (p. 304), edibility unknown, which has a dark amber brown to blackish brown obliquely cup-shaped to spoon-shaped or fan-shaped fruit body, grows on decaying wood of broadleaf trees, and has larger spores, 11–16 × 4.5–6.5 μm.

Exidia glandulosa Bulliard : Fries   Illus. p. 79
COMMON NAME: Black Jelly Roll.
FRUIT BODY: ⅜–¾" (1–2 cm) wide, up to ½" (1.3 cm) high, gland- to blister-like or brain-like, fusing together to form extensive irregular and soft-gelatinous masses up to 5" (18 cm) or more long; surface smooth or somewhat warty, shiny, dark reddish brown to blackish brown or black, drying to a black crust.
MICROSCOPIC FEATURES: spores 10–16 × 4–5 μm, sausage-shaped, smooth, hyaline.
FRUITING: in large clusters on decaying broadleaf logs and fallen branches; April–December; widely distributed in the Southeast; fairly common.
EDIBILITY: edible.
COMMENTS: Also known as Black Witches' Butter. Compare with *Exidia recisa* (p. 305).

Exidia recisa Fries   Illus. p. 80
COMMON NAME: Amber Jelly Roll.
FRUIT BODY: ⅝–1⅜" (1.5–3.5 cm) wide, up to ¾" (2 cm) high, cushion-shaped to brain-like or irregularly lobed, somewhat erect, with a stalk-like base; forming extensive irregular clusters up to 4" (10 cm) or more long; rubbery-gelatinous, yellowish brown to purplish brown or cinnamon brown, shiny, smooth to somewhat roughened and often contorted; margin somewhat thickened and irregular; upper surface typically with scattered, tiny, blackish, wart-like projections (use a hand lens).
MICROSCOPIC FEATURES: spores 11–15 × 3–5.5 μm, sausage-shaped, smooth, hyaline.
FRUITING: in groups or clusters on decaying broadleaf trees; May–October; widely distributed in the Southeast; common.
EDIBILITY: unknown.
COMMENTS: *Exidia glandulosa* (p. 305), edible, forms larger gland-like to blister-like, gelatinous, reddish brown to black clusters on broadleaf logs and branches.

Sebacina incrustans (Fries) Tulasne   Illus. p. 80
COMMON NAME: none.
FRUIT BODY: up to 6" (15.5 cm) or more in diameter; a thin, fibrous-tough to leathery, irregular sheet-like mass, forming small, lateral projections; whitish to pale tan.
MICROSCOPIC FEATURES: spores 11.5–15 × 6–8 μm, elliptic to ovoid, flattened to slightly depressed on one side, smooth, hyaline.
FRUITING: growing on the ground and soon enveloping plant stems, leaves, branches, cones, and debris; May–October; widely distributed in the Southeast; occasional to fairly common.
EDIBILITY: inedible.

Syzygospora mycetophila (Peck) Ginns   Illus. p. 80
  COMMON NAME: Collybia Jelly.
  FRUIT BODY: ⅛–1" (3–25 mm) wide or larger, consisting of brain-like, cup-shaped, tumor-like, or irregular growths that are often fused together; pale yellow to brownish yellow, rubbery-gelatinous.
  MICROSCOPIC FEATURES: spores 6–9 × 1.5–2.5 μm, elliptic to cylindric, smooth, hyaline.
  FRUITING: in dense clusters on the cap, gills, or stalk of *Gymnopus dryophilus* (p. 136); June–November; widely distributed in the Southeast; fairly common.
  EDIBILITY: unknown.
  COMMENTS: Formerly known as *Christiansenia mycetophila* and *Tremella mycetophila*.

Tremella concrescens (Fries) Burt   Illus. p. 80
  COMMON NAME: none.
  FRUIT BODY: a soft, whitish to grayish, irregular, membrane-like, rubbery-gelatinous spreading mass up to 5½" (14 cm) or more in diameter.
  MICROSCOPIC FEATURES: spores variable, 9–14 × 5–8 μm, elliptic and slightly curved to oval or nearly round, smooth, hyaline.
  FRUITING: on the ground and soon enveloping plant stems and leaves; August–October; widely distributed in the Southeast; fairly common.
  EDIBILITY: unknown.

Tremella foliacea Persoon : Fries   Illus. p. 80
  COMMON NAME: Jelly Leaf.
  FRUIT BODY: 2–10" (5–25.5 cm) wide, 2–4" (5–10 cm) high, a lettuce-like cluster composed of numerous leaf-like, rubbery-gelatinous folds; pale to dark reddish brown or rarely pale brownish yellow.
  MICROSCOPIC FEATURES: spores 8–12 × 7–9 μm, oval to nearly round, smooth, hyaline.
  FRUITING: solitary or scattered on decaying wood; July–November; widely distributed in the Southeast; occasional.
  EDIBILITY: edible, but rather bland.
  COMMENTS: Compare with *Auricularia auricula* (p. 303).

Tremella fuciformis Berkeley   Illus. p. 80
  COMMON NAME: Silver Ear.
  FRUIT BODY: up to 2¾" (7 cm) wide, up to 1⅛" (3 cm) high; a delicate, finely leafy, translucent, shiny, silvery white, complexly lobed, jelly-like mass.
  MICROSCOPIC FEATURES: basidiospores 8–14 × 5–8 μm, elliptic, smooth, hyaline; subglobose conidia up to 4.5 × 3 μm may also be present.
  FRUITING: solitary or scattered on decaying broadleaf wood, especially oak; July–November; widely distributed in the Southeast; common.
  EDIBILITY: edible.
  COMMENTS: This mushroom is a tropical and subtropical species that extends northward to the southeastern United States.

Cauliflower Mushrooms

Members of this group form large, rounded, cauliflower-like or lettuce-like clusters that usually grow on the ground at the base of trees. They are choice edibles and when young and fresh are highly prized for their firm texture and excellent flavor.

Sparassis crispa Wulfen : Fries   Illus. p. 81
   COMMON NAME: Rooting Cauliflower Mushroom.
   FRUIT BODY: 4¾–18" (12–46 cm) wide and high; a densely packed, rounded, cauliflower-like or lettuce-like cluster of whitish to pale yellow or tan, flattened, leafy, folded lobes, often resembling ribbon candy; individual branches leaf-like, flexible, darkening upward, typically darkest and broadest at their tips, tapering downward and uniting to form a thick, solid, basal mass above the stalk.
   STALK: 2–5½" (5–14 cm) long, ¾–2" (2–5 cm) thick, tapering toward the base, dark brown to black, deeply rooting.
   FLESH: white, thin, fibrous; odor fragrant or not distinctive; taste not distinctive.
   SPORE PRINT COLOR: white.
   MICROSCOPIC FEATURES: spores 5–6.5 × 3–3.5 μm, oval, smooth, hyaline.
   FRUITING: solitary or in groups on the ground and decaying wood in conifer and mixed conifer and broadleaf forests; August–October; widely distributed in the Southeast; occasional to fairly common.
   EDIBILITY: edible and choice when young, bitter or sour in age.
   COMMENTS: This mushroom is also known as *Sparassis radicata*. The leaf-like branches are typically smaller and more curled than those of *Sparassis herbstii* (p. 307), edible.

Sparassis herbstii Peck   Illus. p. 81
   COMMON NAME: Cauliflower Mushroom.
   FRUIT BODY: 6–15" (15–38 cm) wide and high, rounded; a cauliflower-like or lettuce-like cluster of cream to pale yellow, flattened branches that are wrinkled, curled, and folded, often resembling ribbon candy; individual branches leaf-like, somewhat stiff, typically palest and broadest at their tips, tapering downward and uniting to form a thick, solid, basal mass; lacking a long, rooting stalk.
   FLESH: white, thin, fibrous; odor pleasant; taste mild.
   SPORE PRINT COLOR: white.
   MICROSCOPIC FEATURES: spores 6–10 × 5–7 μm, oval, smooth, hyaline.
   FRUITING: solitary or in groups on the ground, usually at the base of oak and around decaying stumps, sometimes under conifer trees; July–October; widely distributed in the Southeast; occasional to fairly common.
   EDIBILITY: edible and choice when young.
   COMMENTS: Also known as *Sparassis spathulata*. The leaf-like branches are typically larger than those of *Sparassis crispa* (p. 307), edible.

Puffballs, Earthballs, Earthstars, and Allies

At some stage in their development, all members of this group are round to oval, pear- to turban-shaped, or irregularly rounded in outline. Species included here have been placed into six subgroups based on macroscopic features and habit of growth. These subgroups include: earthstars, earthballs, stalked puffballs and allies, gilled puffballs, hypogeous allies, and puffballs.

The earthstars are nearly round at first, often with a short, pointed, beak-like apex. The outer peridium splits open at maturity, forming star-like rays and exposing a thin inner spore case. Earthballs resemble puffballs but have a relatively thick, hard, rind-like spore case. Stalked puffballs have a papery thin spore case and a distinct stalk or a stalk-like base that is often partially or entirely buried. Gilled puffballs form round to oval fruit bodies that contain distorted, sometimes unexposed gills and an internal stalk. Hypogeous allies lack a stalk or stalk-like base and are completely or partially buried in soil or duff. Puffballs lack the combined features of these groups. They have a thin spore case that is often coated with spines and warts, as well as a white spore mass when young.

Arachnion album Schweinitz   Illus. p. 81
    COMMON NAME: none.
    FRUIT BODY: ¼–⅝" (5–15 mm) wide, irregularly globose to subglobose, tapering to a point of attachment and having root-like rhizomorphs, lacking a sterile base.
    SPORE CASE: white when young, becoming yellowish to buff; smooth at first, becoming areolate; thin-walled, fragile, breaking into fragments and exposing peridioles at maturity.
    SPORE MASS: white at first, becoming grayish to brownish; composed of numerous hollow chambers (peridioles) lined with a fertile layer and surrounded by hyphal tissue that eventually breaks down, leaving granular spore-containing particles; odor not distinctive when young, becoming nitrous at maturity.
    MICROSCOPIC FEATURES: spores $4.5–6 \times 3.5–5$ μm, short-elliptic to subglobose with a short pedicel, thick-walled, smooth.
    FRUITING: gregarious on soil in grassy areas, on disturbed ground, along paths, in flowerbeds, and in open groves of broadleaf trees.
    EDIBILITY: unknown.
    COMMENTS: When immature, this species is easily mistaken for other small puffballs, but cutting them in half will reveal the distinctive granular mass of peridioles.

Astraeus hygrometricus (Persoon) Morgan Illus. p. 81
   COMMON NAME: Barometer Earthstar.
      FRUIT BODY: 1½–3½" (4–9 cm) wide, ⅜–1" (1–2.5 cm) high when fully expanded, consisting of a rounded spore case and star-like rays.
      SPORE CASE: ⅜–1" (1–2.5 cm) wide, nearly round to somewhat flattened; whitish to grayish or grayish brown, finely roughened; with a single, irregular, pore-like mouth.
      SPORE MASS: white when young, becoming brown and powdery at maturity.
      RAYS: 6–12, hygroscopic (folding over the spore case when dry and expanding when moistened), 1–2" (2.5–5 cm) long, yellow-brown to reddish brown or grayish to nearly black; interior surface often finely cracked.
      MICROSCOPIC FEATURES: spores 7–11 μm, round, distinctly warted, brown.
      FRUITING: solitary, scattered, or in groups on sandy soil; year-round; widely distributed in the Southeast; infrequent to locally common.
      EDIBILITY: inedible.
      COMMENTS: The opening and closing of the rays in response to humidity explains its common name.

Bovista plumbea Persoon Illus. p. 81
   COMMON NAME: none.
      FRUIT BODY: ¾–3⅛" (2–8 cm) wide, globose to subglobose, attached to the substrate by a clump of fibrous mycelium.
      SPORE CASE: outer peridium thin and white when young, tearing apart and exposing a papery thin, smooth, purplish brown to grayish brown or bluish gray endoperidium, which often has a metallic luster; splitting open at maturity by a rather large apical pore and exposing the spore mass.
      SPORE MASS: white at first, becoming yellowish olive and finally dark brown and powdery at maturity.
      MICROSCOPIC FEATURES: spores 5–7 × 4–5 μm, oval to subglobose with a prominent pedicel that measures 8–14 μm long, minutely roughened, with a double wall, brownish.
      FRUITING: solitary, scattered, or in groups on the ground in pastures, parks, and cemeteries, as well as in old apple orchards, under hawthorns, and in mixed broadleaf woodlands; June–December, often overwintering; widely distributed in the Southeast; fairly common.
      EDIBILITY: edible when the spore mass is white.
      COMMENTS: The Tumbling Puffball, *Bovista pila*, edible when the spore mass is white, has a dark brown to bronze endoperidium at maturity and smaller, spherical spores with a short pedicel (see *MNE*, p. 458).

Bovistella radicata (Mont.) Pat. Illus. p. 81
   COMMON NAME: none.
      FRUIT BODY: 1–3½" (2.5–9 cm) wide, subglobose to broadly top-shaped, attached to the soil by a rooting base.
      SPORE CASE: outer peridium surface overall or at least on the upper half composed of soft, white pyramidal warts that are often fused at their tips to form fasicles; granular to scurfy white material distributed nearly overall, becoming minutely scurfy to nearly smooth at maturity; white at first, dull orange-yellow in age; wearing away irregularly to expose the thin, papery endoperidium; splitting open at maturity by an apical pore or slit that eventually enlarges to expose most of the spore mass.

SPORE MASS: white and spongy when young, becoming yellowish to olive and finally yellow-brown and powdery; subgleba whitish, becoming orange-yellow, forming a well-developed cup-like, sterile base occupying most of the narrowed portion of the fruit body.

MICROSCOPIC FEATURES: spores 4–5 × 3.5–4.5 μm, oval, smooth, hyaline, with one large oil drop and a hyaline pedicel 6–12 μm long; the capillitium of separate units highly branched and entangled.

FRUITING: solitary, scattered, or in groups on soil in open areas and in cultivated fields, pastures, and woodlands, especially with oak and pine; June–December; widely distributed in the Southeast; occasional.

EDIBILITY: unknown.

COMMENTS: Formerly known as *Bovistella ohiensis*. The genus *Bovista* lacks a distinctive, well-developed subgleba, which is always present in *Bovistella*. *Bovistella echinata* (not illustrated), edibility unknown, also has a white fruit body, but it is floccose, is much smaller ¼–⅜" (6–9 mm), and is attached to soil by a short pad of fibrils and not a rooting base.

Calostoma cinnabarina Desvaux   Illus. p. 82

COMMON NAME: Red Slimy-stalked Puffball, Hot Lips.

FRUIT BODY: a spore case supported by a thick, short stalk.

SPORE CASE: ⅜–¾" (1–2 cm) wide and high, oval to nearly round, sometimes collapsed; covered by a thick, gelatinous outer peridium with a thin, bright red inner layer when young; splitting into small seed-like pieces and exposing the smooth, thin-walled, bright reddish orange endoperidium, which fades to orange or orange-yellow in age; with an irregular slit-like mouth surrounded by bright red, elevated ridges.

SPORE MASS: white when young, buff and powdery at maturity.

STALK: ⅝–1½" (1.5–4 cm) long, ⅜–¾" (1–2 cm) thick, nearly equal overall, spongy, coarsely reticulate and pitted, reddish orange to pale reddish brown, covered with a thick gelatinous layer coated with debris.

MICROSCOPIC FEATURES: spores 14–22 × 6–9 μm, oblong-elliptic, pitted, hyaline.

FRUITING: solitary, scattered, or in groups on the ground in woods, frequently in loose or sandy soil, often buried up to the spore case; August–March; widely distributed in the Southeast; occasional.

EDIBILITY: inedible.

COMMENTS: *Calostoma ravenelii* (p. 311), inedible, has a tan to grayish spore case, an elongated stalk up to 2" (5 cm), and elliptic spores, and it lacks a gelatinous outer peridium. Also see *Calostoma lutescens* (p. 340).

Calostoma lutescens (Schweinitz) Burnap   Illus. p. 82

COMMON NAME: Collared Calostoma.

FRUIT BODY: a spore case supported by an elongated stalk.

SPORE CASE: ½–⅞" (1.5–2 cm) wide and high, nearly round, sometimes collapsed; covered by a weakly gelatinous outer peridium that forms a torn ring around the bottom of the yellow endoperidium; with an irregular slit-like mouth surrounded by bright red, elevated ridges.

SPORE MASS: white when young, becoming whitish to buff and powdery at maturity.

STALK: 2–3½" (5–9 cm) long, ⅝–⅞" (1.5–2 cm) thick, nearly equal overall, spongy, coarsely reticulate and pitted, pale yellow, becoming brown, coated with debris.

MICROSCOPIC FEATURES: spores 5.5–8 × 5.5–8 μm, globose, pitted but appearing like reticulate ridges, hyaline.

FRUITING: solitary, scattered, or in groups on the ground in woodlands, usually in loose or sandy soil and on well-decayed stumps, often buried up to the spore case; August–March; widely distributed in the Southeast; frequent.

EDIBILITY: inedible.

COMMENTS: Compare with *Calostoma cinnabarina* (p. 310), inedible, and *Calostoma ravenelii* (p. 311), inedible.

Calostoma ravenelii (Berk.) Massee   Illus. p. 82

COMMON NAME: Ravenel's Calostoma.

FRUIT BODY: a spore case supported by an elongated stalk.

SPORE CASE: ¼–¾" (6–20 mm) wide and high, nearly round, sometimes collapsed; covered by a scurfy grayish to straw-colored outer peridium that lacks a gelatinous layer in all stages and breaks into scales, revealing a whitish to pale yellow endoperidium; with an irregular slit-like mouth surrounded by bright red, elevated ridges.

SPORE MASS: whitish at first, becoming creamy white and powdery at maturity.

STALK: ¾–2" (2–5 cm) long, ¼–½" (6–12 mm) thick, nearly equal overall, spongy, coarsely reticulate and pitted, pale yellow, becoming brown, coated with debris.

MICROSCOPIC FEATURES: spores 10–17 × 6.5–8 μm, oblong-elliptic, minutely pitted but appearing finely punctate, with faint radial lines through the walls, hyaline.

FRUITING: solitary, scattered, or in groups on the ground in woodlands, usually in sandy soil, often buried up to the spore case; August–March; widely distributed in the Southeast; occasional.

EDIBILITY: inedible.

COMMENTS: The absence of a gelatinous layer is a major feature that separates this species from *Calostoma cinnabarina* (p. 310), inedible, and *Calostoma lutescens* (p. 310), inedible.

Calvatia cyathiformis (Bosc) Morgan   Illus. p. 82

COMMON NAME: Purple-spored Puffball.

FRUIT BODY: 2¾–7" (7–18 cm) wide, 3½–8" (9–20 cm) high, round to oval when young, becoming pear-shaped to top-shaped or irregular in age.

SPORE CASE: whitish to pale brown, smooth at first then becoming areolate on the upper portion, breaking into thin, irregular plates that flake off with age.

SPORE MASS: white and solid when immature, yellowish and finally dull purple to purple-brown and powdery at maturity.

MICROSCOPIC FEATURES: spores 4–7 μm, round, weakly echinulate, pale lilac.

FRUITING: solitary, scattered, or in fairy rings on the ground in grassy areas and woodland edges; July–November; widely distributed in the Southeast; fairly common.

EDIBILITY: edible when the spore mass is white.

COMMENTS: *Calvatia craniformis* (not illustrated), edible when the spore mass is white, is very similar, but its spore case surface is often deeply wrinkled to grooved. It often grows on the ground in open grassy areas, but also in woodlands. Its mature spore mass is yellow-brown without purple tones, and it has smaller, round spores that measure 2.5–4 μm.

Calvatia rubroflava (Cragin) Morgan Illus. p. 82
  COMMON NAME: Orange-staining Puffball.
  FRUIT BODY: 1 ⅛–4¾" (3–12 cm) wide, ¾–3⅛" (2–8 cm) high, nearly round to pear-shaped, often flattened somewhat at the apex, with a large sterile base and typically with white rhizomorphs.
  SPORE CASE: nearly white with pinkish or lavender tints when young, becoming yellow to bright yellow or orange when bruised, cut, or rubbed; smooth to finely areolate.
  SPORE MASS: pure white with minute cavities when young, becoming bright yellow-orange to dull orange and powdery in age; no odor at first, but becoming strong and unpleasant as the fruit body matures.
  MICROSCOPIC FEATURES: spores 3–4 μm, round, minutely warted, olive brown.
  FRUITING: scattered or in groups on the ground in gardens, grassy areas, and woodlands; August–December; widely distributed in the Southeast; occasional.
  EDIBILITY: edible when the spore mass is white, but the unpleasant odor of the maturing fruit body soon makes it unpalatable.
  COMMENTS: The bright yellow to orange staining of the peridium is a diagnostic features of this unique puffball.

Chorioactis geaster (Peck) Eckblad Illus. pp. 82, 83
  COMMON NAME: Devil's Cigar.
  FRUIT BODY: up to 4" (10 cm) high, up to 2" (5 cm) wide, elongated, hollow, dark brown, cigar- or spindle-shaped; splitting open at its apex and forming 3–6 earthstar-like rays and a supporting stalk.
  OUTER SURFACE: dry, velvety, covered with a dense coat of appressed, dark brownish hairs.
  INNER SURFACE: moist or dry, smooth, at first white or whitish, becoming yellowish then butterscotch-colored at maturity.
  FLESH: thick, fibrous-tough to leathery.
  MICROSCOPIC FEATURES: asci cylindric or subcylindric, 600–700 × 20–23 μm; spores uniseriate, 50–70 × 12–16 μm, fusoid, often unequal-sided and somewhat allantoid, hyaline or subhyaline; paraphyses strongly thickened above.
  FRUITING: solitary, scattered, or more typically in groups or clusters on the ground attached to decaying cedar elm *(Ulmus crassifolia)* stumps; September–November; reported only from Texas and Japan; rare.
  EDIBILITY: inedible.
  COMMENTS: Previously known as *Urnula geaster*. According to K. C. Rudy, naturalist and education coordinator at the Dallas Zoo, where this unusual fungus has been found, the periodic release of spores is often accompanied by a hiss that is audible from several feet away.

Disciseda candida (Schweinitzii) Lloyd Illus. p. 83
  COMMON NAME: Sand Case Puffball, Acorn Puffball.
  FRUIT BODY: ⅜–1⅜" (1–3.5 cm) wide when fresh, more or less globose to flattened-globose with a single anchoring rhizomorph at the base, becoming acorn-shaped at maturity.
  SPORE CASE: pale tan to grayish or sun-bleached dull white with age; at first covered with a layer of sand and interwoven mycelium, which falls away from the upward facing half at maturity; sterile base absent.

SPORE MASS: white and fleshy at first, becoming yellowish then olive brown before turning to a brown spore powder.

MICROSCOPIC FEATURES: spores 3.5–4.5 μm, globose, with or without a short pedicle, minutely warted, dark brown.

FRUITING: solitary or often in small groups in sandy soil in grasslands, pastures, and old colonized coastal dune habitat; June–May; widely distributed; uncommon.

EDIBILITY: of no culinary interest.

COMMENTS: This small puffball is a member of a unique genus of puffballs that first develop below the surface. At this stage, the eventual pore opening is oriented downward. As the fruit body matures, it breaks free and sheds the sand covering the lower half of the spore case as it emerges and becomes inverted so that the pore opening then faces upward. What now appears to be the base is weighted with adhering sand, leaving a fruit body that resembles an acorn in its cup. The detached spore case spills spores as it is blown about by wind and, owing to the sand-weighted base, always comes to rest with the pore opening facing upward, thus allowing further dispersal of spores by wind and rain.

Elaphomyces granulatus Fries   Illus. p. 83

COMMON NAME: Underground False Truffle, Deer Truffle.

FRUIT BODY: ¾–1⅝" (2–4 cm) wide, round to ovoid, tuber-like.

SPORE CASE: composed of two layers—a pale ochraceous to brownish outer layer with small, somewhat pyramidal warts and a thick, white, compact, inner rind that encloses the spore mass.

SPORE MASS: dark brown to purplish-black, powdery at maturity.

MICROSCOPIC FEATURES: spores 24–32 μm, globose, covered with low irregular warts, blackish brown.

FRUITING: solitary, scattered, or in groups buried in soil and under leaf litter in conifer woods, especially with pine, and in broadleaf woods; August–December; widely distributed in the Southeast; occasional.

EDIBILITY: unknown.

COMMENTS: This fungus is commonly attacked by *Cordyceps capitata* (p. 295), edibility unknown, and *Cordyceps ophioglossoides* (see *MNE*, p. 22).

Endoptychum agaricoides Czerniaier   Illus. p. 83

COMMON NAME: Puffball Agaric.

FRUIT BODY: a large oval to turban-shaped mushroom with a whitish to tan or brownish spore case and distorted, unexposed, white gills that become brown at maturity.

SPORE CASE: ¾–2¾" (2–7 cm) wide, 1–4" (2.5–10 cm) high, oval to turban-shaped or sometimes rounded; surface smooth or finely scaly, whitish at first, becoming tan to brownish at maturity; stalk whitish and internal, sometimes extending slightly below the spore case, attached to the ground by a cord.

SPORE MASS: fleshy; whitish at first, becoming brown at maturity; odor fragrant or not distinctive when fresh; taste sweet or nutty when fresh.

MICROSCOPIC FEATURES: spores 6–9 × 5–7 μm, elliptic, smooth, pale brown.

FRUITING: solitary, scattered, or in groups on the ground and among mosses and grasses in pastures, lawns, gardens, and waste areas; May–November; widely distributed in the Southeast; occasional.

EDIBILITY: edible when young and fresh.

COMMENTS: Spores are dispersed when the fruit body disintegrates or when animals consume them. Compare with *Macowanites arenicola* (p. 317), edibility unknown.

Geastrum fornicatum (Hudson) Fries   Illus. p. 83
COMMON NAME: Arched Earthstar.
FRUIT BODY: star-shaped, up to 3" (7.5 cm) wide and high when fully expanded, consisting of a rounded to compressed spore case and star-like rays.
SPORE CASE: ¾–1" (2–2.5 cm) wide, rounded to compressed, somewhat velvety, brownish to grayish brown, supported by a short stalk up to ⅛" (3 mm) long that is sometimes surrounded by a low basal collar; pore-like mouth large, irregular.
SPORE MASS: white and firm at first, becoming blackish brown and powdery at maturity.
RAYS: 4–5, not hygroscopic, 1–2" (2.5–5 cm) long, brownish, upright, arching, curved backward and downward at maturity, with their tips contacting the edge of a thick mycelial disc.
MICROSCOPIC FEATURES: spores 3.5–4.5 μm, globose, warted, dark brown.
FRUITING: scattered or in groups on the ground among leaves, twigs, mixtures of dung and wood chips or shavings or sawdust, and other organic debris; September–April; widely distributed in the southeast; occasional.
EDIBILITY: inedible.
COMMENTS: *Geastrum quadrifidum* (p. 315), inedible, is similar but smaller, and its pore-like mouth has a sharply defined silky area which is outlined by a circular groove.

Geastrum mirabilis Montagne   Illus. p. 84
COMMON NAME: none.
FRUIT BODY: star-shaped, up to 1⅛" (3 cm) when fully expanded, consisting of a round spore case and star-like rays, arising from a white basal mycelium that binds together debris and soil.
SPORE CASE: ¼–½" (6–12 mm) wide, round, sometimes with a small umbo, conspicuously strigose-tomentose, sometimes matted-tomentose, ochraceous buff to tan or light yellowish brown to reddish brown, stalkless; inner peridium grayish with a delicately felted surface; pore-like mouth distinct, more or less elevated, surrounded by a silky circle.
SPORE MASS: whitish to pale pinkish brown and firm at first, becoming dark grayish brown and powdery at maturity.
RAYS: 6–8, up to ⅜" (1 cm) long, slightly hygroscopic, brownish, the basal half bowl-shaped and holding the sessile spore case.
MICROSCOPIC FEATURES: spores 3–4 μm, globose, minutely warted, grayish brown.
FRUITING: densely caespitose or sometimes scattered or in groups on the ground, usually in sandy soil, in woodlands, and in disturbed areas, often associated with pines; July–December; widely distributed in the Southeast; occasional but easily overlooked.
EDIBILITY: unknown.
COMMENTS: Formerly known as *Geastrum caespitosus*. The small size, caespitose habit, conspicuous strigose-tomentose spore case, and white basal mycelium that binds leaves, twigs, and sandy soil are a most distinctive combination of characters that makes this earthstar easy to identify.

Geastrum quadrifidum Persoon : Persoon Illus. p. 84
  COMMON NAME: none.
  FRUIT BODY: star-shaped, up to 1⅜" (3.5 cm) wide and high when fully expanded, consisting of an ovoid to oblong or somewhat compressed spore case and star-like rays.
  SPORE CASE: ¼–½" (6–12 mm) wide, ovoid to oblong, or somewhat compressed, coated with minute glistening particles, brown to grayish brown, supported by a short stalk up to ⅛" (3 mm) long that is sometimes surrounded by a low basal collar; pore-like mouth elevated, delicate, surrounded by a sharply defined silky area that is outlined by a groove.
  SPORE MASS: white and firm at first, becoming dark smoky purplish and powdery at maturity.
  RAYS: 4–8, not hygroscopic, up to ¾" (2 cm) long, brown, upright, arching, curved backward and downward at maturity with their tips contacting the edge of a thin mycelial disc.
  MICROSCOPIC FEATURES: spores 4–5 μm, globose, distinctly warted, dark smoky purplish.
  FRUITING: scattered or in groups on the ground among leaves and twigs in conifer and broadleaf woodlands; September–May; widely distributed in the Southeast; occasional.
  EDIBILITY: inedible.
  COMMENTS: Formerly known as *Geastrum coronatum*. *Geastrum fornicatum* (p. 314), inedible, is similar but larger, and its pore-like mouth is larger and lacks the sharply defined silky area outlined by a groove.

Geastrum saccatum Fries Illus. p. 84
  COMMON NAME: Rounded Earthstar.
  FRUIT BODY: star-shaped, up to 2" (5 cm) wide and 1" (2.5 cm) high when fully expanded, consisting of a broadly conic to rounded spore case and star-like rays.
  SPORE CASE: ½–¾" (1.2–2 cm) wide, broadly conic to rounded, smooth, brownish to grayish brown or sometimes paler, stalkless; pore-like mouth large, irregular, surrounded by a conspicuous paler disc-like zone.
  SPORE MASS: white and firm at first, becoming brown to purplish brown and powdery at maturity.
  RAYS: 5–7, not hygroscopic, ⅝–1" (1.6–2.5 cm) long, ochre-brown to brownish pink, upright, then curved backward and downward at maturity, forming a distinct sac-like container surrounding the spore case.
  MICROSCOPIC FEATURES: spores 3.5–4.5 μm, globose, finely warted, brown.
  FRUITING: scattered or in groups on the ground among leaf litter in woodlands; June–December; widely distributed in the Southeast; occasional.
  EDIBILITY: inedible.
  COMMENTS: *Geastrum fimbriatum* (not illustrated), inedible, is similar but somewhat smaller, and its pore-like mouth lacks a conspicuous paler disc-like zone. *Geastrum vulgatum* = *Geastrum rufescens* (see photo, p. 84), inedible, is also similar but typically larger, up to 3⅛" (8 cm) in diameter when expanded; its pore-like mouth lacks a conspicuous paler disc-like zone; it has thicker flesh; and the downward-bent star-like lobes often crack transversely.

Lycoperdon acuminatum Bosc Illus. p. 84
  COMMON NAME: none.
  FRUIT BODY: very small, ⅛–⅜" (3–9 mm) wide and tall, shaped like an inverted top or pointed egg.
  SPORE CASE: whitish to pale tan with a yellowish to gray or pale gray-brown endoperidium, thin, delicate, with a thin spongy or granular to short spiny scurf that slowly wears away to expose the endoperidium; eventually forming a pore-like mouth on the apex at maturity; lower portion lacking a stalk-like base.
  SPORE MASS: white and soft at first, becoming pale brown with a slight olivaceous tint and powdery at maturity.
  MICROSCOPIC FEATURES: spores 3.3–4 μm, globose, thick-walled, with a short pedicel, smooth, pale olivaceous to dirty gray.
  FRUITING: scattered or in groups in mossy areas on the bark of living broadleaf and conifer tree trunks, usually 4–15 feet above ground; September–December; widely distributed in the Southeast; occasional to fairly common.
  EDIBILITY: unknown.
  COMMENTS: This species is the smallest known puffball in North America.

Lycoperdon americanum Demoulin Not Illustrated
  COMMON NAME: Spiny Puffball.
  FRUIT BODY: 1–2" (2.5–5 cm) wide, 1–1½" (2.5–4 cm) high, nearly round when young, becoming somewhat flattened and broader than high at maturity, covered with spines.
  SPORE CASE: white at first, becoming brownish in age, covered with clusters of long white spines with fused tips that darken in age and fall off, leaving a net-like pattern on the surface; forming a pore-like mouth at maturity; with a small, sterile, white to grayish base.
  SPORE MASS: white and firm at first, becoming purple-brown and powdery at maturity.
  MICROSCOPIC FEATURES: spores 4–6 μm, globose, warted, purple-brown.
  FRUITING: solitary, scattered, or in groups on the ground, often among leaves in woodlands; June–November; widely distributed in the Southeast; occasional.
  EDIBILITY: edible when the spore mass is white.
  COMMENTS: Formerly known as *Lycoperdon echinatum*. Compare with *Lycoperdon pulcherrimum* (p. 317), edible when the spore mass is white, which also has spines that are fused at their tips, but whose spines do not darken in age and do not leave marks on the surface of the spore case when they fall away.

Lycoperdon marginatum Vittadini Illus. p. 84
  COMMON NAME: Peeling Puffball.
  FRUIT BODY: ⅜–2" (1–5 cm) wide, nearly round when young, becoming somewhat flattened to pear-shaped at maturity and usually broader than tall; usually with a tapering, sterile, stalk-like base.
  SPORE CASE: at first white, covered with short spines or warts that break off in irregular sheets, exposing the nearly smooth, pale to dark olive brown or reddish brown inner surface; forming a pore-like mouth on the apex at maturity.
  SPORE MASS: firm and white at first, becoming olive brown to grayish brown and powdery at maturity.
  MICROSCOPIC FEATURES: spores 3.5–4.5 μm, round, minutely punctate to smooth, sometimes with a broken pedicel, pale brown.

FRUITING: solitary, scattered, or in groups on the ground, often on sandy soil, in oak-pine woods, and in nutrient poor habitats; June–December; fairly common.

EDIBILITY: not recommended; reports vary from edible to poisonous.

COMMENTS: The tendency of the spore case spines and warts to break off in irregular sheets is an important field identification character. The Gem-studded Puffball, *Lycoperdon perlatum*, edible when the spore mass is white, also has a white spore case covered with spines and granules, but the spines do not break off in sheets (see *MNE*, p. 16).

Lycoperdon pulcherrimum Berkeley and Curtis   Illus. p. 84

COMMON NAME: none.

FRUIT BODY: ¾–1¾" (2–4 cm) wide, ⅝–1⅜" (1.8–3.5 cm) high, nearly round when young, becoming somewhat flattened and broader than high at maturity, covered with long spines.

SPORE CASE: white at first, becoming brownish in age, covered with clusters of long spines that have fused tips that fall off and do not leave marks on the smooth, shining, deep brown or purplish brown or silvery brown surface of the exposed endoperidium; forming a pore-like mouth at maturity; tapered below to a slender base.

SPORE MASS: white and firm at first, becoming olivaceous and finally dark purple-brown and powdery at maturity.

MICROSCOPIC FEATURES: spores 4–6 μm, globose, warted, purple-brown.

FRUITING: solitary, scattered, or in groups on the ground; July–November; widely distributed in the Southeast; occasional.

EDIBILITY: edible when the spore mass is white.

COMMENTS: *Lycoperdon americanum* (p. 316), edible when young and the spore mass white, is very similar, but its spines darken in age and leave net-like marks on the surface of the spore case when they fall away.

Lycoperdon pyriforme Schaeffer : Persoon   Illus. p. 85

COMMON NAME: Pear-shaped Puffball.

FRUIT BODY: ⅝–1¾" (1.5–4.5 cm) wide, ¾–2" (2–5 cm) high, pear-shaped to nearly round.

SPORE CASE: whitish when young, becoming yellow-brown to reddish brown with tiny warts and granules or with spines; eventually forming a pore-like mouth on the apex at maturity; lower portion tapering downward, sometimes forming a sterile stalk-like base that is often compressed.

SPORE MASS: white and firm at first, becoming greenish yellow and finally dark olive brown and powdery at maturity.

MICROSCOPIC FEATURES: spores 3–4.5 μm, round, smooth, pale brown.

FRUITING: scattered or in dense clusters on decaying wood, sawdust, and organic debris; July–December, often overwintering; widely distributed in the Southeast; very common.

EDIBILITY: edible when the spore mass is young and completely white.

COMMENTS: This species often fruits in immense quantities on decaying logs and stumps.

Macowanites arenicola S. Miller   Illus. p. 85

COMMON NAME: Sandy Mac.

FRUIT BODY: resembling a gilled mushroom—consisting of a cap, a stalk-like base, and a convoluted gill chamber.

CAP: ¾–1¾" (2–4.5 cm) wide, convex to broadly convex, at times with a slight central depression; surface smooth, tacky or dry, usually with adhering granules of sand, whitish to cream or pale grayish yellow, occasionally with pinkish tones or blotches; margin incurved; flesh thin, watery white to grayish; odor reminiscent of old yogurt; taste not distinctive.

GILL CHAMBER (GLEBA): convoluted and chambered to sublamellate, pale to light yellow or ochraceous.

STALK-COLUMELLA: 5/8–1 1/4" (1.5–3 cm) long, 1/4–1/2" (0.5–1.2 cm) thick, cylindrical or tapering slightly toward the base, extending completely through the gleba, with low, longitudinal ribs or nearly smooth; solid at first, becoming chambered with age; whitish to tinged grayish.

SPORE PRINT COLOR: not obtainable.

MICROSCOPIC FEATURES: spores 8.8–10.4 × 6.8–7.2 μm, broadly elliptic to elliptic, with low warts and connecting lines that form a partial reticulum, hyaline, amyloid.

FRUITING: usually in small groups and often gregarious on sandy soil, especially on old dunes that are colonized with oak and pine; December–March; locally common along the northwest Florida coast, distribution limits yet to be determined.

EDIBILITY: unknown.

COMMENTS: This distinctive mushroom looks similar to species of *Russula*, to which it is closely related. However, unlike normal gilled mushrooms, *Macowanites* species do not forcibly discharge their spores from gills; instead, the spores are passively released as the gleba disintegrates, or possibly they are dispersed by insects. The Sandy Mac's cap often barely rises above the soil surface or remains buried and then is easily overlooked except for a slight hump in the sand. Compare with *Endoptychum agaricoides* (p. 313), edible.

Pisolithus tinctorius (Persoon) Coker and Couch   Illus. p. 85

COMMON NAME: Dye-maker's False Puffball.

FRUIT BODY: 1 3/8–4" (3.5–10 cm) wide, oval to pear-shaped or sometimes club-shaped in age, tapering downward to form a thick, stalk-like rooting base.

SPORE CASE: a thin, smooth, shiny peridium, dingy yellow to yellow-brown, splitting irregularly at maturity to expose numerous tiny yellowish to brownish peridioles embedded in a black gelatinous matrix.

SPORE MASS: reddish brown to dark brown and powdery at maturity, produced by the disintegrating peridioles.

MICROSCOPIC FEATURES: spores 7–12 μm, round, echinulate, brownish.

FRUITING: solitary, scattered, or in groups in sandy soil, typically under oak and pine, commonly with Prickly Pear cactus, often partially buried; July–January; widely distributed in the Southeast; common.

EDIBILITY: inedible.

COMMENTS: This mushroom's common name is a reference to the fact that it can be used to dye wool various shades of brown or black. Also known as *Pisolithus arhizus*.

Rhizopogon nigrescens Coker and Couch   Illus. p. 85

COMMON NAME: none.

FRUIT BODY: 3/4–1 1/2" (1–4 cm) wide, subglobose to top-shaped or irregularly lobed, sometimes partially flattened, stalkless.

SPORE CASE: a thin peridium; surface somewhat sticky when young, becoming tomentose to felty; yellowish at first, becoming brownish orange to grayish brown and finally dark brown to blackish at maturity, staining dull red to reddish brown then brown when bruised.

SPORE MASS: whitish at first, becoming grayish yellow to olive brown at maturity, bruising reddish brown; odor somewhat fragrant.
MICROSCOPIC FEATURES: spores 6–9 × 2–3.5 μm, subfusoid to oblong, smooth, hyaline to pale yellow.
FRUITING: scattered or in groups, usually partially buried in sandy soil under pines; August–January; widely distributed in the Southeast; fairly common.
EDIBILITY: unknown.
COMMENTS: The name *nigrescens* means "blackening," a reference to the color of the mature spore case. This species is thought to have evolved from a species of *Suillus*. Spore case stains brownish violet to grayish red with the addition of KOH and slowly dark olive progressing to black with $FeSO_4$. *Rhizopogon atlanticus* (see photo, p. 85), edibility unknown, is similar, but it has a spore case that is whitish when young, becoming pinkish cinnamon to yellowish brown in age or on drying; and its spores measure 7–8.5 × 3–4 μm, are bluntly elliptic, and appear truncate. *Elaphomyces granulatus* (p. 313), inedible, forms nearly round to oval dark yellow-brown to reddish brown fruiting bodies that are covered with small hard warts; it grows buried in soil or humus, usually 2–4 inches below the surface in conifer and sometimes broadleaf woods; and it is often attacked by *Cordyceps capitata* (p. 295), edibility unknown, a parasitic fungus that produces an above-ground dark reddish brown to olive black head on a yellow stalk.

Rhopalogaster transversarium (Bosc) Johnston — Illus. p. 85
COMMON NAME: none.
FRUIT BODY: ⅝–1¾" (1.5–4.5 cm) wide, 1⅛–3¾" (3–9.5 cm) high, narrowly to broadly club-shaped.
SPORE CASE: the enlarged upper portion granular-roughened to scurfy with scattered matted and brownish fibers when young and fresh, becoming nearly smooth in age; pale reddish brown to yellow-brown or olive brown, becoming pale brownish orange as the ground color is exposed; rupturing irregularly to expose the spore mass.
STALK: the narrowed lower portion covered with a dense coat of matted fibers and colored like the spore case when young and fresh, becoming nearly smooth and pale brownish orange then whitish as the fibers wear away in age; flesh watery-gelatinous, meat-like, and conspicuously marbled or veined when fresh young specimens are sectioned longitudinally; becoming dry, fibrous, and forming a whitish columella that extends upward and branches through the spore mass to the top of the spore case.
SPORE MASS: distinctly chambered like a honeycomb, consisting of elongated peridioles formed by the branching columella, gelatinous; pale to dark reddish brown, slowly staining blackish when exposed, becoming olive brown to dark brown and eventually powdery in age.
MICROSCOPIC FEATURES: spores 5.5–7.5 × 3–4.5 μm, elliptic, smooth, pale brown.
FRUITING: solitary, scattered, or in groups on decaying organic matter such as wood chips, logs, limbs, leaf litter, and mulch in pine-oak woods; July–February; widely distributed from North Carolina to Florida, west to Texas; occasional.
EDIBILITY: unknown.
COMMENTS: Young and fresh specimens, as well as mature ones, are shown in the two illustrations provided. Application of $FeSO_4$ to the surface of the fruit body produces a green staining reaction.

Scleroderma bovista Fries   Illus. p. 86
- COMMON NAME: none.
- FRUIT BODY: ⅝–1¾" (1.5–4.5 cm) wide, more or less spherical, becoming flattened on the upper surface in age, attached to the soil by a thick stalk-like base.
- SPORE CASE: firm, dry, smooth when young, soon developing fine cracks and typically divided into small patches 1/16–¼" (2–6 mm) wide, slightly to markedly scaly in age; straw yellow to pale orange-yellow when young, becoming orange-yellow to reddish brown with olive gray tints at maturity; 1/32–1/16" (0.8–2 mm) thick, splitting irregularly over the top and sides at maturity; staining dark reddish with the application of KOH.
- STALK-LIKE BASE: ¼–1⅜" (5–35 mm) long, ½–1⅛" (1.2–3 cm) thick, straw yellow to pale orange-yellow, composed of a dense mass of entangled mycelioid cords with trapped sand.
- SPORE MASS: dark blackish brown.
- MICROSCOPIC FEATURES: spores 10–16 μm, globose, with a reticulum that is usually well developed but may be imperfectly developed on some; the reticulum measures 1/16–⅛" (1.5–3 mm) high; clamp connections present.
- FRUITING: solitary, scattered, or in groups on soil in grassy areas, waste areas, and woodlands, especially oak-pine; July–November; widely distributed in the Southeast; occasional.
- EDIBILITY: poisonous.
- COMMENTS: The more or less spherical, yellow to reddish brown fruiting body, yellow to orange-yellow stalk-like base, and dark blackish brown spore mass are the distinctive features. *Scleroderma meridionale* (p. 320), edibility unknown, is somewhat similar but has a prominent stalk-like base and thicker spore case wall that splits into star-like rays at maturity. *Scleroderma texense* (see photo, p. 86), poisonous, is similar, but its surface is whitish to pale brown and coated with small and flat to pyramid-shaped scales; it stains yellowish to reddish with the application of KOH; and it has smaller spores that measure 8–12 μm.

Scleroderma meridionale Demoulin and Malençon   Illus. p. 86
- COMMON NAME: none.
- FRUIT BODY: ¾–2⅜" (2–6 cm) wide, globose to subglobose or irregular in outline, tapered downward to form a thick, stalk-like rooting base.
- SPORE CASE: moderately thick, up to 1/16" (2 mm) thick; surface dry, roughened, conspicuously areolate and warted at maturity, ochraceous tan to bright ochraceous yellow, with dull grayish to yellowish brown warts, slowly splitting into irregular lobes at maturity to expose the spore mass.
- STALK-LIKE BASE: 1–3½" (2.5–9 cm) long, ¾–1¾" (2–4.5 cm) thick, tapered in either direction, with coarse, irregular, blunt projections; scurfy-roughened, dull ochraceous orange to ochre-yellow or brownish, typically coated with sand.
- SPORE MASS: dark gray to brownish gray or blackish gray and coarsely powdery at maturity.
- MICROSCOPIC FEATURES: spores 12–20 μm, globose, echinulate, reticulate, dark brown.
- FRUITING: solitary, scattered, or in groups partially buried in sand; September–March; widely distributed in the Southeast; occasional to fairly common.
- EDIBILITY: unknown.
- COMMENTS: Also known as *Scleroderma macrorrhizon*. The Pigskin Poison Puffball = Common Earthball, *Scleroderma citrinum*, poisonous, has a nearly round to somewhat flattened, thick, rind-like, pale wood brown to yellow-brown or golden brown spore

case that is covered with coarse warts; it lacks a stalk; and it grows on the ground and on decaying wood (see *MNE,* p. 16).

Scleroderma polyrhizon (Gmelin) Lévielle   Illus. p. 86
  COMMON NAME: Earthstar Scleroderma.
  FRUIT BODY: 1½–4½" (4–12 cm) wide, round to oval or irregular when closed, expanding up to 6" (15.5 cm) and resembling a giant earthstar when open.
  SPORE CASE: wall ⅛–⅜" (3–10 mm) thick, hard, rind-like, rough, areolate to somewhat scaly, dingy white to straw-colored or pale yellow-brown; splitting open at maturity into 4–8 star-like rays and exposing the spore mass.
  SPORE MASS: firm when young, becoming powdery, brown to purplish brown, becoming blackish brown at maturity.
  MICROSCOPIC FEATURES: spores 5–10 μm, globose, coated with short spines, sometimes forming a partial reticulum, purple-brown.
  FRUITING: solitary or in groups, often partially buried in sandy soil; August–December; widely distributed in the Southeast; occasional to fairly common.
  EDIBILITY: poisonous.
  COMMENTS: Formerly known as *Scleroderma geaster. Scleroderma floridanum* (see photo, p. 86), edibility unknown, is very similar, but it has larger spores, 9–12 μm, and its fruit body is tan to pale grayish brown and peels open at maturity, forming 5–8 thick rays with a dark brown inner surface and spore mass.

Scleroderma verrucosum (Bull.: Pers.) Persoon   Illus. p. 86
  COMMON NAME: none.
  FRUIT BODY: ¾–2½" (2–7 cm) wide, more or less globose above a narrowed and furrowed stalk-like base with attached white rhizomorphs.
  SPORE CASE: wall thin, less than 1/32" (1 mm) thick, dry, leathery, coated with flattened scales, becoming smoother in age, reddish brown to ochraceous.
  SPORE MASS: whitish to cream and firm when young, soon becoming purplish black and marbled, finally becoming powdery grayish brown at maturity.
  MICROSCOPIC FEATURES: spores 9–11 μm, globose, spiny, brown.
  FRUITING: usually in groups on the ground, especially in sandy soil in woods and along roadsides and woodland margins; July–December; widely distributed in the Southeast; occasional.
  EDIBILITY: poisonous.
  COMMENTS: *Scleroderma bovista* (p. 320), poisonous, has a shorter stalk-like base and reticulated spores. *Scleroderma citrinum,* poisonous, has a thicker and more distinctly scaly spore case (see *MNE,* p. 16). *Scleroderma polyrhizon* (p. 321), poisonous, is typically larger, up to 5" (12 cm) wide when expanded, and as it matures, its outer spore case splits into irregular ray-like segments similar to an earthstar.

Vascellum pratense (Pers. em. Quél.) Kreisel   Illus. p. 86
  COMMON NAME: Western Lawn Puffball.
  FRUIT BODY: ¾–2" (2–5 cm) wide and high, turban- to pear-shaped, with a rounded to somewhat depressed center, outer surface granular to finely spiny, spore mass supported by a sterile stalk-like base.
  SPORE CASE: at first white to pale yellowish and covered with granules and fine spines that wear away to expose a shiny brownish endoperidium; forming a pore-like mouth on

the apex at maturity that disintegrates down to the sterile base, causing the fruit body to appear bowl-like.

SPORE MASS: white and firm at first, becoming olive then dark olive brown and powdery at maturity.

MICROSCOPIC FEATURES: spores 3.5–5.5 μm, subglobose, thick-walled, finely warted, brownish.

FRUITING: solitary, scattered, or in groups on the ground in grassy areas; July–November; along the Gulf Coast from Florida to Texas; occasional, more common in the western United States.

EDIBILITY: edible when the spore mass is white.

COMMENTS: Also known as *Vascellum depressum*.

Zelleromyces cinnabarinus Singer and A. H. Smith   Illus. p. 86

COMMON NAME: Milky False Truffle.

FRUIT BODY: ⅜–1¼" (1–3 cm) wide, subglobose with a small, white, basal knob of mycelium at the point of attachment; often flattened or irregularly compressed when growing in compact clusters; stalkless; surface dry, somewhat velvety or cottony at first, becoming smooth and whitish to mottled orangish then orange to brown or reddish brown at maturity.

INTERIOR: chambered or marbled with sterile tissue and a distinct branching columella.

FLESH: firm at first, nearly crumbly when old, orangish white, becoming brownish in age, exuding a white latex when fresh specimens are cut; odor pungent, like rubber, when the mushroom is old, or not distinctive; taste not distinctive.

MICROSCOPIC FEATURES: spores 12–17.5 × 11.5–15.5 μm, globose to broadly elliptic, thick-walled, surface ornamented with warts and ridges that form a broken to nearly complete reticulum, hyaline, pale yellow in KOH, amyloid; basidia 2-spored.

FRUITING: usually clustered or in groups on the ground beneath pines in hard-packed soil, on roadsides, and in waste areas; August–April; widespread in the Southeast; common.

EDIBILITY: unknown.

COMMENTS: This hypogeous relative of the milk mushrooms, the genus *Lactarius*, is fairly easy to recognize by its association with pines and its exudation of latex when fresh fruit bodies are cut. Previously known as *Octaviania ravenelii*. Some taxonomists suggest that the correct name for this species is *Zelleromyces ravenelii*. Species of *Rhizopogon* are similar and also form mycorrhizae with pines, but they lack latex and a columella.

Carbon and Cushion Fungi

Members of this group belong to the class Pyrenomycetes, commonly called the flask fungi. They are closely related to the *Cordyceps, Claviceps,* and Allies (p. 295). Their fertile surfaces are finely roughened like sandpaper owing to the protruding necks of numerous flask-shaped reproductive structures called *perithecia,* which produce ascospores. These fungi occur on decaying wood, usually broadleaf trees, and sometimes on the surrounding ground.

Although some species may be gelatinous or powdery, most are fibrous-tough to woody, and several are hard, black, and carbonaceous at maturity. Some resemble crust fungi, but differ by having a finely roughened, sandpaper-like surface. Most members are cushion-shaped, and a few have distinct stalks.

Daldinia concentrica (Bolt. : Fr.) Cesati and de Notaris   Illus. p. 87
  COMMON NAME: Carbon Balls, Crampballs.
  FRUIT BODY: ¾–2" (2–5 cm) wide, cushion-shaped to nearly round or irregular; surface uneven and furrowed, reddish brown, becoming black, somewhat shiny, finely roughened, often with minute pores; flesh concentrically zoned when cut vertically, fibrous to powdery and carbon-like, dark purplish brown alternating with darker or sometimes whitish zones; perithecia embedded in a single layer near the surface.
  MICROSCOPIC FEATURES: spores 12–17 × 6–9 μm, irregularly elliptic with one side flattened, smooth, dark brown.
  FRUITING: solitary or in clusters on decaying broadleaf trees; year-round; widely distributed in the Southeast; common.
  EDIBILITY: inedible.
  COMMENTS: The concentrically zoned interior tissue is a diagnostic feature of this species.

Diatrype stigma (Hoffmann : Fries) Fries   Illus. p. 87
  COMMON NAME: none.
  FRUIT BODY: a wide-spreading crust, highly variable in shape and size, up to 9" (23 cm) or more long and ¹⁄₃₂–¹⁄₁₆" (1–1.5 mm) thick; surface shiny, finely roughened to nearly smooth, sometimes minutely cracked, purplish to reddish brown when young, soon blackish brown to black; flesh whitish, fibrous-tough; perithecia embedded just beneath the surface in a single layer.
  MICROSCOPIC FEATURES: spores 6–9 × 1.5–2 μm, sausage-shaped, smooth, hyaline.
  FRUITING: on decaying broadleaf branches, especially beech, often surrounding them; year-round; widely distributed in the Southeast; common.
  EDIBILITY: inedible.
  COMMENTS: The nearly smooth, shiny surface and hyaline spores are characteristic of this

species. Other similar black, spreading crusts have more coarsely roughened surfaces or dark brown spores.

Hypoxylon fragiforme (Persoon : Fries) Kickx Illus. p. 87
  COMMON NAME: Red Cushion Hypoxylon.
  FRUIT BODY: 1/16–5/8" (1.5–16 mm) wide, round to cushion-shaped, dry, hard, conspicuously roughened with raised dots, grayish white at first, then salmon pink, becoming cinnamon to brick red at maturity and finally blackening in age; interior a dark brown to black layer of embedded perithecia surrounding a hollow center.
  MICROSCOPIC FEATURES: spores 10–15 × 5–8 μm, elliptic with one side flattened and a narrow longitudinal groove, sometimes with one to three oil drops, smooth, dark brown.
  FRUITING: in dense clusters or sometimes scattered on the bark of decaying broadleaf trees; year-round; widely distributed in the Southeast; common.
  EDIBILITY: inedible.
  COMMENTS: This common fungus is restricted to beech trees. *Hypoxylon fuscum* (not illustrated), inedible, is similar but has rounded surface bumps rather than raised dots; grows on birch, hazel, and alder; and has spores that measure 12–13 × 5–6 μm. *Hypoxylon rubiginosum* (p. 324), inedible, also has a brick red fruit body, but it forms crust-like spreading patches up to 4" (10 cm) or more long.

Hypoxylon rubiginosum (Persoon : Fries) Fries Illus. p. 87
  COMMON NAME: none.
  FRUIT BODY: variable in size and shape but often cushion-shaped; individual fruiting bodies usually fusing and forming crust-like spreading patches up to 4" (10 cm) or more long; surface uneven, often furrowed, brick red to reddish brown or purplish brown, in age wearing away and exposing a black interior.
  MICROSCOPIC FEATURES: spores 9–12 × 4–5.5 μm, irregularly elliptic, with one side flattened, smooth, dark brown.
  FRUITING: in groups or confluent patches on broadleaf trunks, logs, and stumps; year-round; widely distributed in the Southeast; fairly common.
  EDIBILITY: inedible.
  COMMENTS: *Hypoxylon fragiforme* (p. 324), inedible, forms a more cushion-shaped fruit body that is grayish white at first, then salmon pink, then brick red, then black in age; it does not form crust-like spreading patches. Some color forms of *Hypoxylon rubiginosum* are similar to *Hypoxylon fuscum* (not illustrated), inedible, which has larger spores that measure 12–13 × 5–6 μm.

Peridoxylon petersii (Berk. and Curt.) Shear Illus. p. 87
  COMMON NAME: Split-skin Carbon Cushion.
  FRUIT BODY: 3/4–2 1/4" (2–6 cm) wide, cushion-shaped or oval or irregular; surface grayish brown to yellowish brown, splitting open to expose a shiny, black, slightly roughened, sandpaper-like interior; flesh reddish brown to blackish brown, thick, firm, containing numerous tiny, oval, seed-like perithecia arranged in five or six layers; odor unpleasant, creosote-like; taste unpleasant.
  MICROSCOPIC FEATURES: spores 6–8 × 3.5–4 μm, elliptical, smooth, bluish black.
  FRUITING: solitary, scattered, or in groups on decaying broadleaf stumps, logs, and fallen branches; June–October; widely distributed in the Southeast; occasional.
  EDIBILITY: inedible.
  COMMENTS: Formerly known as *Camarops petersii*.

Ustulina deusta (Fries) Petrak Illus. p. 87
  COMMON NAME: Carbon Cushion.
  FRUIT BODY: 1¾–3½" (4.5–9 cm) wide, ⅛–¼" (3–6 mm) thick, spreading, crust-like, forming extensive sheets up to 15" (38.5 cm) or more that are irregular in outline; surface at first grayish white, soft, powdery, with multiple lobes (asexual stage), becoming hard, black, finely roughened and resembling burned wood at maturity (sexual stage); flesh soft and white at first, becoming black and brittle in age; perithecia embedded just beneath the surface in a single layer.
  MICROSCOPIC FEATURES: spores 28–36 × 7–10 μm, irregularly elliptic with one side flattened, smooth, dark brown.
  FRUITING: on stumps, roots, and decaying trunks of broadleaf trees; June–December, often overwintering; widely distributed in the Southeast; common.
  EDIBILITY: inedible.
  COMMENTS: Mature specimens are easily overlooked. Our photograph shows a young, fresh specimen, which represents the commonly observed asexual stage of this fungus.

Xylaria magnoliae J. D. Rogers Illus. p. 87
  COMMON NAME: none.
  FRUIT BODY: 1–3" (2.5–7.5 cm) high, up to 3/16" (2 mm) thick, upright, slender, filamentous to spindle-shaped, sometimes branched, tough, coated with white asexual spores when immature, becoming black at maturity when coated with sexual spores produced in embedded perithecia.
  MICROSCOPIC FEATURES: sexual spores 11–17 × 3–6 μm, spindle-shaped or irregularly elliptic, smooth, yellowish.
  FRUITING: several to many on decaying magnolia cones; May–November; widely distributed in the Southeast; common.
  EDIBILITY: inedible.
  COMMENTS: The exclusive association with magnolia cones makes this fungus easy to identify.

Xylaria oxyacanthae Tulasne Illus. p. 88
  COMMON NAME: none.
  FRUIT BODY: at the asexual stage 2–4" (5–10 cm) high, 1/16–⅛" (1–3 mm) wide at the base, upright, slender, filamentous to spindle-shaped, sometimes branched, tough, coated with white asexual spores on the upper half and blackish below; at the sexual stage up to 2¾" (7 cm) high, up to ¼" (5 mm) wide, black overall, roughened by embedded perithecia.
  MICROSCOPIC FEATURES: asexual spores 3.5–4 × 3–3.5 μm, globose to ellipsoid, smooth, hyaline; sexual spores 10–13 × 4.5–6 μm, irregularly elliptic with one side flattened and a straight, spore-length germ slit, smooth, dark brown.
  FRUITING: usually solitary but sometimes in pairs or groups, arising from hickory and pecan nut shells; asexual stage May–June; sexual stage July–October; widely distributed in the Southeast; fairly common.
  EDIBILITY: inedible.
  COMMENTS: The specific relationship with hickory and pecan nuts makes this fungus easy to identify.

Xylaria persicaria (Schw.: Fr.) Berk. and Curtis Illus. p. 88
  COMMON NAME: none.
  FRUIT BODY: at the asexual stage ¾–2" (2–5 cm) high, about 1/16" (1–2 mm) thick at the

base, upright, slender, filamentous to spindle-shaped, tough, coated with white asexual spores on the upper half and blackish below; at the sexual stage up to 2" (5 cm) high, about 1/16" (1–2 mm) thick, black overall, roughened by embedded perithecia.

MICROSCOPIC FEATURES: asexual spores 7.5–10 × 3–4 μm, almond-shaped to ellipsoid, smooth, hyaline; sexual spores 10–16 × 4–6 μm, irregularly elliptic with one side flattened and a long spiraling germ slit, smooth, dark brown.

FRUITING: one to several arising from sweet gum fruits and peach pits; asexual stage May–June; sexual stage July–October; widely distributed in the Southeast; fairly common.

EDIBILITY: inedible.

COMMENTS: The specific association with sweetgum fruits and peach pits makes this fungus easy to identify.

Xylaria polymorpha (Persoon : Mérat) Greville

Illus. p. 88

COMMON NAME: Dead Man's Fingers.

FRUIT BODY: 3/4–3 1/2" (2–9cm) high, 3/8–1 1/8" (1–3 cm) thick, unbranched, irregularly clavate or spindle-shaped, exterior powdery and white at first (asexual stage), becoming hard, black, carbonaceous, finely roughened, often wrinkled and minutely cracked at maturity (sexual stage); stalk short, indistinct, cylindric; flesh white, fibrous-tough; perithecia embedded just beneath the surface in a single layer.

MICROSCOPIC FEATURES: sexual spores 22–30 × 5–9 μm, spindle-shaped or irregularly elliptic with one side flattened, smooth, dark brown.

FRUITING: densely clustered or sometimes solitary on or near decaying broadleaf stumps and elsewhere on the ground growing up from buried wood; June–November, sometimes overwintering; widely distributed in the Southeast; common.

EDIBILITY: inedible.

COMMENTS: *Xylaria tentaculata* (p. 326), inedible, forms a fragile, gray to blackish erect stalk and a head with a crown of spreading tentacles.

Xylaria tentaculata Berkeley and Broome

Illus. p. 88

COMMON NAME: Fairy Sparklers.

FRUIT BODY: 3/4–1 3/8" (1.6–3.5 cm) high, consisting of a stalk and head, with a crown of tentacles; stalk 5/8–1" (1.6–2.5 cm) long, 1/16–1/8" (1.5–3 mm) thick, erect, nearly equal to irregular and twisted, frequently curved near the base; surface granular-scurfy, gray to blackish, often with bluish tints; flesh white; head 1/8–3/8" (3–10 mm) long, 1/8–1/4" (3–6 mm) wide, cylindric, distinctly scurfy, opening at the apex by a prominent ostiole and crowned by up to eight or more tentacles, gray to pinkish gray; tentacles 1/2–1 1/2" (1.3–4 cm) long, up to 1/32" (0.75 mm) thick, partially erect then spreading, tapering toward the tips, sometimes branched, pale gray to pinkish gray, coated with pale gray to whitish powder at maturity.

MICROSCOPIC FEATURES: spores 14–19 × 6–8.5 μm, elliptic, smooth, with two large oil drops, hyaline.

FRUITING: solitary, scattered, or in groups, sometimes clustered, on organic debris such as decaying wood and humus and among mosses; July–October; widely distributed in the Southeast; fairly common.

EDIBILITY: inedible.

COMMENTS: This mushroom is very fragile and easily overlooked.

Crust and Parchment Fungi and Allies

This group is a very large and highly variable collection of species that form thin, spreading, crust-like to papery growths, usually on decaying wood. Some species are nearly flat, whereas others have small, projecting, shelf-like caps usually formed by being bent backward at the margin. Their fertile surfaces may be warted, wrinkled, cracked, tooth-like, or smooth, but they lack true pores and are not finely roughened like sandpaper. Crust-like species with pores on their fertile surfaces are included in the polypores.

Several hundred species of crust and parchment fungi have been identified. We have included some of the more common and conspicuous species likely to be encountered.

Aleurodiscus oakesii (Berkeley and Curtis) Hoehnel and Litschauer Illus. p. 88
  COMMON NAME: Hophornbeam Disc.
  FRUIT BODY: 1/32–1/8" (1–3 mm) wide, disc-shaped, flattened to slightly elevated at the margin, attached at a single point at the center, fleshy to leathery; fertile (upper) surface concave, glabrous, smooth to finely wrinkled, pale brownish pink to pale brown; sterile (lower) surface whitish and tomentose.
  FLESH: thin, whitish, firm or leathery.
  SPORE PRINT COLOR: white.
  MICROSCOPIC FEATURES: spores 18–21 × 12–13 μm, ovate to broadly elliptic, smooth, hyaline.
  FRUITING: scattered or confluent on broadleaf trees, especially hophornbeam, hickory, and oak; June–December, sometimes overwintering; widely distributed in the Southeast; fairly common.
  EDIBILITY: inedible.
  COMMENTS: *Aleurodiscus amorphus* (not illustrated), inedible, is very similar, but it has minutely echinulate spores and grows on conifer trees.

Cotilydia diaphana (Schw.) Lentz Illus. p. 88
  COMMON NAME: Stalked Stereum.
  FRUIT BODY: 1/4–1 3/8" (6–3.5 cm) wide, 1/2–1 3/4" (1.3–4.5 cm) high, deeply vase-shaped or split into flattened spatula-like sections with an entire or torn margin; sterile (upper) surface moist or dry, faintly to strongly zoned, streaked and wrinkled or folded, translucent when fresh, white to grayish or pinkish buff; fertile (lower) surface faintly zoned, slightly to conspicuously folded, minutely roughened, whitish to buff or grayish; stalk present as a downward tapered extension of the fertile surface, with a white basal mycelium.

FLESH: thin, fibrous, whitish; odor and taste not distinctive.
SPORE PRINT COLOR: white.
MICROSCOPIC FEATURES: spores 4–6 × 2.5–4 μm, elliptic to oval, smooth, hyaline.
FRUITING: scattered or in groups on the ground in broadleaf woodlands; August–December; widely distributed in the Southeast; occasional.
EDIBILITY: inedible.
COMMENTS: Formerly known as *Stereum diaphanum*.

Hydnochaete olivaceum (Schweinitz) Banker   Illus. p. 89

COMMON NAME: Brown-toothed Crust.
FRUIT BODY: a spreading, resupinate, crust-like mass up to 4" (10 cm) wide or wider, up to 8" (20 cm) long; surface dry, leathery, olive brown to cinnamon brown, with jagged teeth.
FLESH: up to ⅛" (3 mm) thick, leathery, brown; odor and taste not distinctive.
SPORE PRINT COLOR: white.
MACROCHEMICAL TESTS: all parts stain dark brown to black with the application of KOH.
MICROSCOPIC FEATURES: spores 4.5–7 × 1.2–1.5 μm, sausage-shaped, smooth, hyaline; setae present on the teeth, 30–150 × 9–14 μm, sharp and tapered at both ends, thick-walled, reddish brown.
FRUITING: a spreading crust on the underside of decaying branches of broadleaf trees, especially oak; May–December, often overwintering; widely distributed in the Southeast; common.
EDIBILITY: inedible.
COMMENTS: Formerly known as *Irpex cinnamomeus*. *Hymenochaete tabacina* = *Hymenochaete badio-ferruginea*, inedible, is similar but forms a thin spreading crust with small, shelf-like, and stalkless caps that grow in overlapping clusters (see *MNE*, p. 15).

Phlebia incarnata (Schrader : Fries) Nakasone and Burdsall   Illus. p. 89

COMMON NAME: none.
FRUIT BODY: ¾–1⅝" (2–4 cm) wide, 1⅛–3⅛" (3–8 cm) long, fan-shaped to semicircular, somewhat cartilaginous to leathery, stalkless caps; sterile (upper) surface moist or dry, finely pubescent to subglabrous, coral pink when young and fresh, becoming salmon buff in age; fertile (lower) surface poroid, consisting of a network of radiating, branched folds, pinkish ochre to salmon buff.
FLESH: about ⅛" (2–4 mm) thick, spongy to leathery, whitish to buff.
SPORE PRINT COLOR: white.
MICROSCOPIC FEATURES: spores 4–5 × 2–3 μm, elliptic, smooth, hyaline.
FRUITING: in overlapping clusters on logs and stumps of broadleaf trees; July–December; widely distributed in the Southeast; occasional.
EDIBILITY: unknown.
COMMENTS: Also known as *Merulius incarnatus*. Compare with *Phlebia tremellosa* (p. 329), edibility unknown, which has a hairy to wooly, white to pale yellow upper surface, a yellowish to brownish orange or pinkish orange fertile surface, and spores that measure 3–4 × 0.5–1.5 μm. *Phlebia radiata* (not illustrated), edibility unknown, forms cartilaginous discs or patches with a radiating wrinkled reddish orange to pink fertile surface, usually on the underside of fallen trunks and branches of broadleaf and conifer trees.

Phlebia tremellosa (Schrader: Fries) Nakasone and Burdsall   Illus. p. 89
  COMMON NAME: Trembling Merulius.
  FRUIT BODY: ¾–1½" (2–4 cm) wide, 2–4" (5–10 cm) long, fan-shaped to semicircular, spongy to fibrous, stalkless caps or spreading sheets; sterile (upper) surface dry or moist, hairy to woolly, white to pale yellow; fertile (lower) surface with radiating to wrinkled ridges and crossveins, often forming pore-like depressions on mature specimens, yellowish orange to brownish orange or pinkish orange.
  FLESH: about 1/16" (2 mm) thick, fleshy and gelatinous, white to yellowish.
  SPORE PRINT COLOR: white.
  MICROSCOPIC FEATURES: spores 3–4 × 0.5–1.5 μm, sausage-shaped, smooth, hyaline.
  FRUITING: in overlapping clusters, often laterally fused, usually on decaying broadleaf wood but sometimes on conifer logs and stumps; July–January; widely distributed in the Southeast; occasional.
  EDIBILITY: unknown.
  COMMENTS: Also known as *Merulius tremellosus*. Compare with *Phlebia incarnata* (p. 328), edibility unknown, which has a finely pubescent to subglabrous, coral pink to salmon buff upper surface, a pinkish ochre to salmon buff fertile surface, and spores that measure 4–5 × 2–3 μm.

Pulcherricium caeruleum (Persoon) Parmasto   Illus. p. 89
  COMMON NAME: Velvet Blue Spread.
  FRUIT BODY: 1/8–¾" (3–20 mm) wide, thin, crust-like, rounded to irregular, becoming confluent and spreading to form thin patches 1½–6" (4–15.5 cm) or more in diameter; surface dry, velvety, dark blue to blackish blue, paler toward the margin.
  FLESH: soft and membranous when fresh, becoming hard and crust-like when dry; odor and taste not distinctive.
  SPORE PRINT COLOR: white.
  MICROSCOPIC FEATURES: spores 6–10 × 4–5 μm, elliptic, smooth, hyaline.
  FRUITING: usually on the underside of decaying broadleaf logs and branches, especially oak; August–December, sometimes overwintering; widely distributed in the Southeast; occasional.
  EDIBILITY: inedible.
  COMMENTS: Also known as *Corticum caeruleum*.

Stereum complicatum (Fries) Fries   Illus. p. 89
  COMMON NAME: Crowded Parchment.
  FRUIT BODY: 3/8–1¼" (1–3 cm) wide, fan-shaped; upper surface dry, with stiff hairs at the base and fine silky hairs near the margin, typically concentrically zoned and variously colored from orange to orange-yellow, ochraceous gray, or reddish brown, sometimes radially sulcate; fertile lower surface smooth, pale orange to tan, becoming grayish when dry, often exuding a red juice if cut when fresh and moist; margin often crimped or wavy.
  FLESH: thin, leathery, tough, whitish to pinkish buff; odor and taste not distinctive.
  SPORE PRINT COLOR: white.
  MICROSCOPIC FEATURES: spores 5–6.5 × 2–2.5 μm, slightly curved or nearly cylindric, smooth, hyaline.
  FRUITING: in groups or overlapping clusters, often laterally fused to form rows and completely covering twigs and branches of conifers and broadleaf trees, especially oaks; year-round; widely distributed in the Southeast; common.

EDIBILITY: inedible.

COMMENTS: Also known as *Stereum rameale*. *Stereum hirsutum* (not illustrated), inedible, is very similar but has a dense coating of stiff hairs over all of its variously colored but typically creamy buff upper surface, and its lower fertile surface does not exude red juice if cut when fresh and moist. *Stereum striatum* (not illustrated), inedible, has a silvery to pale gray upper surface with tiny, radiating, silky fibers, and it grows on decaying branches of American hornbeam *(Carpinus caroliniana)*. Also compare with *Stereum ostrea* (p. 330), which is typically larger and has a multicolored, zoned upper surface and a reddish brown lower fertile surface.

Stereum ostrea (Blume and Nees : Fries) Fries Illus. p. 89

COMMON NAME: False Turkey-tail.

FRUIT BODY: ⅜–2¾" (1–7 cm) wide, shell- to petal-shaped, thin, leathery, overlapping, sometimes laterally fused; upper surface coated with fine silky hairs, typically concentrically zoned with reddish brown and various other colors, especially gray, yellow, and orange, often whitish at the margin; fertile lower surface smooth, lacking pores (use a hand lens), reddish brown to reddish buff or buff.

FLESH: thin, leathery, tough, whitish to buff; odor and taste not distinctive.

SPORE PRINT COLOR: white.

MICROSCOPIC FEATURES: spores 5–7.5 × 2–3 μm, cylindric, smooth, hyaline.

FRUITING: on decaying broadleaf branches, logs, and stumps; June–December, often persisting through winter; widely distributed in the Southeast; common.

EDIBILITY: inedible.

COMMENTS: The False Turkey-tail is often misidentified as the Turkey-tail, *Trametes versicolor* (p. 271), inedible, a polypore with many tiny pores on its white, fertile undersurface.

Xylobolus frustulatus (Persoon : Fries) Boidin Illus. p. 89

COMMON NAME: Ceramic Parchment.

FRUIT BODY: ¼–1" (6–25.5 mm) wide, a crust-like layer of numerous many-sided plates resembling broken pieces of dull ceramic tile, aggregated into irregular patches, pinkish white to reddish brown when young, becoming tan to grayish tan in age.

FLESH: up to 1 mm thick, very hard, whitish; odor and taste not distinctive.

SPORE PRINT COLOR: white.

MICROSCOPIC FEATURES: spores 3.5–5 × 2.5–3 μm, oval, smooth, hyaline.

FRUITING: in clusters up to 6" (15.5 cm) wide on barkless broadleaf stumps and logs; year-round; widely distributed in the Southeast; fairly common.

EDIBILITY: inedible.

COMMENTS: Previously known as *Stereum frustulosum*. Two-tone Parchment, *Laxitextum bicolor* (not illustrated), inedible, forms a crust-like and spreading patch of spongy pliant overlapping caps with a coffee-colored felty upper surface and a white lower surface.

Hypomyces

Members of this group are fungi that parasitize and disfigure other fungi; such fungi are also known as *hyperparasites*. About forty-two species of *Hypomyces* are known to attack various gilled mushrooms, boletes, and polypores, and another half dozen are known to parasitize certain Ascomycetes, including *Leotia lubrica* and *Humaria hemisphaerica*. Species of *Hypomyces* belong to a group of Ascomycetes commonly called flask fungi. They produce flask-shaped to rounded sexual structures called *perithecia* in which asci produce ascospores. The perithecia are often partially embedded in the host tissue, and their protruding necks are responsible for the sandpaper-like texture that covers them.

Many also have an asexual stage that precedes the sexual stage and that typically resembles a white mold or a yellow powder. The asexual spores produced are called *conidia* and *aleuriospores*. *Hypomyces* species are closely related to other flask fungi, including the Carbon and Cushion Fungi, and to *Cordyceps, Claviceps,* and Allies.

Identification of most species requires microscopic examination. A complete treatment of *Hypomyces* is beyond the scope of this work; therefore, we have included only those species that are commonly encountered or that have conspicuous features.

Hypomyces chrysospermus Tulasne   Illus. p. 90
   COMMON NAME: Bolete Mold.
   FRUIT BODY: the asexual stage appears first, is white and moldy, partially or completely covers its host, and produces conidia; it becomes yellow to golden yellow and powdery as yellow to yellow-brown aleuriospores are produced; the sexual stage consists of globose to flask-shaped, orange-yellow to red-brown perithecia containing asci and ascospores, producing a roughened, sandpaper-like texture.
   MICROSCOPIC FEATURES: conidia 10–30 × 5–12 μm, elliptic, 1-celled, smooth, hyaline; aleuriospores 10–25 μm, globose, thick-walled, prominently warted, yellow to golden yellow or yellow-brown; ascospores 15–30 × 4–6 μm, 2-celled, spindle-shaped, hyaline.
   FRUITING: on various boletes; June–November; widely distributed in the Southeast; common.
   EDIBILITY: inedible; the parasitized boletes should not be eaten.
   COMMENTS: The sexual stage is seldom encountered.

Hypomyces hyalinus (Schweinitz : Fries) Tulasne   Illus. p. 90
   COMMON NAME: Amanita Mold.
   FRUIT BODY: 4–12" (10–30 cm) high, up to 2¾" (7 cm) thick, firm, solid, column- to club-shaped, usually quite phallic, chalky white, yellowish or pinkish to pale orange,

roughened like sandpaper; host cap, gills, and other structures rarely discernible; odor and taste not distinctive.

MICROSCOPIC FEATURES: spores 15–20 × 4.5–6.5 μm, spindle-shaped, 2-celled, prominently warted, hyaline.

FRUITING: on several species of *Amanita,* especially frequent on *Amanita rubescens* (p. 112), edible; June–November; widely distributed in the Southeast; frequent and often abundant.

EDIBILITY: reportedly edible if parasitizing an edible species such as *Amanita rubescens,* but certainly not recommended.

COMMENTS: The host mushroom's identity can only be presumed; reddish-staining specimens are most likely *Amanita rubescens.* No asexual stage is known.

Hypomyces lactifluorum (Schweinitz : Fries) Tulasne   Illus. p. 90

COMMON NAME: Lobster Fungus, Lobster Mushroom.

FRUIT BODY: orange to reddish orange, sometimes with whitish areas, roughened like sandpaper, growing over the surface of funnel-shaped to irregular caps, stalks, and deformed gills of host mushrooms; parasitized caps are typically dense, often partially buried in conifer debris, measure 2–7⅞" (5–20cm) wide, and are white and firm within.

MICROSCOPIC FEATURES: spores 35–40 × 4.5–7 μm, spindle-shaped, 2-celled, prominently warted, hyaline.

FRUITING: solitary, scattered, or in groups on species of *Lactarius* and *Russula*; July–October; widely distributed in the Southeast; infrequent to fairly common.

EDIBILITY: edible with caution; generally considered to be edible because none of the known hosts is poisonous.

COMMENTS: The Lobster Fungus is a popular edible fungus, even though the identity of the host species is usually undetermined. No asexual stage is known.

Hypomyces luteovirens (Fries : Fries) Tulasne   Illus. p. 90

COMMON NAME: Russula Mold.

FRUIT BODY: yellowish green to dark green, roughened like sandpaper, growing on the gills of host species.

MICROSCOPIC FEATURES: spores 28–35 × 4.5–5.5 μm, spindle-shaped, 1-celled, nearly smooth to prominently warted, hyaline.

FRUITING: solitary, scattered, or in groups on various species of *Russula*; July–November; widely distributed in the Southeast; occasional.

EDIBILITY: unknown.

COMMENTS: Because this parasite rarely attacks the cap surface of the host mushroom, it can be detected only by picking and inspecting the underside of specimens of *Russula.* No asexual stage is known.

Cedar-Apple Rust

Cedar-apple Rust is a fungal pathogen of both cedar and apple trees. To be successful, this fungus must spend part of its life cycle in both host plants. The formation on cedar trees of orange, jelly-like horns that later contain mature spores is a diagnostic feature of this fungus.

Gymnosporangium juniperi-virginiana Schweinitz　　　　　　　　　　Illus. p. 90
- COMMON NAME: Cedar-apple Rust.
- FRUIT BODY: a gall measuring 1–4" (2.5–10 cm) in diameter at maturity, at first a small greenish brown swelling on the upper surface of a cedar branches, enlarging rapidly to form an overwintering stage that turns reddish brown to dark brown and forms small circular depressions on the surface, during the spring producing conspicuous orange (immature) to orange-brown (mature) jelly-like horns that arise from the small circular depressions and measure ⅜–¾" (1–2 cm) long.
- MICROSCOPIC FEATURES: teliospores 15–21 × 42–65 μm, 2-celled, rhombic-oval to elliptic, thick-walled, brown, on long stalks.
- FRUITING: solitary or in groups on cedar branches; May–December, sometimes overwintering; widely distributed in the Southeast; occasional to fairly common; conspicuous stage on cedar appears only for a short period in spring.
- EDIBILITY: of no culinary interest.
- COMMENTS: This parasitic fungus divides its life cycle between two hosts, cedar and apple, and causes considerable damage to both plants. Galls eventually die but often remain attached to the tree for a year or more. The jelly-like horns give rise to spores that infect apple leaves.

APPENDIXES

GLOSSARY

RECOMMENDED READING

INDEXES

APPENDIX A

MICROSCOPIC EXAMINATION OF MUSHROOMS

Macroscopic characters, also known as field characters, are features that can be detected without the use of a microscope. Examples include shape, size, color, smell, taste, and spore print. With practice, many mushroom species can easily be identified using only macroscopic characters; the Yellow Morel, Chanterelle, and Sulphur Shelf are three examples.

Accurate identification or confirmation of some mushrooms requires the use of a microscope, however. Black earth tongues, species of *Geoglossum* and *Trichoglossum*, are difficult to identify to genus and nearly impossible to identify to species using only macroscopic characters. With a microscope, however, the identification becomes a relatively simple task. Two mushrooms easily distinguished microscopically are the Purple-gilled Laccaria, *Laccaria ochropurpurea,* and its look-alike, the Sandy Laccaria, *Laccaria trullisata*. The spores of *Laccaria ochropurpurea* are mostly globose and spiny. The spores of *Laccaria trullisata* are mostly ellipsoid and smooth.

If you own a microscope, are thinking of purchasing one, have access to one, or are curious about its use in mycology, consider the following information. Proper microscopic examination of mushrooms requires an instrument in good condition with a substage condenser, an artificial light source (preferably built in) instead of a mirror, and quality lenses. The microscope should be equipped with at least three objective lenses: 10×, 40×, and 100× (oil-immersion), each of which produces a sharp image. An ocular micrometer should be inserted into the ocular lens to facilitate the measurement of spores, cystidia, hyphae, and other structures. Before you can calculate measurements, the ocular micrometer must be calibrated using a stage micrometer, a tiny glass ruler with known increments. This step is important because the magnification of each combination of lenses is different. Most microscope dealers and college and university biology departments have stage micrometers and may assist you with the calibration.

The microscope plays a key role in fungal taxonomy. Numerous microscopic structures are commonly used to differentiate species and genera: spores, cystidia, basidia, clamp connections, hyphae, asci, paraphyses, and others. A thorough discussion of microscopic structures and techniques is beyond the scope of this work. However, for those who have a microscope, we have included spore details in the descriptions. Additional microscopic information has been provided when necessary.

Spores from a spore print or from fresh or dried material may be used when determining shape, size, and other characters. If a spore print is available, scrape off a minute sample using a razor or knife blade, emulsify it in a small drop of water or in a mounting medium such as 3 percent KOH, and carefully place a coverslip over the specimen. Using a pencil eraser, gently apply pressure on the coverslip to distribute the spores and remove excess fluid and air bubbles. Place a drop of immersion oil on the coverslip and carefully lower the 100× objective lens into the oil. Adjust the lighting and examine the spores.

If you use fresh material, remove a small section of gill, place it in the mounting fluid, add the coverslip, gently apply pressure, and examine it as described in the previous paragraph. Pieces of dried material may also be used, but they should be soaked in wetting agents such as 70–95 percent ethyl alcohol for two or three minutes before being placed in the mounting medium. This wetting helps hyphae absorb water and regain their original appearance.

APPENDIX B

CHEMICAL REAGENTS AND MUSHROOM IDENTIFICATION

Chemical reagents are also used as taxonomic tools in the identification and classification of mushrooms. Two macroscopically similar genera or species may be differentiated according to their reaction to an applied chemical reagent. For example, *Lentinus* and *Lentinellus,* two wood-inhabiting genera with serrated gills, are differentiated according to the reaction of their spores in Melzer's reagent: the spores of *Lentinellus* species stain bluish gray to bluish black, a reaction called *amyloid;* the spores of *Lentinus* species do not stain bluish gray to bluish black and are thus *inamyloid.* Two commonly collected boletes, *Boletus auriporus* and *Boletus innixus,* can be very difficult to differentiate. Adding a drop of ordinary household ammonia to the cap surface provides a rapid distinction between the two species: the cap surface of *Boletus auriporus* stains burgundy red to dull blood red or dark vinaceous; *Boletus innixus* immediately produces a green flash, then stains dull orange-red.

The following chemical reagents and reactions are mentioned in the text where pertinent and can be very useful in mushroom identification:

NH_4OH = ammonium hydroxide: 3–10 percent solution
   Household ammonia is perfectly adequate. Ammonium hydroxide is used as a mounting medium for microscopic examination and is ideal for dried material after it is wetted in 70–95 percent ethyl alcohol. It is also used to produce macroscopic color reactions such as the green flash of *Boletus innixus.*

$FeSO_4$ = iron sulfate: 10 percent solution
   Iron sulfate produces macroscopic color reactions when applied to

cap, stalk, and flesh tissue. For example, the cap surface of *Boletus luridiformis* stains dark olive green with this reagent.

Melzer's Reagent

Melzer's reagent is a special solution of iodine, potassium iodide, chloral hydrate, and water. It is used to separate genera and groups of species that produce white or lightly colored spores. It also serves as a valuable test for identifying some species. Spores, hyphae, paraphyses, and other structures produce specific color reactions when mounted in Melzer's reagent. Spores or other cells that stain bluish gray to bluish black are called *amyloid;* those that stain reddish brown are called *dextrinoid;* and those that stain yellow or remain colorless are called *inamyloid*. It is possible to determine whether spores are amyloid, dextrinoid, or inamyloid without using a microscope. First, obtain a spore print on glass or any other nonporous surface. Place a small drop of Melzer's reagent on a glass slide. Using a razor blade or sharp knife, carefully scrape off some spores and transfer them to the glass slide *next* to the reagent. Hold the slide over a piece of white paper in a well-lighted area. Tilt the slide to mix the spores and reagent and observe any color change (*note:* color changes are usually rapid). If a color change is not evident, check again after a few minutes.

KOH = potassium hydroxide: 3–5 percent solution

Potassium hydroxide is used to produce color reactions when applied to cap, stalk, and flesh tissue. For example, the flesh of the Shiny Cinnamon Polypore, *Coltricia cinnamomea*, instantly stains black with the addition of this reagent. Potassium hydroxide is also used as a mounting medium for fresh and dried specimens; it stains various microscopic structures, including certain cystidia and hyphae.

APPENDIX C

CLASSIFICATION

Species included in this book are members of the kingdom Fungi and the division Eumycota or true Fungi. They are classified in three subdivisions: Ascomycotina—the sac fungi, Basidiomycotina—the club fungi, and Zygomycotina—the bread molds and allies. The subdivision Ascomycotina is the second largest group represented in this book and includes cup fungi, morels, carbon fungi, *Cordyceps,* and others. All Ascomycetes are able to reproduce sexually, and many are able to reproduce both sexually and asexually. These fungi produce sexual spores called *ascospores,* which are formed internally in a closed sac-like container called an *ascus,* and asexual spores called *conidia,* which are produced externally. Some species' sexual stage is often conspicuous and commonly collected. The sexual-stage genera are also included. The subdivision Basidiomycotina is the largest group represented in this work and includes gilled mushrooms, puffballs, polypores, and many others. This group reproduces only sexually and produces *basidiospores,* which are formed externally, usually on a club-shaped structure, called a *basidium*. The subdivision Zygomycotina includes mold-like fungi, which often occur on other host fungi. Some are saprobes, and others are parasites. They reproduce both sexually and asexually. They form thick-walled sexual spores called *zygospores* and asexual spores called *sporangiospores*. The sporangiospores are produced internally in tiny, round to oval containers called *sporangia*. These fungi are sometimes collected because they occur on host mushrooms.

Fungal taxonomy is currently undergoing dramatic changes as new discoveries are made, so that extensive information about fungal classification is not included in this work. The following lists contain the Ascomycotina, Basidiomycotina, and Zygomycotina genera included in this book.

GENERA OF ASCOMYCOTINA

Aleuria
Ascocoryne
Bisporella
Bulgaria
Camarops
Chlorociboria
Chorioactis
Chlorosplenium
Claviceps
Cordyceps
Cudonia
Daldinia
Diatrype
Elaphomyces
Galiella

Gyromitra
Helvella
Humaria
Hypomyces
Hypoxylon
Isaria
Leotia
Macroscyphus
Microglossum
Morchella
Nomuraea
Peridoxylon
Peziza
Plicaria

Podostroma
Ptychoverpa
Spadicioides
Spathularia
Spathulariopsis
Sphaerosporella
Trichoglossum
Underwoodia
Urnula
Ustulina
Verpa
Wolfina
Xylaria
Xylocoremium

GENERA OF BASIDIOMYCOTINA

Agaricus
Agrocybe

Albatrellus
Aleurodiscus

Amanita
Anellaria

Anthracophyllum
Antrodia
Arachnion
Armillaria
Arrhytidia
Aseroe
Asterophora
Astraeus
Aurantioporus
Aureoboletus
Auricularia
Austroboletus
Baeospora
Bjerkandera
Blumenavia
Bolbitus
Boletellus
Boletinellus
Boletus
Bondarzewia
Bovista
Bovistella
Callistosporium
Calocera
Calostoma
Calvatia
Camarophyllus
Cantharellus
Catathelasma
Cerrena
Chalciporus
Chlorophyllum
Christiansenia
Chroogomphus
Chrysomphalina
Clathrus
Claudopus
Clavaria
Clavariadelphus
Clavicorona
Clavulina
Clavulinopsis

Clitocybe
Clitocybula
Clitopilus
Collybia
Coltricia
Conocybe
Copelandia
Coprinus
Corticum
Cortinarius
Cotilydia
Craterellus
Crepidotus
Crinipellis
Crucibulum
Cyathus
Cyptotrama
Cystoderma
Dacrymyces
Dacryopinax
Daedalea
Daedaleopsis
Dentinum
Dictyophora
Disciseda
Endoptychum
Entoloma
Exidia
Favolus
Fistulina
Flammulina
Fomes
Fomitopsis
Galerina
Ganoderma
Geastrum
Gerronema
Gloeophyllum
Gloeoporus
Gomphus
Grifola
Gymnopilus

Gymnopus
Gymnosporangium
Gyrodon
Gyroporus
Hapalopilus
Hebeloma
Hericium
Heterobasidion
Hexagonia
Hohenbuehelia
Humidicutis
Hydnellum
Hydnocharte
Hydnopolyporus
Hydnum
Hygrocybe
Hygrophorus
Hymenochaete
Hypholoma
Hypsizygus
Inocybe
Inonotus
Irpex
Ischnoderma
Laccaria
Lachnocladium
Lactarius
Laetiporus
Laternea
Laxitextum
Leccinum
Lentinellus
Lentinula
Lentinus
Lenzites
Lepiota
Leucoagaricus
Leucocoprinus
Leucopaxillus
Limacella
Linderia
Lycoperdon

Lysurus
Macowanites
Macrocybe
Macrolepiota
Marasmiellus
Marasmius
Melanoleuca
Meripilus
Merulius
Microporellus
Mucilopilus
Mutinus
Mycena
Mycorrhaphium
Naematoloma
Nigrofomes
Nigroporus
Nolanea
Octaviania
Omphalina
Omphalotus
Panaeolus
Panellus
Panus
Paxillus
Phaeolus
Phallogaster
Phallus
Phellinus
Phellodon

Phlebia
Pholiota
Phylloporus
Phyllotopsis
Pisolithus
Pleurotus
Plicaturopsis
Pluteus
Polyporus
Poria
Porphyrellus
Pouzarella
Psathyrella
Pseudocoprinus
Pseudofavolus
Pseudofistulina
Psilocybe
Pulcherricium
Pulveroboletus
Pycnoporus
Ramaria
Rhizopogon
Rhodocollybia
Rhodocybe
Rhopalogaster
Ripartitella
Russula
Schizophyllum
Schizopora

Scleroderma
Sebacina
Simblum
Sparassis
Spongipellis
Squamanita
Stereum
Strobilomyces
Strobilurus
Stropharia
Suillellus
Suillus
Syzygospora
Thelephora
Trametes
Tremella
Tremellodendropsis
Trichaptum
Tricholoma
Tricholomopsis
Trogia
Tylopilus
Vascellum
Volvariella
Xanthoconium
Xerocomus
Xeromphalina
Xylobolus
Zelleromyces

GENERA OF ZYGOMYCOTINA

Spinellus

APPENDIX D

MYCOPHAGY

COLLECTING MUSHROOMS FOR THE TABLE

The same general collecting tips discussed in the introduction apply to collecting mushrooms for the table. If you are unsure of the identification of your specimens, be certain to collect them carefully in order to preserve all diagnostic structures, and keep collections wrapped separately to avoid mixing species. If you are confident of the identification of the mushrooms you are collecting for eating, then you may choose to "field dress" them: cut them close to the ground or other substrate upon which they are growing; cut away dried bits or parts that may house unwanted slugs, insects, or larvae; and brush off dirt, leaves, pine needles, and other debris.

As with any food item, select only mushrooms that are fresh. Be alert for obvious signs of decay: foul odors, spongy stalks, or discoloration. Use common sense to determine the ratio of insect-infested to edible mushroom tissue (the "bug-to-bolete" ratio). Leaving a few "very mature" specimens in the field to help ensure continued spore dispersal is always a good idea. Most important, collect gently; disturb the substrate and the habitat as little as possible. In this way, you minimize damage to the mycelium, help ensure continued fungal growth, and preserve the environment for others.

Finally, don't use mushrooms from contaminated habitats (chemically treated lawns, landfills, crop fields, and toxic waste sites, or along major roadways and near railroad tracks) for the table. Mushrooms may concentrate toxic substances.

A multitude of gustatory delights are out there for the picking for those who exercise a bit of common sense and caution.

GUIDELINES FOR EATING WILD MUSHROOMS

The first and most vital rule regarding the consumption of wild mushrooms is: if in doubt, throw it out! *Never eat any mushroom unless you are certain of its identification as an edible species!* It is not worth the risk of illness or death.

Perhaps no where else is superstition and myth as common as it is in regard to eating wild mushrooms. "Old wives' tales" abound; every family has its "secret"; and each ethnic group seems to have its ways of telling the "good ones" from the bad. However, none of the myths or secrets or formulas is reliable. They might even be dangerous. The only valid "rule of thumb" for safely separating edible and poisonous species is that there isn't one. If you are interested in gathering wild mushrooms to eat, you must learn them species by species.

Once you are certain that the mushrooms you have collected are edible, be aware of some simple considerations. Like with many other foods, certain popular edible mushrooms may cause adverse or allergic responses in some individuals—not much different from somebody being sensitive to wheat or peanut butter, for instance. Whenever eating a mushroom species for the first time, eat a small portion, sautéed lightly in butter or margarine. Store the rest of your collection in the refrigerator for at least forty-eight hours before eating more of it. This practice serves two purposes. First, you will discover the taste and texture of that mushroom (so you can use it more creatively in your cooking). Second, you will discover whether or not you are sensitive to that particular mushroom. Most adverse reactions consist of mild gastrointestinal upset requiring no medical attention. In the rare instance of a serious reaction indicating possible misidentification of the species, you then have fresh specimens on hand to take to the hospital because treatment is conditional on the type of toxins. Such a reaction should not be a concern, however, if proper care is taken. It is also comforting to know that if you have an adverse reaction to mushrooms in a particular genus, it does not mean you are sensitive to others. A good friend of ours is allergic to the genus *Agaricus,* including the common grocery store white button mushroom. However, he enjoys eating several other wild genera with impunity and great gusto!

Gastrointestinal upsets related to mushroom consumption are occasionally owing to sheer overindulgence. Some individuals have adverse reactions if they consume alcohol between forty-eight hours before and forty-eight hours after eating certain mushrooms (the Alcohol Inky and Black Morel, for example). Eating infested or spoiled mushrooms can cause unpleasant reactions. Always thoroughly cook your mushrooms. Many contain certain proteins that may cause adverse reactions if not denatured by heat. In addition, some edible mushrooms are covered with sticky, slimy gluten. This glutenous layer causes gastric distress for some individuals. To be on the safe side, remove the glutenous layer prior to cooking.

CLEANING, COOKING, AND PRESERVING THE HARVEST

Once you are back home with the rewards of a day spent collecting, it is best to remove clumps of dirt and debris from the mushrooms before storing them. Wrap the mushrooms in waxed paper or in brown paper bags and store them in the refrigerator. A final cleaning should be done just prior to cooking or preserving. Never store mushrooms in plastic bags or plastic wrap, both of which trap moisture and hasten spoilage.

Mushrooms retain water, so use as little of it as needed when cleaning them. A sharp knife and a soft brush or damp cloth or paper towel are all that is needed. Many culinary shops sell special mushroom brushes for this purpose. If water must be used, allow the mushrooms to drain on paper towels for about ten minutes before cooking. Because a great deal of a mushroom's flavor is located in the outer skin of the mushroom cap, peeling mushrooms is generally not recommended. Exceptions include the Slippery Jack boletes.

Most mushrooms, if fresh, will keep for up to a week in the refrigerator. There are exceptions, such as the Inky Caps, which become a black inky liquid within hours, even while refrigerated. You can also preserve mushrooms by drying, freezing, salting, pickling, and canning. Many fine mushroom cookbooks give specific information on preserving various mushroom species. Some are listed in the "Recommended Reading" section.

Mushrooms lend themselves to a variety of cooking styles. They can be sautéed, baked, grilled, and used in soups, stews, and casseroles. As you become familiar with each mushroom's culinary characteristics, experiment with what you like best. Again, referring to a reputable cookbook is a fine way to begin.

MUSHROOM RECIPES

We have included the following recipes as examples of some of the ways mushrooms may be prepared.

Baked Brie Pastries with Boletes

8 ounces Brie cheese, room temperature, cut into cubes with rind
½ teaspoon dried rosemary, crumbled
pinch cayenne pepper
1 egg, lightly beaten
1 cup coarsely chopped boletes
1 tablespoon olive oil
15 phyllo pastry sheets
1 cup (2 sticks) unsalted butter, melted

Sauté boletes in olive oil over moderately high heat until golden brown.

Using food processor, blend cheese until smooth. Add rosemary, cayenne, egg, and sautéed boletes. Blend until smooth.

Butter large baking sheets. Place 1 phyllo sheet on work surface (keep remainder covered with slightly damp towel). Brush phyllo lightly with melted butter. Top with second phyllo sheet. Brush lightly with butter. Repeat with third phyllo sheet. Cut stacked, buttered phyllo lengthwise into 3½-inch-wide strips. Then cut crosswise into 3½-inch-wide squares. Place 1 teaspoon cheese filling in center of each square, gather corners together over center, and crimp firmly. Transfer to prepared sheets, spacing 1 inch apart. Brush tops lightly with butter.

Repeat process using phyllo sheets, butter, and filling. Refrigerate at least 1 hour (can be prepared 1 day ahead).

Preheat oven to 350° F. Bake until crisp and golden brown, about 20–25 minutes. Cool 5 minutes. Serve warm.

Makes about sixty appetizers.

Chanterelles with Pasta and Greens

    3 cloves garlic, minced
    ¼ cup olive oil
    3 tablespoons butter
    1 pound chanterelles, cleaned and thickly sliced
    ½ teaspoon dried rosemary
    ½ pound sweet Italian sausage links, cut into 1-inch pieces
    1 pound penne rigate or pasta of your choice
    6 ounces fresh spinach (or greens of your choice), cleaned and coarsely chopped
    ¼ cup freshly grated Pecorino Romano cheese, plus more to accompany the meal
    freshly ground black pepper

In a deep skillet, melt butter with olive oil. Add garlic and sauté over medium-high heat until golden. Add chanterelles and continue cooking, stirring occasionally, until tender and most of the liquid has evaporated, about 10 minutes. Add rosemary. Remove from heat.

While cooking garlic and mushrooms, fry sausage pieces in a separate pan until well done. Remove from pan, drain, and add to mushroom/garlic mixture.

Bring a large pot of salted water to a boil, and cook the penne rigate until al dente. Drain well.

Add pasta and spinach to mushroom/sausage mixture. Add ¼ cup grated cheese and mix well. Add up to 1 tablespoon additional olive oil if needed. Add black pepper to taste.

Serve with additional black pepper and grated cheese. Goes nicely with fresh tomato-basil salad and a full bodied red wine.

Serves four.

Wild Mushrooms and Asparagus Frittata

    4 cups sliced wild mushrooms (morels, honey mushrooms, boletes)
    2 tablespoons butter
    1 small onion, finely chopped
    2 large yellow bell peppers, cut into ¼-inch strips
    3 pounds fresh asparagus, trimmed and cut into ¼-inch pieces
    10 large eggs
    ½ cup heavy cream
    1 tablespoon dried parsley
    1 cup shredded Colby or Monterrey Jack cheese
    1½ teaspoons salt
    ¼ teaspoon freshly ground black pepper

Butter a 13-by-9-by-2-inch baking dish. Preheat oven to 350° F.

Melt 1 tablespoon butter in a large skillet and cook bell peppers and chopped onion over medium-low heat until tender, about 10 minutes.

In a separate skillet, melt remaining 1 tablespoon butter, and sauté mushrooms over medium heat, stirring frequently, until tender and all liquid has evaporated.

Meanwhile, blanch asparagus pieces in boiling water for 1 minute. Drain in a colander, rinsing under cold water.

Beat eggs in a large bowl. Mix in cream, cheese, parsley, salt, and pepper.
Add the asparagus, bell pepper mixture, and the mushrooms.

Pour into prepared baking pan. Bake for 30–35 minutes in the middle of the oven until golden and set. Cool.

Serves ten to twelve if accompanying a meal or six to eight as a main course.

New Red Potatoes and Honey Mushrooms

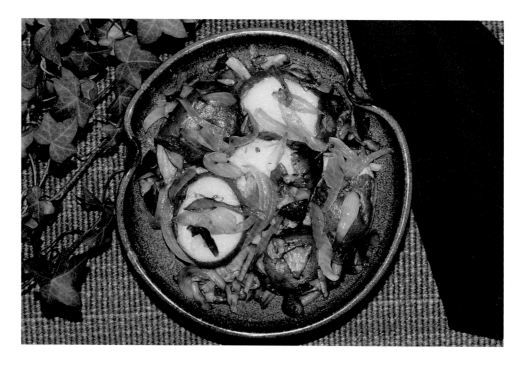

1½ lbs. (about 12 medium-small) new red potatoes
4 strips bacon
1 large onion, diced
2–4 cups sliced honey mushrooms
salt and freshly ground black pepper

Wash potatoes and quarter or half the larger ones, so they all are of a uniform size. Boil them in salted water until fork tender.

While potatoes are cooking, fry bacon in skillet until crisp, and drain bacon strips on paper towels.

Sauté onion in bacon fat until lightly browned. Add mushrooms and sauté until all mushroom liquid is expressed and evaporated.

Drain potatoes in colander. Return to pot. Toss gently with onion-mushroom mixture. Crumble bacon and add to potatoes.

Season to taste with salt and pepper.

Serves four.

Curried Chicken Mushroom

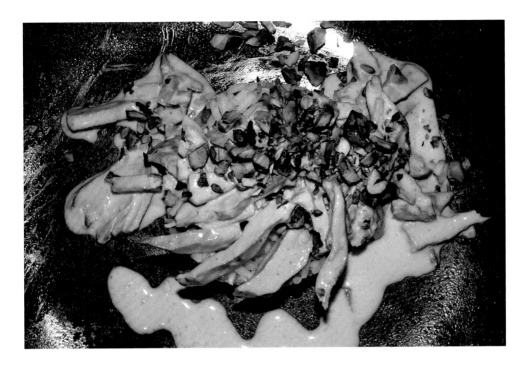

    3 tablespoons butter
    3 tablespoons extra-virgin olive oil
    3–4 shallots, minced
    3 cups very young Chicken Mushroom sliced into strips 2–3 inches long and ¼ inch wide
    ¼ teaspoon curry powder
    1 cup frozen baby green peas (optional)
    ¼ cup dry white wine
    1 cup (8 ounces) heavy cream
    salt
    roasted pistachios for garnish

In a large skillet melt butter with olive oil over medium-high heat.

Add shallots and sauté until golden brown.

Add mushroom strips and continue cooking approximately 5 minutes, stirring frequently.

Add curry powder and mix well.

Add peas, if desired, and wine. When peas are heated through, or, if not using peas, when mixture is bubbling, add the cream. Simmer until thickened.

Salt to taste.

Serve over polenta, rice pilaf, or pasta of your choice, with pistachios sprinkled on top.

Serves four.

Gnocchi Carbonara with Wild Mushrooms and Ham

>1 package Knorr Carbonara sauce mix
>1 cup diced cooked ham
>2 cups sliced mushrooms of your choice
>2 tablespoons olive oil
>1 pound gnocchi or pasta of your choice
>4 scallions, green tops only, minced
>garlic salt
>freshly ground black pepper

Prepare carbonara sauce according to package directions.

Add ham; cover and keep warm over low heat.

Sauté mushrooms in olive oil over moderately high heat until all liquid has evaporated and mushrooms are golden brown. Season to taste with garlic salt.

While mushrooms are cooking, bring a large pot of water to a boil and cook gnocchi until al dente (follow timing on package). Drain well.

Serve gnocchi with carbonara and ham sauce, spooning mushrooms on top and garnish with minced scallions.

Add freshly ground black pepper to taste.

Serves four as a main course.

Shrimp and Shrooms with Creamy Pesto Sauce

    1 package Knorr Creamy Pesto pasta sauce mix
    2 medium-large King Boletes
    2 tablespoons olive oil
    1 dozen jumbo shrimp shelled, deveined, and cooked, approx. 20
    garlic salt
    fresh snow pea pods for garnish

Prepare Creamy Pesto sauce according to package directions. Keep warm.

Clean and trim boletes. Slice lengthwise into ½-inch slices. Heat olive oil in large skillet. Sauté mushrooms over medium-high heat until golden brown on both sides and all liquid has evaporated. Season to taste with garlic salt. Keep warm.

Blanch pea pods for 1 minute in boiling water. Arrange on four plates with shrimp and mushrooms. Top with creamy Pesto sauce.

Serves four as an appetizer or side salad.

Glossary

acrid: producing a burning sensation when tasted
agglutinated-fibrillose: having fibers that are glued together
aleuriospore: an asexual spore supported by a tiny stalk and not enclosed in a container; also known as a *conidium*
allantoid: sausage-shaped
ampullaceous: flask or bottle-shaped
amygdaliform: almond-shaped
amyloid: staining grayish to blue-black in Melzer's reagent
anastomosing: fusing to form a network
angular-lobate: divided into angular lobes
annular zone: a poorly defined ring
annual: completing growth in one season
apex (pl. **apices**): the uppermost portion of the stalk, or the portion of the spore closest to the point of attachment to the basidium
apical: pertaining to the apex
apical pore: a small opening or thin area in the wall at the apex of a spore; also known as a *germ pore*
apiculus: a short projection at or near the apex of a spore
appendiculate: hung with fragments of the partial veil
appressed: flattened onto the surface
arachiform: peanut-shaped
arcuate: curved like a bow
areolate: marked out into small areas by cracks or crevices
areolate-scaly: scales marked out into small areas by cracks or crevices
ascocarp: the fruiting body of an Ascomycete
Ascomycetes: a major group of fungi that includes those species that produce ascospores in asci
ascospore: a sexual spore formed within an ascus
ascus (pl. **asci**): a sac-like cell in which ascospores are formed

asexual: not resulting from fertilization
attached: joined to the stalk
attenuate: gradually narrowed
avellaneous: pale gray with a pinkish tint
azonate: lacking zones
basal: located at the base
base: the lowest portion of the stalk
Basidiomycetes: a major group of fungi that includes those species that produce spores borne on a basidium
basidiospore: a spore formed on a basidium
basidium (pl. **basidia**): typically a club-shaped cell on which basidiospores are formed
biennial: occurring or maturing every two years
bloom: a dull, thin coating that is typically whitish
boletinoid: having radially arranged and elongated pores
broadleaf: referring to any non-cone-bearing deciduous tree or shrub, such as maple or birch
buff: dull white to very pale yellow
bulb: a swollen portion, especially at the base of a stalk
bulbous: having a bulb-like base
button: the young developing stage of a mushroom
caespitose: occurring in groups
campanulate: bell-shaped
canescence: a whitish to grayish dust-like bloom
cap: the upper part of a mushroom, which supports gills, tubes, spines, or a smooth surface on its underside
capillitium: sterile, thread-like filaments mixed with spores
carbonaceous: hard and blackened
cartilaginous: fibrous tough, often splitting lengthwise in strands
caulocystidia: cystidia found on the stalk

central: attached to the middle of the cap
cheilocystidia: cystidia that occur on the edge of a gill; compare with *pleurocystidia*
chlamydospores: thick-walled asexual spores that are often nearly round
clamp connection: a semicircular bridge-like structure that connects two adjoining cells in some Basidiomycetes
clavate: club-shaped
collarium: a collar or ring on the stalk apex into which the inner edges of the gills are inserted; the gills do not touch the stalk
columella: a persistent sterile column within a spore case
compressed: flattened longitudinally
concolorous: having the same colors
concrescent: growing together
confluent: becoming continuous together, merging
conic: shaped more or less like an inverted cone
conical: pertaining to an inverted cone
conidium (pl. **conidia**): an asexual spore supported by a tiny stalk and not enclosed in a container; also known as an *aleuriospore*
conifer: a cone-bearing tree; one that has needle-like leaves, such as spruce, fir, hemlock, or pine
context: inner tissue of a fruit body
convex: curved or rounded like the exterior of a circle
coprine: an amino acid found in some mushrooms, which when consumed before, with, or after alcohol may cause nausea, vomiting, a flushed feeling, rapid breathing, and other signs and symptoms of poisoning
corrugated: coarsely wrinkled or ridged
cortina: a spiderweb-like partial veil
cortinate: appearing spiderweb-like
crenate: finely scalloped
crenulate: very finely scalloped
crested: having a showy tuft or projection
crossveined: having tiny veins that connect adjoining gills or vein-like ridges
cuticle: the outermost tissue layer of the cap; more correctly called a *pileipellis*
cylindric: pertaining to a cylinder
cystidia: sterile cells that project between and usually beyond the basidia or on the cap or stalk surface
daedaleoid: resembling a maze
decurrent: descending or running down the stalk; a form of gill attachment
decurved: bent downward
delimited: marked or bordered by
deliquesce: liquify, as in the gills of the genus *Coprinus*
deliquescent: having the ability to deliquesce
deliquescing: dissolving into a fluid
depressed: sunken
dermatocystidia: sterile cells located on the outer surface of the stalk or cap
dextrinoid: staining orange to orange-brown or pinkish red to dark red or reddish brown in Melzer's reagent
disc: the central area of the surface of a mushroom cap
distant: spaced widely apart
doglegged: having an abrupt angle or sharp bend
duff: decaying plant matter on the ground in a forest
eccentric: away from the center, off-center
echinulate: having spines
effused-reflexed: spread out along the substrate then projecting outward at the margin to form a cap
elevated: raised upward above the plane
ellipsoid: resembling an elongated oval with similarly curved ends
elliptic: pertaining to an elongated oval with similarly curved ends
elliptic-cylindric: more or less cylindric with similarly curved ends
elongated-ellipsoid: a somewhat stretched-out oval
endoperidium: innermost layer of a spore case
entire: even; not broken, serrated, or lacerated
equal: having the same thickness over the entire length
eroded: partially worn away and appearing ragged
evanescent: slightly developed and soon disappearing
exoperidium: outermost layer of a spore case
expanded: enlarged and elongated
faces: the sides of a gill
farinaceous: having an odor of fresh meal or somewhat resembling watermelon
ferruginous: rust-colored
fertile surface: the spore-bearing surface
FeSO$_4$: iron sulfate, usually a 10 percent solution
fetid: having an offensive odor
fiber: a hair-like structure present on the cap or stalk of some mushrooms

fibril: a tiny fiber
fibrillose: composed of fibrils
fibrillose-cottony: composed of fibrils resembling cotton balls
fibrillose-punctate: marked with tiny points, dots, scales, or spots that are composed of fibrils
fibrillose-scaly: having fibrils that appear scale-like
fibrous: composed of fibers
filamentous: composed of filaments
filiform: thread-like
fimbriate: minutely fringed
flesh: the inner tissue of a fruiting body
floccose: tufted like cotton balls
floccose-fibrillose: composed of tufted fibers
floccose-lacerate: appearing torn into tufted cotton-like balls
floccose-membranous: having a tufted cottony membrane
floccose-squamose: having tufted fibers that form tiny scales
flocculence: minute floccose decorations
forked: divided into two or more branches
free: not attached to the stalk
fruit body: the reproductive stage of a fungus, commonly called a mushroom
fulvous: reddish cinnamon; colored like a red fox
furfuraceous: covered with scurfy particles
furrowed: marked by grooves
fuscous: dark brownish gray to brownish black
fusiform: spindle-shaped and narrowing at both ends
fusiform-ellipsoid: elliptical but somewhat spindle-shaped
fusiform-elliptic: primarily elliptic but somewhat spindle-shaped
fusoid: somewhat spindle-shaped
Gasteromycetes: a group of Basidiomycetes, such as puffballs, which produce spores in closed chambers within the fruiting body
genus (pl. **genera**): taxonomic grouping of closely related species
germ pore: a thin portion of the spore wall through which the hypha passes during germination; also known as an *apical pore*
gills: thin to thick, knifeblade-like, spore-bearing structures on the cap undersurface of some mushrooms
glabrous: bald and smooth
glancing: appearing shiny when viewed at a tilted angle

glandular dots: sticky spots on the surface of the stem
gleba: spore-bearing tissue of stinkhorns, puffballs, and other Gasteromycetes
globose: round
gluten: a sticky, glue-like, pectinous material
glutinous: having gluten
granular: resembling tiny grains
granules: tiny grains
granulose: composed of tiny grains
gregarious: closely scattered over a small area
guttules: oil-like drops
H_2SO_4: hydrochloric acid; used in concentrated form to determine colorfastness of the spores of *Panaeolus* species
habit: the manner of growth, such as solitary or in fused groups
habitat: the substrate from which the mushroom grows, such as among sphagnum mosses, on wood, or on soil
hemispheric: shaped like half of a sphere
hilar appendix: a small projection on a spore where it was attached to supporting tissue
hoary: covered with very fine, silky down or having a whitish to grayish sheen
homogeneous: composed of uniform cells or tissue
humus: decaying organic plant material in soil
hyaline: transparent; clear and nearly colorless
hydrazine: a colorless, corrosive liquid released from some mushrooms during cooking; used in rocket fuel
hygrophanous: appearing water-soaked when fresh, fading to a paler color as water is lost
hygroscopic: readily absorbing water
hymenial cystidia: sterile cells arising on gills, spines, or pores
hyperparasites: fungal parasites that attack other fungi
hypha (pl. **hyphae**): thread-like filaments of fungal cells
hypogeous: occurring below the ground surface
inamyloid: unchanging or staining pale yellow in Melzer's reagent; neither amyloid nor dextrinoid
incurved: bent inward toward the stalk
inflorescence: the flowering portion of a stem
infundibuliform: deeply funnel-shaped
inrolled: bent inward toward the stalk and upward
intervenose: having veins on the gill faces that often extend between the gills or from gill to gill

KOH: potassium hydroxide, usually made up in a 3–5 percent concentration in water; used to test color reactions
labyrinthine: resembling a maze
lacerated: appearing torn
lacerate-scaly: appearing torn into scales
lacrymoid: shaped like a teardrop
lamella (pl. **lamellae**): a gill on the underside of a mushroom cap
lamellate: having gills
lamellulae: short gills that do not reach the stalk
lanceolate: of much greater length than width and tapering
lateral: attached to the margin of a cap
latex: a watery or milk-like fluid that exudes from some mushrooms when they are cut or bruised
lobe: subdivision of the whole
longitudinal: oriented along the vertical axis of the stalk
lorchel: a common name for *Gyromitra esculenta* and other false morels
lubricous: smooth and slippery
margin: the edge of a mushroom cap
marginate: having gill edges that are darker-colored than the faces; having a circular ridge on a bulb
marginate-depressed: provided with a narrow, circular, horizontal platform on the upper side
matted-fibrillose: having fibers that are appressed to the surface median: reference to a ring located midway down on a stalk
Melzer's reagent: a solution containing iodine used for testing fungal spores and tissues for color reactions
membranous: having a membrane
mesoperidium: middle layer of a spore case
mucronate: having a short, sharp point
multiseptate: having several to many cross-walls
mycelium: a mass of hyphae, typically hidden in a substrate
mycophagist: a person who eats mushrooms
mycorrhiza: a mutually beneficial association between a fungus and a tree or other plant
naked: bald, smooth
napiform: turnip-shaped, bulbous above and tapering below
NH$_4$OH: ammonium hydroxide; used to test color reactions on mushroom tissues
nitrous: a reference to nitrogen-containing compounds, which are pungent and unpleasant
oblanceolate: reversely lanceolate
oblong: longer than wide and with somewhat flattened ends
obovate: ovate, with the broader end opposite to the point of attachment
obovoid: ovoid, with the broader end opposite to the point of attachment
obtuse: rounded or blunt
ochraceous: pale brownish orange-yellow
ochre: brownish orange-yellow
ostiole: a small pore-like opening
ovate: shaped like an egg
ovoid: somewhat egg-shaped
palisade: a perpendicular arrangement of cells in close proximity to each other
papilla: a small nipple-shaped elevation
paraphysis (pl. **paraphyses**): a distinctive sterile cell in the spore-producing layer of Ascomycetes that keeps the asci erect
parasite: an organism that obtains its nutrients from a living host
partial veil: a layer of fungal tissue that covers the gills or pores of some immature mushrooms
patches: flattened remnants of the universal veil
pedicel: a slender stalk
pendant: hanging or draping
perennial: continuing growth from year to year
perforate: pierced through
peridiole: a tiny, egg-like structure that contains spores
peridium (pl. **peridia**): a layer of the spore case of puffballs and other Gasteromycetes
perithecium (pl. **perithecia**): a minute, flask-shaped structure containing asci
pileipellis: the outermost layers of the outer surface of the cap, also known as a *cuticle*
pileocystidia: cystidia located on the surface of the cap
pileotrama: supporting tissue found in the cap
plane: flat
pleurocystidia: cystidia that occur on the gill faces; compare with *cheilocystidia*
plicate: deeply grooved, sometimes pleated or folded
plicate-striate: having deeply grooved, nearly parallel lines
polar: located at the most distant part

pores: the open ends of the tubes of a bolete or polypore
pore surface: the undersurface of the cap of a bolete or polypore, where the open ends of the tubes are visible
poroid: resembling pores or composed of pores
pruina: powdery particles, flakes, or dots
pruinose: appearing finely powdered
pseudoreticulate: having a false reticulum that is usually formed by anastomosing ridges; not formed from the attachment of the tubes to the stalk during early stages of development
pubescent: having short, soft, downy hairs
pulverulent: powdery
pulvinate: cushion-shaped, strongly convex
punctae: tiny points, dots, scales, or spots
punctate: marked with tiny points, dots, scales, or spots
pyramidal: pyramid-shaped
pyriform: pear-shaped
radial: pointed away from a common central point, like the spokes of a wheel
radiating: spreading away from a common central point
radicating: forming a root-like extension in the ground
recurved: curved backward or downward
resupinate: having the fruit body attached to the substrate and facing outward
reticulate: covered with a net-like pattern of ridges
reticulation: raised, net-like ridges
reticulum: a system of raised, net-like ridges found on the stalk surface or spores of some mushrooms
revolute: rolled up toward the disc, then toward the margin; opposite of *inrolled*
rhizomorph: a group of thick, cord-like strands of hyphae growing together as a single organized unit
rimose: having distinct cracks or crevices
rimose-areolate: cracked and forming tiny patches
ring: remnants of a partial veil that remains attached to the stalk after the veil ruptures
saccate: sheath-like or cup-shaped
saprobe: an organism that lives off dead or decaying matter
scabers: small, stiff, granular points on the surface of the stalks of some mushrooms; a distinctive feature of the bolete genus *Leccinum*

scabrous: with small, short, rigid projections
scale: an erect, flattened, or recurved projection or torn portion of the cap or stalk surface
scaly-punctate: marked with tiny dots or points composed of scales
sclerotium (pl. **sclerotia**): a small, rounded to irregular body composed of dormant hyphae and often soil components
scrobiculate: pitted or smeared; having flat, shiny areas
scurfy: roughened by tiny flakes or scales
scurfy-fibrous: tiny fibers arranged as flakes or scales
seceding: attached at first and later separating
septate: divided by crosswalls
serrate: jagged or toothed like a saw blade
sessile: lacking a stalk
setae: sharply pointed sterile cells that are usually brown or yellow and project on the surface of the stalk or other portion of the fruiting body of some mushrooms
sexual: pertaining to fertilization involving two compatible cells
sinuate: gradually narrowed and becoming concave near the stalk
sinuous-daedaloid: elongated and wavy
sordid: dingy or dull
sphaeropedunculate: having a large round tip on a short stalk
spines: tapered, typically downward-pointing projections on a mushroom cap's undersurface; also ornamentation of spores or outer walls of certain puffballs
sporangiospore: asexual spore formed in a sporangium
sporangium: a sac-like microscopic structure in which asexual spores are produced
spore: a microscopic reproductive cell with the ability to germinate and form hyphae
spore case: a structure containing the spore mass in species of Gasteromycetes
spore mass: a dense layer of spores
spore print: a deposit of spores on a piece of paper or glass from a mushroom's gills, tubes, or other spore-producing structures
squamules: small scales
squamulose: composed of small scales
stalk: the structure that arises from the substrate and supports the cap or spore case of a mushroom
sterile surface: a portion that lacks reproductive structures

sterile tissue: tissue not directly involved with the reproductive process
stipitipellis: the outer layers of the stalk
striate: having small and more or less parallel lines or furrows
striatulate: having very fine lines or furrows
strigose: coated with long, coarse, stiff hairs
stuffed: containing a soft tissue that usually disappears in age, leaving a hollow space
subapical: slightly below the apex
subattenuate: abruptly narrowed, but not truncate
subdecurrent: extending slightly down the stalk
subdistant: gill spacing halfway between close and distant
subellipsoid: nearly ellipsoid
subelliptic: nearly elliptic
subfusiform: nearly spindle-shaped
subfusoid: somewhat spindle-shaped; tapered slightly at both ends
subglabrous: nearly bald and smooth
subglobose: nearly round
subhyaline: nearly transparent
subhygrophanous: appearing water-soaked when fresh, fading to a slighty paler color when dry
sublamellate: nearly gill-like
submarginate: having a somewhat distinctly marked border on the bulb portion of the stalk
submembranous: somewhat resembling a membrane
subnavicular: somewhat boat-shaped when a boat is viewed from above
suboblong: nearly oblong
subovoid: slightly egg-shaped
substrate: organic matter that serves as a food source for a fungal mycelium
subvelutinous: somewhat velvety
subventricose: slightly swollen in the middle and somewhat tapered in both directions
subviscid: slightly sticky or tacky
sulcate: grooved; deeper than striate, less than plicate
superior ring: a ring located on the upper stalk surface
tawny: dull yellowish brown
teeth: spines that point downward
teliospore: thick-walled spores formed at the terminal stage of the life cycle of rusts and smuts
terete: rounded like a broom handle
terrestrial: growing on the ground

tiers: nearly parallel rows
tomentose: coated with a thick, matted covering of hairs
trama: supporting tissue of the fruit body
tramal: pertaining to the trama
translucent-striate: appearing to have lines or furrows when gill edges are viewed through a moist, nearly transparent cap tissue
trilobate: having three lobes
truncate: appearing cut off at the end
tuber: a fleshy underground structure that gives rise to reproductive fruit bodies
tuberculate: having small warts or bumps
tuberculate-striate: having lines or furrows that are roughened by small warts or bumps
tubes: narrow, parallel, spore-producing cylinders on the undersurface of the cap of a bolete or polypore
turbinate: shaped like a top
twisted-striate: having parallel lines that spiral
umbilicate: having a central funnel-shaped depression
umbo: a pointed or rounded elevation at the center of a mushroom cap
umbonate: having an umbo
uniseriate: arranged in a single row
universal veil: a layer of fungal tissue that completely encloses immature stages of some mushrooms
uplifted: elevated toward the plane
veil: a layer of fungal tissue that covers all or part of some immature mushrooms (see *universal veil* and *partial veil*)
ventricose: swollen in the middle and tapering to somewhat of a point
verruculose: minutely warted
villose: coated with long, tiny, soft hairs
vinaceous: wine-colored; pinkish red, pale purplish red, brown, cinnamon, etc.
virgate: streaked with fibrils
viscid: sticky or tacky
volva: a typically cup-like sac that remains around the base of a mushroom stalk when the universal veil ruptures
warts: small patches of tissue that remain on the top of a mushroom cap when the universal veil ruptures
zones: concentric bands of different colors on the surface of the cap or stalk of some mushrooms

Recommended Reading

TECHNICAL PUBLICATIONS

Adaskaveg, J. E., and R. L. Gilberton. 1988. Basidiospores, Pilocystidia, and Other Basidiocarp Characters in Several Species of the *Ganoderma lucidum* Complex. *Mycologia* 80: 433–507.

Banik, M. T., H. H. Burdsall Jr., and T. J. Volk. 1998. Identification of Groups within *Laetiporus sulphureus* in the United States Based on RFLP Analysis of the Nuclear Ribosomal DNA. *Folia Cryptog. Estonica. Fasc.* 33: 9–14.

Bérubé, J. A., and M. Dessureault. 1988. Morphological Characterization of *Armillaria ostoyae* and *Armillaria sinapina* sp. nov. *Can. J. Bot.* 66: 2027–2034.

———. 1989. Morphological Studies of the *Armillaria mellea* Complex: Two New Species, *A. gemina* and *A. calvescens*. *Mycologia* 81: 216–225.

Bessette, A. E., A. R. Bessette, D. P. Lewis, and S. H. Metzler. 1993. A New Substrate for *Strobilurus conigenoides* (Ellis) Singer. *Mycotaxon* 68: 299–300.

Bessette, A. E., and R. L. Homola. 1986. *Lentinus adhaerens*, a Species New to North America. *Mycologia* 78: 296–298.

Bigelow, H. E. 1973. The Genus *Clitocybula*. *Mycologia* 65: 1101–1116.

———. 1982. *North American Species of Clitocybe*. Part 1. J. Cramer, Berlin. 280 pp.

———. 1985. *North American Species of Clitocybe*. Part 2. J. Cramer, Berlin. 240 pp.

Bird, C. J., and D. W. Grund. 1979. Nova Scotian Species of *Hygrophorus*. *Proc. Nova Scotia Inst. of Sci.* 29: 1–131.

Both, E. E. 1993. *The Boletes of North America: A Compendium*. Buffalo Museum of Science, Buffalo. 436 pp.

Coker, W. C., and A. H. Beers. 1943. *The Boletaceae of North Carolina*. Univ. of North Carolina Press, Chapel Hill. 96 pp.

Coker, W. C., and J. N. Couch. 1928. *Gasteromycetes of the Eastern United States and Canada*. Univ. of North Carolina Press, Chapel Hill. 201 pp.

Gilbertson, R. L., and L. Ryvarden. 1986. *North American Polypores*. Vol. 1. Fungiflora, Oslo. 433 pp.

———. 1987. *North American Polypores*. Vol. 2. Fungiflora, Oslo. 451 pp.

Gilliam, M. S. 1975. *Marasmius* Section Chordales in the Northeastern United States and Adjacent Canada. *Contr. Univ. Mich. Herb.* 11: 25–40.

———. 1976. The Genus *Marasmius* in the Northeastern United States and Adjacent Canada. *Mycotaxon* 4: 1–144.

Grund, D. W., and K. A. Harrison. 1976. *Nova Scotian Boletes*. Bibliotheca Mycologica, band 47. J. Cramer, Vaduz. 283 pp.

Guzmán, G. 1983. *The Genus* Psilocybe. Nova Hedwigia, heft 74. J. Cramer, Vaduz. 439 pp.

Halling, R. E. 1983. *The Genus Collybia (Agaricales)*. Mycologia Memoir no. 8. New York Botanical Garden. J. Cramer, Vaduz. 148 pp.

Harrington, F. A. 1990. *Sarcoscypha* in North America (Pezizales, Sarcocyphaceae). *Mycotaxon* 38: 417–458.

Harrison, K. A. 1961. *Stipitate Hydnums of Nova Scotia*. Canadian Dept. of Agriculture, Ottawa. 60 pp.

Henderson, D. M., P. D. Orton, and R. Watling, eds. 1979. *Coprinaceae Part 1: Coprinus*. Her Majesty's Stationary Office, Edinburgh. 149 pp.

Hesler, L. R. 1965. *North American Species of Crepidotus.* Hafner, New York. 168 pp.
——. 1967. Entoloma in Southeastern North America. Nova Hedwigia, heft 23. Verlag von J. Cramer, Lehre, Germany. 196 pp.
——. 1969. *North American Species of Gymnopilus.* Hafner, New York. 117 pp.
Hesler, L. R., and A. H. Smith. 1963. *North American Species of Hygrophorus.* Univ. of Tennessee Press, Knoxville. 416 pp.
——. 1979. *North American Species of Lactarius.* Univ. of Michigan Press, Ann Arbor. 841 pp.
Jenkins, D. T. 1986. *Amanita of North America.* Mad River Press, Eureka, Calif. 198 pp.
Kauffman, C. H. 1971a. *The Gilled Mushrooms (Agaricales) of Michigan and the Great Lakes Region.* Vol. 1. Reprint. Dover, New York. 442 pp.
——. 1971b. *The Gilled Mushrooms (Agaricales) of Michigan and the Great Lakes Region.* Vol. 2. Reprint. Dover, New York. 481 pp.
Kibby, G., and R. Fatto. 1990. *Keys to the Species of Russula in Northeastern North America.* Kibby-Fatto Enterprises, Somerville, N.J. 61 pp.
Kirk, P. M., P. F. Cannon, J. C. David, and J. A. Staplers, eds. 2002. *Ainsworth and Bisby's Dictionary of the Fungi.* 9th. ed. CABI and Oxford Univ. Press, Oxford. 672 pp.
Largent, D. L. 1977. *The Genus Leptonia on the Pacific Coast of the United States.* Bibliotheca Mycologica, band 55. J. Cramer, Vaduz. 286 pp.
——. 1986. *How to Identify Mushrooms to Genus I: Macroscopic Features.* Rev. ed. Mad River Press, Eureka, Calif. 166 pp.
Largent, D. L., and T. J. Baroni. 1988. *How to Identify Mushrooms to Genus VI: Modern Genera.* Mad River Press, Eureka, Calif. 277 pp.
Largent, D. L., D. Johnson, and R. Watling. 1977. *How to Identify Mushrooms to Genus III: Microscopic Features.* Mad River Press, Eureka, Calif. 148 pp.
Largent, D. L., and H. D. Thiers. 1977. *How to Identify Mushrooms to Genus II: Field Identification of Genera.* Mad River Press, Eureka, Calif. 32 pp.
Mazzer, S. J. 1976. *A Monographic Study of the Genus* Pouzarella. Bibliotheca Mycologica, band 46. J. Cramer, Vaduz. 191 pp.
Miller, O. K. 1964. Monograph of *Chroogomphus* (Gomphidiaceae). *Mycologia* 56: 526–549.
——. 1968. A Revision of the Genus *Xeromphalina. Mycologia* 60: 156–188.
——. 1970. The Genus *Panellus* in North America. *Michigan Botanist* 9: 17–30.
——. 1971. The Genus *Gomphidius* Fries, with a Revised Description of the Gomphidiaceae and Keys to the Genera. *Mycologia* 53: 1129–1163.
Miller, O. K., and L. Stewart. 1971. The Genus *Lentinellus. Mycologia* 63: 333–369.
Miller, O. K., T. J. Volk, and A. E. Bessette. 1995. A New Genus, *Leucopholiota*, in the Tricholomataceae (Agaricales) to Accommodate an Unusual Taxon. *Mycologia* 88: 137–139.
Moser, M. 1983. *Keys to Agarics and Boleti.* Gustav Fischer Verlag, Stuttgart. 535 pp.
Motta, J. J., and K. Korhonen. 1986. A Note on *Armillaria mellea* and *Armillaria bulbosa* from the Middle Atlantic States. *Mycologia* 78: 471–474.
Mueller, G. M. 1992. Systematics of *Laccaria* (Agaricales) in the Continental United States and Canada, with Discussions on Extralimital Taxa and Descriptions of Extant Types. *Fieldiana* Pub. 1435, n.s. no. 30. 158 pp.
Overholts, L. O. 1953. *The Polyporaceae of the United States, Alaska, and Canada.* Univ. of Michigan Press, Ann Arbor. 466 pp.
Ovrebo, C. L. 1989. *Tricholoma*, Subgenus *Tricholoma*, Section *Albidogrisea:* North American Species Found Principally in the Great Lakes Region. *Can. J. Bot.* 67: 3134–3152.
Pegler, D. N. 1983. *The Genus Lentinus—A World Monograph.* Her Majesty's Stationary Office, London. 281 pp.
Pfister, D. H., and A. E. Bessette. 1985. More Comments on the Genus *Acervus. Mycotaxon* 22: 435–438.
Redhead, S. A. 1986. Mycological Observations 15–16: On *Omphalia* and *Pleurotus. Mycologia* 87: 522–528.
Redhead, S. A., J. Ginns, and R. A. Shoemaker. 1987. The *Xerula (Collybia, Oudemansiella) radicata* Complex in Canada. *Mycotaxon* 30: 357–405.

Seaver, F. J. 1928. *The North American Cup-fungi (Operculates)*. Hafner, New York. 284 pp.
——. 1942. *The North American Cup-fungi (Operculates) Supplement*. Lancaster Publishing Company, Lancaster, Penn. 90 pp.
——. 1951. *The North American Cup-fungi. (Inoperculates)*. Lancaster Publishing Company, Lancaster, Penn. 428 pp.
Shaffer, R. L. 1957. *Volvariella* in North America. *Mycologia* 49: 545–579.
Singer, R., and A. H. Smith. 1943. A Monograph on the Genus *Leucopaxillus* Boursier. *Papers of the Mich. Academy of Sciences, Arts, and Letters* 28: 85–132.
——. 1947. Additional Notes on the Genus *Leucopaxillus*. *Mycologia* 39: 725–736.
Smith, A. H. 1947. *The North American Species of Mycena*. Univ. of Michigan Press, Ann Arbor. 521 pp.
——. 1951. The North American Species of *Naematoloma*. *Mycologia* 43: 467–521.
——. 1972. The North American Species of *Psathyrella*. *Memoirs of the New York Botanical Garden* 24: 1–633.
Smith, A. H., and L. R. Hesler. 1963. *North American Species of Hygrophorus*. Univ. of Tennessee Press, Knoxville. 416 pp.
——. 1968. *The North American Species of Pholiota*. Hafner, New York. 402 pp.
Smith, A. H., and R. Singer. 1945. A Monograph of the Genus *Cystoderma*. *Papers of the Mich. Academy of Science, Arts, and Letters* 30: 71–124.
——. 1964. *A Monograph on the Genus Galerina Earl*. Hafner, New York. 168 pp.
Smith, A. H., and H. D. Thiers. 1964. *A Contribution toward a Monograph of North American Species of Suillus*. Univ. of Michigan Press, Ann Arbor. 116 pp.
——. 1971. *The Boletes of Michigan*. Univ. of Michigan Press, Ann Arbor. 428 pp.
Snell, W. H., and E. A. Dick. 1957. *A Glossary of Mycology*. Harvard Univ. Press, Cambridge, Mass. 181 pp.
——. 1970. *The Boleti of Northeastern North America*. J. Cramer, Lehre. 115 pp.
Stuntz, D. E. 1977. *How to Identify Mushrooms to Genus IV: Keys to Families and Genera*. Mad River Press, Eureka, Calif. 94 pp.
Watling, R. 1977. *How to Identify Mushrooms to Genus V: Cultural and Developmental Features*. Mad River Press, Eureka, Calif. 169 pp.
——. 1982. *3 Bolbitiaceae: Agrocybe, Bolbitius, and Conocybe*. British Fungus Flora Agarics and Boleti. Her Majesty's Stationary Office, Edinburgh. 139 pp.
Weber, N. S. 1972. The Genus *Helvella* in Michigan. *Michigan Botanist* 11: 183–186.
Wells, V. L., and P. E. Kempton. 1968. A Preliminary Study of *Clavariadelphus* in North America. *Michigan Botanist* 7: 35–57.
Zeller, S. M., and A. H. Smith. 1964. The Genus *Calvatia* in North America. *Lloydia* 27: 148–186.

NONTECHNICAL PUBLICATIONS

Ammirati, J. F., J. A. Traquir, and P. A. Horgen. 1985. *Poisonous Mushrooms of the Northern United States and Canada*. Univ. of Minnesota Press, Minneapolis. 396 pp.
Barron, G. 1999. *Mushrooms of Northeast North America*. Lone Pine, Vancouver. 336 pp.
Bessette, A. E. 1985. *Guide to Some Edible and Poisonous Mushrooms of New York*. Canterbury Press, Rome, N.Y. 24 pp.
——. 1988. *Mushrooms of the Adirondacks: A Field Guide*. North Country Books, Utica, N.Y. 145 pp.
Bessette, A. E., A. R. Bessette, and D. W. Fischer. 1997. *Mushrooms of Northeastern North America*. Syracuse Univ. Press, Syracuse, N.Y. 584 pp.
Bessette, A. E., O. K. Miller, A. R. Bessette, and H. H. Miller. 1995. *Mushrooms of North America in Color—A Field Guide Companion to Seldom-Illustrated Fungi*. Syracuse Univ. Press, Syracuse, N.Y. 188 pp.
Bessette, A. E., W. C. Roody, and A. R. Bessette. 2000. *North American Boletes—A Color Guide to the Fleshy Pored Mushrooms*. Syracuse Univ. Press, Syracuse, N.Y. 400 pp.

Bessette, A. E., and W. J. Sundberg. 1987. *Mushrooms: A Quick Reference Guide to Mushrooms of North America*. Macmillan, New York. 174 pp.

Bessette, A. R., and A. E. Bessette. 1993. *Taming the Wild Mushroom: A Culinary Guide to Market Foraging*. Univ. of Texas Press, Austin. 113 pp.

———. 2001. *The Rainbow Beneath My Feet: A Mushroom Dyer's Field Guide*. Syracuse Univ. Press, Syracuse, N.Y. 176 pp.

Bessette, A. R., A. E. Bessette, and W. J. Neill. 2001. *Mushrooms of Cape Cod and the National Seashore*. Syracuse Univ. Press, Syracuse, N.Y. 174 pp.

Bigelow, H. E. 1974. *Mushroom Pocket Field Guide*. MacMillan, New York. 117 pp.

Breitenbach, J., and F. Kranzlin. 1986. *Fungi of Switzerland*. Vol. 2, *Non-gilled Fungi*. Verlag Mykologia, Lucerne, Switzerland. 412 pp.

Evenson, V. S. 1997. *Mushrooms of Colorado and the Southern Rocky Mountains*. Westcliffe, Denver. 208 pp.

Fischer, D. W., and A. E. Bessette. 1992. *Edible Wild Mushrooms of North America: A Field-to-Kitchen Guide*. Univ. of Texas Press, Austin, 254 pp.

Groves, J. W. 1979. *Edible and Poisonous Mushrooms of Canada*. 2d rev. ed. Research Branch, Agriculture Canada, Ottawa. 326 pp.

Horn, B., R. Kay, and D. Abel. 1993. *A Guide to Kansas Mushrooms*. Univ. Press of Kansas, Lawrence. 297 pp.

Katsaros, P. 1990. *Familiar Mushrooms*. Knopf, New York. 192 pp.

Kendrick, B. 1992. *The Fifth Kingdom*. 2d ed. Focus Information Group, Newburyport, Mass. 406 pp.

Kibby, G. 1992. *Mushrooms and Other Fungi*. Smithmark, New York. 192 pp.

Kimbrough, J. 2000. *Common Florida Mushrooms*. Univ. of Florida, Gainesville. 342 pp.

Lincoff, G. H. 1981. *The Audubon Society Field Guide to North American Mushrooms*. Knopf, New York. 498 pp.

Lincoff, G. H., and D. H. Mitchell. 1977. *Toxic and Hallucinogenic Mushroom Poisoning*. Van Nostrand Reinhold, New York. 267 pp.

Metzler, S., V. Metzler, and O. K. Miller Jr. 1992. *Texas Mushrooms*. Univ. of Texas Press, Austin. 350 pp.

Miller, O. K. 1973. *Mushrooms of North America*. E. P. Dutton, New York. 360 pp.

Miller, O. K., and H. H. Miller. 1980. *Mushrooms in Color*. E. P. Dutton, New York. 286 pp.

Phillips, R. 1981. *Mushrooms and Other Fungi of Great Britain and Europe*. Ward Lock Limited, London. 288 pp.

———. 1991. *Mushrooms of North America*. Little, Brown, Boston. 319 pp.

Roody, W. C. 2003. *Mushrooms of West Virginia and the Central Appalachians*. Univ. Press of Kentucky, Lexington. 520 pp.

Sicard, M., and Y. Lamoureux. 1999. *Les Champignons Sauvages du Québec*. Fides, Québec. 399 pp.

Smith, A. H. 1949. *Mushrooms in Their Natural Habitats*. Sawyer's, Portland, Ore. 626 pp.

Smith, A. H., H. V. Smith, and N. S. Weber. 1979. *How to Know the Gilled Mushrooms*. William C. Brown, Dubuque, Iowa. 334 pp.

Smith, A. H., and N. S. Weber. 1980. *The Mushroom Hunter's Field Guide*. Univ. of Michigan Press, Ann Arbor. 316 pp.

Smith, H. V., and A. H. Smith. 1973. *How to Know the Non-gilled Fleshy Fungi*. William C. Brown, Dubuque, Iowa. 402 pp.

Stamets, P. 1993. *Growing Gourmet and Medicinal Mushrooms*. Ten Speed Press, Berkeley, Calif. 552 pp.

Sundberg, W. J., and J. A. Richardson. 1980. *Mushrooms and Other Fungi of the Land Between the Lakes*. Tennessee Valley Authority, Knoxville. 60 pp.

Thorn, R. G. 1991. *Mushrooms of Algonquin Provincial Park*. Friends of Algonquin Park, Whitney, Ontario. 32 pp.

Weber, N. S. 1988. *A Morel Hunter's Companion*. Two Peninsula Press, Lansing, Mich. 209 pp.

Weber, N. S., and A. H. Smith. 1985. *A Field Guide to Southern Mushrooms*. Univ. of Michigan Press, Ann Arbor. 280 pp.

Index to Common Names

Abortive Entoloma, 131
Abruptly-bulbous Agaricus, 101
Acorn Puffball, 312
Alcohol Inky, 123
Amanita Mold, 331
Amber Jelly Roll, 305
Apple Bolete, 1
Apricot Milk Cap, 162
Arched Earthstar, 314
Artist's Conk, 253
Ash-tree Bolete, 230

Ballou's Russula, 191
Barometer Earthstar, 309
Bearded Tooth, 202
Bear Lentinus, 163
Bear's-head Tooth, 202
Beefsteak Morel, 280
Beefsteak Polypore, 252
Beetle Cordyceps, 295
Berkeley's Polypore, 250
Big Laughing Gym, 134
Bitter Bolete, 214
Blackening Russula, 193
Black-footed Marasmius, 174
Black Jelly Roll, 305
Black Morel, 282
Black-staining Polypore, 263
Black Trumpet, 96
Black Velvet Bolete, 239
Black Witches' Butter, 305
Bladder Cup, 301
Bleeding Mycena, 176, 177
Blewit, 121
Blue Stain, 300
Bluing Bolete, 232
Blusher, The, 112
Bolete Mold, 331
Bradley, 162
Brain Mushroom, 279
Brick Cap, 144
Brick Tops, 144
Brown-haired White Cup, 301
Brownish Chroogomphus, 119
Brown-toothed Crust, 328
Bubble Gum Fungus, 294
Buff Fishy Milk Cap, 155

Buff Fishy Milky, 155
Burnsite Pholiota, 182
Burnt-orange Bolete, 240

Canary Trich, 198
Carbon Balls, 323
Carbon Cushion, 325
Carolina False Morel, 279
Carrot-foot Amanita, 106
Cauliflower Mushroom, 307
Cedar-apple Rust, 333
Celandine Lactarius, 149
Cep, 213
Ceramic Parchment, 330
Chalky-white Bolete, 209
Changeable Melanoleuca, 175
Chanterelle, 93
Charred Pholiota, 182
Chestnut Bolete, 231
Chicken Mushroom, 262
Chlorine Amanita, 105
Cinnabar-red Chanterelle, 93
Citron Amanita, 105
Club-footed Clitocybe, 120
Club-like Tuning Fork, 304
Club-shaped Stinkhorn, 277
Clustered Dune Hygrocybe, 139
Clustered Psathyrella, 187
Coker's Amanita, 106
Collared Calostoma, 310
Collybia Jelly, 136, 306
Columned Stinkhorn, 274
Comb Tooth, 202
Common Bird's-nest, 298
Common Brown Cup, 301
Common Earthball, 320
Common Laccaria, 146
Common Morel, 282, 283
Common Psathyrella, 186
Common Split Gill, 99, 100
Conifer-cone Baeospora, 117
Conifer False Morel, 280
Cornsilk Cystoderma, 130
Corpse Finder, 138
Corrugated-cap Milky, 150
Corrugated Milk Cap, 150
Crampballs, 323

Creamy Maze Crust, 269
Crested Coral, 287
Crested Polypore, 247
Crimped Gill, 99
Crowded Parchment, 329
Crown-tipped Coral, 287

Deadly Galerina, 132
Dead Man's Fingers, 326
Deceptive Milk Cap, 151
Deceptive Milky, 151
Deer Mushroom, 185
Deer Truffle, 313
Dense-gilled Russula, 192
Destroying Angel, 113
Devil's Cigar, 312
Devil's Urn, 302
Dog Stinkhorn, 276
Double-gilled Stropharia, 197
Dune Stinkhorn, 277
Dune Witch's Hat, 140
Dye-maker's False Puffball, 318
Dye Polypore, 265

Early Blackish-red Russula, 196
Early Spring Entoloma, 179
Earthstar Scleroderma, 321
Elegant Stinkhorn, 276
Elfin Cup, 300
Elm Oyster, 144
Enoki-take, 131
Enotake, 131

Fairy Sparklers, 326
False Turkey-tail, 330
Family Collybia, 122
Fat-footed Clitocybe, 120
Fawn Mushroom, 185
Firm Russula, 191
Flame-colored Chanterelle, 94
Flat-toped Coral, 291
Florida Inky Cap, 124
Fluted-stalked Fungus, 294
Fluted White Helvella, 280
Fragile Leucocoprinus, 170
Fragrant Chanterelle, 97
Frost's Amanita, 107

Frost's Bolete, 1
Fungus Flower, 273
Fused Marasmius, 174
Fuzzy Foot, 201

Gem-studded Puffball, 317
Giggle Mushroom, 188
Goat's Foot, 248
Gold Drop Milk Cap, 150
Golden Chanterelle, 93, 180
Golden Scruffy Collybia, 130
Golden Spreading Polypore, 266
Golden Trumpets, 201
Graceful Bolete, 207
Gray Coral, 287
Green-headed Jelly Club, 292
Green Quilt Russula, 192
Green-spored Lepiota, 119
Green-spored Parasol, 119
Green Stain, 300

Hairy Panus, 165
Hairy Rubber Cup, 300
Hairy-stalked Entoloma, 186
Half-free Morel, 282, 283
Hated Amanita, 113
Head-like Cordyceps, 295
Hedgehog, 204
Hemispheric Agrocybe, 103
Hemlock Varnish Shelf, 255
Hesler's Amanita, 108
Honeycomb, 282
Honey Mushroom, 115, 131
Hophornbeam Disc, 327
Hot Lips, 310

Indigo Milk Cap, 155
Indigo Milky, 155

Jack O'Lantern, 93, 95, 180
Japanese Umbrella Inky, 126
Jelly Babies, 292
Jelly Leaf, 306

Kidney-shaped Tooth, 204
King Bolete, 213
Knobbed Squamanita, 196

Land Fish, 282
Lawyer's Wig, 123
Leaf-like Oyster, 138
Lemon-yellow Lepiota, 169
Lilac Fiber Head, 145
Ling Chih, 254
Little Helmets, 124
Little Red Stinkhorn, 276
Lizard's Claw Stinkhorn, 275
Lobster Fungus, 332
Lobster Mushroom, 332

Lorchel, 280
Luminescent Panellus, 181

Magenta Coral, 286
Magic Mushroom, 188
Magnolia-cone Mushroom, 197
Malodorous Lepiota, 166
Meadow Mushroom, 102
Mica Cap, 125
Milk-white Toothed Polypore, 269
Milky False Truffle, 322
Mossy Maze Polypore, 250
Multicolor Gill Polypore, 263
Mustard-yellow Polypore, 266

Netted Stinkhorn, 274
Non-inky Coprinus, 124

Oak-loving Collybia, 136
Oak-loving Gymnopus, 136
Ochre Jelly Club, 292
Onion-stalk Lepiota, 170
Orange Earth Tongue, 293
Orange-gilled Waxy Cap, 142
Orange Jelly, 304
Orange-mat Coprinus, 125
Orange Mock Oyster, 183
Orange Peel, 299
Orange Rough-cap Tooth, 203
Orange-staining Puffball, 312
Orange Witches' Butter, 304
Ornate-stalked Bolete, 221
Oyster Mushroom, 184

Parasol, 172
Pear-shaped Puffball, 317
Peck's Milk Cap, 157
Peck's Milky, 157
Peeling Puffball, 316
Pestle-shaped Coral, 291
Pig's Ear Gomphus, 97
Pigskin Poison Puffball, 320
Pineapple Bolete, 208
Pine Cone Mushroom, 282
Pine Spike, 119
Pine Varnish Conk, 254
Pink Bottom, 102
Pink Mycena, 177
Pinwheel Marasmius, 175
Plums and Custard, 200
Pointed-stalked Marasmiellus, 174
Poison Paxillus, 182
Poison Pie, 136
Porcini, 213
Powder Cap, 116
Powdery Sulfur Bolete, 235
Puffball Agaric, 313

Pungent Cystoderma, 130
Purple Club Coral, 287
Purple-gilled Laccaria, 146, 337
Purple Jelly Drops, 299
Purple-spored Puffball, 311

Ravenel's Calostoma, 311
Ravenel's Mutinus, 276
Ravenel's Stinkhorn, 277
Recurved Cup, 301
Red-and-Yellow Bolete, 211
Red Cushion Hypoxylon, 324
Reddening Lepiota, 168
Red-gilled Cort, 128
Red-juice Tooth, 203
Red-mouth Bolete, 227
Red Slimy-stalked Puffball, 310
Resinous Polypore, 261
Ribbed-stalked Cup, 300
Ringless Honey Mushroom, 116
Rooting Cauliflower Mushroom, 307
Rose Coral, 289
Rose-red Russula, 194
Rosy Brick-red Bolete, 224
Rosy-brown Waxy Cap, 143
Rough-stalked Hebeloma, 137
Rounded Earthstar, 315
Ruddy Panus, 165
Russula Mold, 332

Saddle-shaped False Morel, 280
Salmon Unicorn Entoloma, 178
Salmon Waxy Cap, 143
Sand Case Puffball, 312
Sandy Laccaria, 147, 337
Sandy Mac, 317, 318
Satyr's Beard, 202
Scaly Vase Chanterelle, 98
Semi-ovate Panaeolus, 180
Shaggy Mane, 123
Shaggy Parasol, 119
Shaggy-stalked Bolete, 206
Shaggy-stalked Lepiota, 166
Shiny Cinnamon Polypore, 251
Silky-shining Russula, 194
Silver-blue Milk Cap, 157
Silver-blue Milky, 157
Silver Ear, 306
Silvery Agaricus, 103
Slender Red-pored Bolete, 218
Slippery Jill, 238
Smaller Indigo Milk Cap, 154
Smoky Polypore, 249
Smooth Chanterelle, 95
Smooth-stalked Helvella, 281
Smooth Thimble-cap, 283
Spindle-shaped Yellow Coral, 288
Spiny Puffball, 316

Splash Cups, 298
Split-pore Polypore, 269
Split-skin Carbon Cushion, 324
Sponge Mushroom, 282
Spotted Bolete, 244
Spotted Collybia, 189
Spotted Cort, 127
Stalked Stereum, 327
Starfish Stinkhorn, 273
Steinpilz, 213
Straight-stalked Entoloma, 179
Straw-colored Fiber Head, 145
Suburban Psathyrella, 186
Sulphur Shelf, 262
Sulphur Tuft, 144
Sweet Tooth, 204
Swollen-stalked Cat, 118

Tender Nesting Polypore, 258
Thin-maze Flat Polypore, 251
Tree-Ear, 303
Trembling Merulius, 329
Trooping Cordyceps, 296

Trumpet Chanterelle, 96
Tufted Collybia, 135
Tufted Gymnopus, 135
Tumbling Puffball, 309
Turkey-tail, 271, 330
Two-colored Bolete, 211
Two-tone Parchment, 330

Underground False Truffle, 313

Variable Pholiota, 183
Vase Thelephore, 285
Veiled Oyster, 184
Velvet Blue Spread, 329
Velvet Foot, 131
Velvet-footed Pax, 181
Violet-branched Coral, 287
Violet Collybia, 136
Violet-gray Bolete, 241
Violet Gymnopus, 136
Violet Toothed Polypore, 271
Viscid Violet Cort, 127
Voluminous-latex Milky, 162

Walnut Mycena, 177
Weeping Milk Cap, 162
Western Lawn Puffbowl, 321
White Dunce Cap, 123
White Marasmius, 173
White-pored Sulphur Shelf, 261
White Worm Coral, 286, 287
Wine-colored Agaricus, 103
Winter Mushroom, 131
Witches' Butter, 304
Witch's Egg, 277
Witch's Hat, 140
Wood Clitocybe, 180
Wrinkled Thimble-cap, 282

Yellow Blusher, 107
Yellow Bolbitius, 117
Yellow Morel, 149, 282
Yellow-red Gill Polypore, 255
Yellow-tipped Coral, 289
Yellow Unicorn Entoloma, 178

Index to Scientific Names

Page number in *italic* denotes illustration.

Agaricus
  abruptibulbus, *16*, 101
  argenteus, 103
  arvensis, 101, 102
  campestris, *16*, 102
  placomyces, 102
  pocillator, *16*, 102
  porphyrocephalus, *16*, 102
  silvicola, 101
  subrutilescens, *16*, 103
Agrocybe
  pediades, 104
  semiorbicularis, *16*, 103
Albatrellus
  confluens, 249
  cristatus, *58*, 247
  ovinus, 249
  pes-caprae, *58*, 248
  subrubescens, *58*, 248
Aleuria aurantia, *77*, 299
Aleurodiscus
  amorphus, 327
  oakesii, *88*, 327
Amanita
  abrupta, *16*, 104
  abruptiformis, 110
  anisata, 110
  atkinsoniana, *17*, 104
  bisporigera, 114
  caesarea, 110
  chlorinosma, *17*, 105
  citrina f. *citrina*, *17*, 105
  citrina f. *lavendula*, 106
  cokeri, *17*, 106
  daucipes, *17*, 106
  excelsa var. *alba*, 104
  flavoconia, 107
  flavorubescens, *17*, 107
  frostiana, *18*, 107
  hesleri, *18*, 108
  jacksonii, 110
  komarekensis, *18*, 108
  longipes, *18*, 109
  muscaria var. *flavivolvata*, *18*, 109
  muscaria var. *formosa*, 109
  muscaria var. *persicina*, 109
  mutabilis, *18*, 110

  onusta, 105
  pantherina var. *multisquamosa*, 104
  parcivolvata, *19*, 110
  peckiana, 114
  pelioma, 111
  phalloides, 113
  polypyramis, *19*, 111
  ravenelii, *19*, 112
  rhopalopus f. *rhopalopus*, *19*, 112
  rhopalopus f. *turbinata*, 112
  roseotincta, 108, 109
  rubescens, *19*, 112
  spreta, *19*, 113
  submutabilis, 110
  verna, 114
  virosa, *19*, 113
  volvata, *20*, 114
Anellaria
  phalaenarum, 181
  semiovatus, 181
  separatus, 181
  sepulchralis, *20*, 114
Anthracophyllum lateritium, *20*, 115
Antrodia
  albida, *58*, 249
  juniperina, 249
  serialis, 249
Arachnion album, *81*, 308
Armillaria
  dryina, 184
  gemina, 116
  mellea, *20*, 115, 131
  ostoyae, *20*, 116
  tabescens, *20*, 116
  umbonata, 197
Arrhytidia involuta, 303
Ascocoryne
  cylichnium, 300
  sarcoides, *77*, 299
Aseroe rubra, *65*, 273
Asterophora
  lycoperdoides, *20*, 116
  parasitica, 117
Astraeus hygrometricus, *81*, 309
Aurantiosporus croceus, 257
Aureoboletus innixus, 217

Auricularia
  auricula, *79*, 303
  polytricha, *79*, 303
Austroboletus
  betula, *48*, 206
  gracilis var. *gracilis*, *48*, 207
  subflavidus, *48*, 207

Baeospora
  myosura, *21*, 117
  myriadophylla, 117
Bisporella citrina, 299
Bjerkandera
  adusta, *58*, 249
  fumosa, 250
Blumenavia angolensis, *66*, 274
Bolbitius
  variicolor, 118
  vitellinus, *x*, *21*, 117
Boletellus
  ananas, *48*, 208
  betula, 207
  fallax, 208
  pictiformis, 208
Boletinellus merulioides, 231
Boletus
  abruptibulbus, *49*, 208
  affinis var. *maculosus*, 244
  albisulphureus, *49*, 209
  ananus, 208
  atkinsonii, 220, 229
  aurantiosplendens, 210
  auriflammeus, *49*, 209
  auriporus, *49*, 210, 338
  bicolor var. *bicolor*, *49*, 211
  bicolor var. *subreticulatus*, 211
  carminiporus, *49*, 212
  coccineus, 208
  communis, 222
  curtisii, *49*, 212
  dupainii, *49*, 213
  edulis, *50*, 213
  erythropus, 218
  fairchildianus, 224, 226
  firmus, *50*, 214
  flammans, *50*, 215
  floridanus, *50*, 215
  frostii, 216

Boletus (continued)
  frostii ssp. *floridanus*, 216
  griseus, 50, 216
  hortonii, 50, 216
  hypocarycinus, 50, 218, 219
  illudens, 228
  inedulis, 230
  innixus, 50, 217, 338
  longicurvipes, 51, 217
  luridellus, 51, 224
  luridiformis, 51, 218, 339
  luridus, 51, 219
  mahagonicolor, 51, 219
  miniato-olivaceus, 227
  miniato-pallescens, 227
  nobilis, 51, 220
  oliveisporus, 220
  ornatipes, 51, 221
  pallidoroseus, 211
  pallidus, 52, 221
  patrioticus, 52, 222
  piedmontensis, 214
  pinophilus, 214
  pseudorubinellus, 230
  pulverulentus, 52, 222
  purpureorubellus, 52, 223
  retipes, 221
  roseolateritius, 52, 224
  roseopurpureus, 52, 224
  roxanae, 210
  rubellus, 224
  rubellus var. *purpureus*, 223
  rubricitrinus, 52, 225
  rubroflammeus, 212, 215
  rubropunctus, 218
  rufomaculatus, 53, 226
  satanas var. *americanus*, 214
  sensibilis, 53, 211, 226
  separans, 245
  speciosus, 225
  stramineus, 209, 246
  subglabripes, 53, 227
  subglabripes var. *corrugis*, 217
  subluridellus, 53, 215
  subvelutipes, 53, 227
  tenax, 53, 228
  variipes, 53, 229
  viridiflavus, 211
  weberi, 53, 229
Bondarzewia berkeleyi, 58, 250
Bovista
  pila, 309
  plumbea, 81, 309
Bovistella
  echinata, 310
  ohiensis, 310
  radicata, 81, 309
Bulgaria rufa, 300

Callistosporium
  luteo-olivaceum, 21, 118
  purpureomarginatum, 118
Calocera
  cornea, 79, 304
  viscosa, 304
Calostoma
  cinnabarina, 82, 310
  lutescens, 82, 310
  ravenelii, 82, 311
Calvatia
  craniformis, 311
  cyathiformis, 82, 311
  rubroflava, 82, 312
Camarophyllus pratensis, 143
Camarops petersii, 324
Cantharellus
  cibarius, 14, 93
  cinnabarinus, 14, 93
  confluens, 14, 94
  ignicolor, 14, 94
  infundibuliformis, 96
  lateritius, 14, 95
  odoratus, 97
  persicinus, 14, 95
  tabernensis, 14, 96
  tubaeformis, 14, 96
Catathelasma ventricosa, 21, 118
Cerrena unicolor, 58, 250
Chalciporus
  pseudorubinellus, 54, 230
  rubinellus, 230
Chlorociboria
  aeruginascens, 77, 300
  aeruginosa, 300
Chlorophyllum molybdites, 21, 119
Chlorosplenium aeruginascens, 300
Chorioactis geaster, 82, 83, 312
Christiansenia mycetophila, 306
Chroogomphus
  rutilus, 21, 119
  vinicolor, 119
Chrysomphalina
  chrysophylla, 132
  strombodes, 132
Clathrus
  angolensis, 274
  columnatus, 275
Claudopus vinaceocontusus, 120
Clavaria
  purpurea, 287
  vermicularis, 72, 286
  zollingeri, cover photo, 72, 286
Clavariadelphus
  pistillaris, 73, 291
  truncatus, 73, 291
Claviceps purpurea, 295
Clavicorona pyxidata, 72, 287

Clavulina
  amethystina, 287
  cinerea, 287
  cristata, 72, 287
Clavulinopsis
  aurantio-cinnabarina, 72, 288
  fusiformis, 73, 288
  helveola, 288
  laeticolor, 288
  pulchra, 288
Clitocybe
  clavipes, 21, 120
  ectypoides, 132, 180
  gibba, 120
  illudens, 180
  irina, 121
  nuda, 22, 121
  robusta, 121
  subconnexa, 22, 121
Clitocybula
  familia, 22, 122
  lacerata, 122
Clitopilus
  prunulus, 190
  roseiavellaneus, 190
Collybia
  acervata, 174, 189
  butyracea, 189
  cirrhata, 122
  confluens, 135
  cookei, 22, 122
  dryophila, 136
  familia, 122
  iocephala, 136
  luteo-olivaceus, 118
  luxurians, 189
  maculata, 189
  spongiosa, 135
  subnuda, 135
  tuberosa, 122
  velutipes, 131
Coltricia
  cinnamomea, 59, 251
  perennis, 251
Conocybe
  lactea, 22, 123
  tenera, 123
Copelandia westii, 22, 115
Coprinus
  americanus, 22, 124, 126
  atramentarius, 23, 123
  comatus, 23, 123
  disseminatus, 23, 124
  floridanus, 23, 124
  laniger, 23, 125
  micaceus, 23, 125
  plicatilis, 24, 126
  radians, 125

sterquilinus, 124
variegatus, 124, 126
Cordyceps
capitata, 75, 295
longisegmentis, 295
melolonthae, 75, 295
militaris, 296
olivaceo-virescens, 296
olivascens, 75, 296
ophioglossoides, 295, 313
sphecocephala, 75, 296
Corticum caeruleum, 329
Cortinarius
cinnabarinus, 128
delibutus, 24, 126
iodes, 24, 127
lewisii, 24, 127
marylandensis, 24, 128
semisanguineus, 24, 128
sublilacina, 129
torvus, 127
violaceus, 127
Cotilydia diaphana, 88, 327
Craterellus
cornucopioides, 97
fallax, 15, 96
odoratus, 15, 97
Crepidotus
fulvotomentosus, 129
mollis, 24, 129
Crinipellis
scabella, 130
stipitaria, 129
zonata, 130
Crucibulum
laeve, 76, 298
vulgare, 298
Cudonia
circinans, 74, 292
lutea, 74, 292
Cyathus striatus, 76, 298
Cyptotrama
asprata, 25, 130
chrysopeplum, 130
Cystoderma
amianthinum var.
   amianthinum, 130
amianthinum var.
   rugosoreticulatum, 25, 130
fallax, 130
granosum, 130
terrei, 130

Dacrymyces
corticioides, 303
involutus, 303
palmatus, 79, 304
Dacryopinax
elegans, 79, 304

spathularia, 79, 304
Daedalea
confragosa, 251
quercina, 251
unicolor, 251
Daedaleopsis confragosa, 59, 251
Daldinia concentrica, 87, 323
Dentinum repandum, 204
Diatrype stigma, 87, 323
Dictyophora duplicata, 66, 274
Disciseda candida, 83, 312

Elaphomyces granulatus, 83, 313
Endoptychum agaricoides, 83, 313
Entoloma
abortivum, 25, 131
cuspidatum, 178
murraii, 178
salmoneum, 179
strictius, 179
vernum, 179
Exidia
glandulosa, 79, 305
recisa, 80, 305

Favolus
brasiliensis, 267
cucullatus, 268
Fistulina
hepatica, 59, 252
radicata, 268
Flammulina velutipes, 25, 131
Fomes
fasciatus, 59, 252
fomentarius, 252
Fomitopsis
meliae, 253
nivosa, 59, 253
palustris, 253

Galerina
autumnalis, 132
marginata, 25, 132
Galiella rufa, 77, 300
Ganoderma
applanatum, 59, 253
colossum, 254
curtisii, 254
lucidum, 59, 254
meredithae, 254
tsugae, 59, 255
zonatum, 60, 253, 254
Geastrum
caespitosus, 314
coronatum, 315
fimbriatum, 315
fornicatum, 83, 314
mirabilis, 84, 314
quadrifidum, 84, 315

rufescens, 315
saccatum, 84, 315
vulgatum, 84, 315
Gerronema strombodes, 26, 132
Gloeophyllum
sepiarium, 60, 255
striatum, 60, 256
trabeum, 256
Gloeoporus dichrous, 60, 257
Gomphus
clavatus, 15, 97
floccosus, ii, 15, 98
Grifola frondosa, 263
Gymnopilus
fulvosquamulosus, 26, 133
liquiritiae, 26, 133
luteofolius, 135
palmicola, 26, 134
penetrans, 26, 134
rufosquamulosus, 133
sapineus, 133, 134
spectabilis, 26, 134
Gymnopus
acervatus, 174, 189
alliaceus, 164
confluens, 26, 135
dryophilus, 26, 136
iocephalus, 27, 136
luxurians, 189
spongiosus, 135
subnudus, 135
*Gymnosporangium juniperi-
   virginiana*, 90, 333
Gyrodon
merulioides, 54, 230
rompelii, 231
Gyromitra
caroliniana, 69, 279
esculenta, 69, 279
infula, 69, 280
Gyroporus
castaneus, 54, 231
cyanescens var. *cyanescens*, 54, 232
cyanescens var. *violaceotinctus*, 232
phaeocyanescens, 232
subalbellus, 54, 232
umbrinosquamosus, 232

Hapalopilus
albo-citrinus, 257
croceus, 60, 257
nidulans, 60, 258
rutilans, 258
salmonicolor, 257
Hebeloma
crustuliniforme, 27, 136
mesophaeum, 137

Hebeloma (continued)
  sinapizans, 27, 137
  syriense, 27, 138
Helvella
  acetabulum, 77, 300
  albella, 69, 280
  crispa, 70, 280
  elastica, 70, 281
  griseoalba, 300
  lacunosa, 281
  macropus, 70, 281
Hericium
  americanum, 202
  coralloides, 47, 202
  erinaceus, 47, 202
  erinaceus ssp. *erinaceo-abietis*, 203
  ramosum, 202
Heterobasidion annosum, 253
Hexagonia
  hydnoides, 61, 258
  taxodii, 268
Hohenbuehelia
  geogenia, 138
  petaloides, 27, 138
Humaria hemisphaerica, 77, 301
Humidicutis marginata, 143
Hydnellum
  aurantiacum, 47, 203
  peckii, 203
  pineticola, 204
  spongiosipes, 47, 203
Hydnochaete olivaceum, 89, 328
Hydnopolyporus
  fimbriatus, 61, 251
  palmatus, 259
Hydnum
  adustum, 205
  repandum, 47, 204
  repandum var. *album*, 47, 204
  rufescens, 204
Hygrocybe
  acutoconica, 27, 138
  andersonii, 27, 139
  chlorophana, 27, 139
  coccinea, 140
  conica, 28, 140
  conica var. *conicoides*, 141
  conicoides, 28, 140
  flavescens, 140
  marginata, 143
  nitida, 28, 141
  ovina, 28, 141
  persistens, 139
  pratensis, 143
Hygrophorus
  acutoconicus, 139
  cantharellus, 141
  chlorophanus, 140

coccineus, 140
conicus, 140
hypothejus, 28, 142
marginatus var. *concolor*, xvi, 143
marginatus var. *marginatus*, 28, 142
marginatus var. *olivaceus*, 143
nitidus, 141
ovinus, 142
pratensis, 28, 143
roseibrunneus, 29, 143
subovinus, 142
Hymenochaete
  badio-ferruginea, 328
  tabacina, 328
Hypholoma
  capnoides, 144
  fasciculare, 29, 144
  sublateritium, 29, 144
  subviride, 144
Hypomyces
  chrysospermus, 90, 331
  hyalinus, 90, 113, 331
  lactifluorum, 90, 332
  luteovirens, 90, 195, 332
Hypoxylon
  fragiforme, 87, 324
  fuscum, 324
  rubiginosum, 87, 324
Hypsizygus
  elongatipes, 145
  marmoreus, 145
  tessulatus, 29, 144
  ulmarius, 145

Inocybe
  calamistrata, 186
  fastigiata, 146
  geophylla var. *geophylla*, 145
  geophylla var. *lilacina*, 29, 145
  rimosa, 29, 145
  rimosa var. *microsperma*, 146
Inonotus
  cuticularis, 260
  dryadeus, 61, 259
  hispidus, 61, 260
  quercustris, 61, 260
  rickii, 261
  texanus, 260
  tomentosus, 266
Irpex
  cinnamomeus, 328
  lacteus, 269, 272
Isaria flabelliformis, 294
Ischnoderma
  benzoinum, 261
  resinosum, 61, 261

Laccaria
  bicolor, 146
  laccata, 30, 146
  ochropurpurea, 30, 146, 337
  trullisata, 30, 147, 337
Lachnocladium semivestitum, 290
Lactarius
  agglutinatus, 30, 151
  alachuanus var. *alachuanus*, 30, 147
  alachuanus var. *amarissimus*, 148
  allardii, 30, 148
  atroviridis, 30, 148
  caeruleitinctus, 152
  carolinensis, 156
  chelidonium var. *chelidonium*, 30, 149
  chelidonium var. *chelidonioides*, 149, 157
  chrysorheus, 31, 150
  corrugis, 31, 150
  croceus, 31, 151
  deceptivus, 31, 151
  dunfordii, 156
  floridanus, 31, 152
  glaucescens, 31, 152
  hygrophoroides var. *hygrophoroides*, 153
  hygrophoroides var. *lavendulaceus*, 153
  hygrophoroides var. *rugatus*, 31, 153
  imperceptus, 31, 154
  indigo var. *diminutivus*, 32, 154
  indigo var. *indigo*, 32, 155
  louisii, 163
  luteolus, 32, 155
  maculatipes, 32, 156
  moschatus, 150
  paradoxus, 32, 157
  peckii var. *glaucescens*, 158
  peckii var. *lactolutescens*, 158
  peckii var. *peckii*, 32, 157
  piperatus var. *piperatus*, 153
  plumbeus, 149
  proximellus, 32, 158
  psammicola f. *glaber*, 159
  psammicola f. *psammicola*, 32, 158
  pseudodeliciosus var. *paradoxiformis*, 159
  pseudodeliciosus var. *pseudodeliciosus*, 33, 159
  quietus var. *incanus*, 33, 154
  salmoneus var. *curtisii*, 160
  salmoneus var. *salmoneus*, 33, 159
  similis, 148
  subplinthogalus, 33, 160

subvelutinus, 162
subvernalis var. *cokeri*, 160
sumstinei, 160
tomentoso-marginatus, 33, 161
vinaceorufescens, 150
volemus var. *flavus*, 33, 161
volemus var. *volemus*, 33, 162
xanthydrorheus, 33, 162
yazooensis, 34, 151
Laetiporus
  cincinnatus, 62, 261
  persicinus, 62, 262
  sulphureus, 62, 262
  sulphureus var. *semialbinus*, 262
Laternea angolensis, 274
Laxitextum bicolor, 330
Leccinum
  albellum, 54, 233
  crocipodium, 234
  nigrescens, 54, 233
  rugosiceps, 234
  snellii, 54, 234
Lentinellus
  cochleatus, 163
  omphalodes, 163
  ursinus, 34, 163
  vulpinus, 163
Lentinula
  boryana, 164
  detonsa, 164
  raphanica, 34, 163
Lentinus
  crinitis, 34, 164
  detonsus, 164
  lepideus, 165, 186
  levis, 34, 164
  siparius, 166
  strigosus, 34, 165
  suavissimus, 164
  tephroleucus, 34, 165
  tigrinus, 164
  torulosus, 165
Lenzites betulina, 62, 263
Leotia
  atrovirens, 292
  lubrica, 74, 292
  viscosa, 74, 292
Lepiota
  acutesquamosa, 166, 167
  americana, 168
  besseyi, 35, 168
  cepaestipes, 170
  clypeolaria, 35, 166
  cortinarius, 35, 168
  cristata, 35, 166
  humei, 169
  lutea, 170
  meleagris, 35, 167
  phaeosticta, 168

  phaeostictiformis, 35, 167
  viridiflava, 169
Leucoagaricus
  americanus, 35, 168
  hortensis, 35, 168
  meleagris, 167
  viridiflavoides, 169
Leucocoprinus
  birnbaumii, 36, 169
  cepaestipes, 36, 170
  fragilissimus, 36, 170
  lilacinogranulosus, 36, 170
  magnicystidiosus, 171
  meleagris, 167
Leucopaxillus albissimus, 122
Limacella
  glioderma, 171
  illinita var. *illinita*, 36, 171
  illinita var. *argillacea*, 171
Linderia columnata, 66, 274
Lycoperdon
  acuminatum, 84, 316
  americanum, 316
  echinatum, 316
  marginatum, 84, 316
  perlatum, 317
  pulcherrimum, 84, 317
  pyriforme, 85, 317
Lysurus
  borealis, 275
  gardneri, 66, 275
  periphragmoides, 67, 275

Macowanites arenicola, 85, 317
Macrocybe titans, 37, 171
Macrolepiota
  gracilenta, 172
  procera, 37, 172
  prominens, 172
  rachodes, 119, 168, 172, 173
  subrachodes, 37, 173
Macroscyphus macropus, 282
Marasmiellus
  albuscorticis, 37, 173
  nigripes, 174
  praeacutus, 174
Marasmius
  candidus, 174
  capillaris, 175
  cohaerens, 37, 174
  fulvoferrugineus, 38, 174
  magnisporus, 174
  pulcherripes, 175
  rotula, xii, 38, 175
  scorodonius, 174
  siccus, 175
Melanoleuca
  alboflavida, 175
  melaleuca, 38, 175

Meripilus
  giganteus, 263
  sumstinei, 62, 263
Merulius
  incarnatus, 328
  tremellosus, 329
Microglossum rufum, 74, 293
Microporellus
  dealbatus, 62, 264
  obovatus, 62, 264
Morchella
  angusticeps, 282
  conica, 282
  elata, 70, 282
  esculenta, 12, 70, 149, 282
  semilibera, 70, 282
Mucilopilus
  conicus, 241
  conicus var. *reticulatus*, 241
Mutinus
  bovinus, 276
  caninus, 276, 279
  curtisii, 276
  elegans, 67, 276
  ravenelii, 68, 276
Mycena
  epipterygia var. *epipterygioides*, 38, 176
  epipterygia var. *viscosa*, 38, 176
  haematopus, 38, 176
  luteopallens, 39, 177
  pura, 39, 177
  sanguinolenta, 177
Mycorrhaphium adustum, 48, 204

Naematoloma
  fasciculare, 144
  sublateritium, 144
Nigrofomes melanoporus, 265
Nigroporus vinosus, 63, 265
Nolanea
  murraii, 39, 178
  nodospora, 186
  quadrata, 39, 178
  strictia, 39, 179
  strictior, 179
  verna, 39, 179
Nomuraea atypicola, 76, 296

Octaviania ravenelii, 322
Omphalina ectypoides, 39, 180
Omphalotus
  illudens, 180
  olearius, 40, 180

Panaeolus
  phaleanarum, 115, 181
  semiovatus, 40, 180

Panaeolus (continued)
   separatus, 181
   solidipes, 115
Panellus
   stipticus, 40, 181
Panus
   crinitis, 164
   levis, 165
   rudis, 165
   siparius, 166
Paxillus
   atrotomentosus, 40, 181
   involutus, 40, 182
   panuoides, 182
Peridoxylon petersii, 87, 324
Peziza
   ammophila, 77, 302
   phyllogena, 301
   repanda, 77, 301
   vesiculosa, 301
Phaeolus schweinitzii, 63, 265
Phallogaster saccatus, 68, 277
Phallus
   hadriani, 68, 277
   impudicus, 277, 278
   ravenelii, 68, 277
Phellinus
   chrysoloma, 63, 266
   gilvus, 63, 266
   pini var. abietus, 266
Phellodon
   alboniger, 205
   niger, 48, 205
Phlebia
   incarnata, 89, 328
   radiata, 328
   tremellosa, 89, 329
Pholiota
   carbonaria, 183
   highlandensis, 41, 182
   lenta, 183
   polychroa, 41, 183
Phylloporus
   boletinoides, 55, 234
   leucomycelinus, 55, 235
   rhodoxanthus, 235
Phyllotopsis nidulans, 41, 183
Pisolithus
   arhizus, 318
   tinctorius, 85, 318
Pleurotus
   cystidiosus, 185
   dryinus, 41, 184
   ostreatus complex, 41, 184
   populinus, 185
   sapidus, 185
Plicaria anthracina, 78, 301
Plicaturopsis crispa, 15, 99

Pluteus
   atricapillus, 185
   cervinus, 41, 185
   magnus, 185
   pellitus, 185
   petasatus, 42, 185
Podostroma alutaceum, 295
Polyporus
   alveolaris, 268
   elegans, 63, 267
   gilvus, 266
   hydnoides, 258
   mori, 268
   tenuiculus, 63, 266
   varius, 63, 267
Poria versipora, 269
Porphyrellus
   gracilis, 207
   subflavidus, 208
Pouzarella nodospora, 42, 186
Psathyrella
   candolleana, 42, 186
   hydrophila, 42, 187
   piluliformis, 187
   umbonata, 42, 187
   velutina, 187
Pseudocoprinus
   disseminatus, 124
   plicatilis, 126
Pseudofavolus cucullatus, 64, 267
Pseudofistulina radicata, 64, 268
Psilocybe cubensis, 42, 188
Ptychoverpa bohemica, 283
Pulcherricium caeruleum, 89, 329
Pulveroboletus
   melleoluteus, 236
   ravenelii, 55, 235
Pycnoporus
   cinnabarinus, 269
   sanguineus, 64, 268

Ramaria
   concolor, 287
   conjunctipes, 289
   fennica, 289
   formosa, 289
   murrillii, 73, 288
   subbotrytis, 73, 289
Rhizopogon
   atlanticus, 85, 319
   nigrescens, 85, 318
Rhodocollybia
   butyracea, 42, 188
   maculata, 43, 189
   maculata var. scorzonerea, 189
Rhodocybe
   mundula, 190
   roseoavellanea, 43, 189
Rhopalogaster transversarium, 85, 319

Ripartitella brasiliensis, 43, 190
Russula
   adusta, 193
   aeruginea, 192
   amoenolens, 43, 190
   anomala, 196
   atropurpurea, 196
   ballouii, 43, 191
   compacta, 43, 191
   crustosa, 44, 192
   cyanoxantha, 195
   densifolia, 44, 192
   emetica, 195
   foetentula, 44, 193
   fragilis, 195
   fragrantissima, 193
   krombholzii, 196
   laurocerasi, 193
   nigricans, 193
   ornaticeps, 195
   pectinatoides, 191
   perlactea, 44, 193
   rosacea, 194
   rubripurpurea, 194
   sanguinea, 44, 194
   sericeonitans, 44, 194
   silvicola, 44, 195
   subalbidula, 45, 195
   subfoetens, 193
   subgraminicolor, 192
   variata, 192
   vinacea, 45, 196
   virescens, 192

Schizophyllum commune, 15, 99
Schizopora paradoxa, 64, 269
Scleroderma
   bovista, 86, 320
   citrinum, 320, 321
   floridanum, 86, 321
   geaster, 321
   macrorrhizon, 320
   meridionale, 86, 320
   polyrhizon, 86, 321
   texense, 86, 320
   verrucosum, 86, 321
Sebacina incrustans, 80, 305
Simblum
   sphaerocephalum, 276
   texense, 276
Spadicioides clavariae, 287
Sparassis
   crispa, 81, 307
   herbstii, 81, 307
   radicata, 307
   spathulata, 307
Spathularia
   flavida, 74, 293
   velutipes, 293

Spathulariopsis
  flavida, 293
  velutipes, 293
Sphaerosporella brunnea, 78, 301
Spinellus fusiger, 177
Spongipellis
  pachyodon, 64, 269
  unicolor, 270
Squamanita umbonata, 45, 196
Stereum
  complicatum, 89, 329
  diaphanum, 328
  frustulosum, 330
  hirsutum, 330
  ostrea, 89, 330
  rameale, 330
  striatum, 330
Strobilomyces
  confusus, 236
  dryophilus, 55, 236
  floccopus, 236
Strobilurus conigenoides, 45, 197
Stropharia
  bilamellata, 45, 197
  coronilla, 198
Suillellus pictiformis, 208
Suillus
  brevipes, 55, 239
  cothurnatus, 238
  decipiens, 55, 237
  granulatus, 238, 239
  hirtellus, 55, 237
  hirtellus ssp. *cheimonophilus*, 238
  pictus, 237
  pinorigidus, 238
  salmonicolor, 56, 238
  subalutaceus, 56, 238
  subaureus, 238
  subluteus, 238
Syzygospora mycetophila, 80, 306

Thelephora
  anthocephala var. *americana*, 71, 284
  anthocephala var. *anthocephala*, 284
  palmata, 71, 284
  terrestris f. *concrescens*, 71, 284
  terrestris f. *terrestris*, 285
  vialis, 71, 285

Trametes
  cubensis, 270
  elegans, 64, 270
  hirsuta, 271
  menziesii, 65, 270
  pubescens, 271
  versicolor, 65, 271
Tremella
  concrescens, 80, 306
  foliacea, 80, 306
  fuciformis, 80, 306
  lutescens, 304
  mycetophila, 306
Tremellodendropsis semivestitum, 73, 289
Trichaptum
  abietinum, 271-72
  biforme, 65, 271
Trichoglossum walteri, 74, 293
Tricholoma
  caligatum, 45, 198
  equestre, 198
  flavovirens, 45, 198
  inamoenum, 199
  odorum, 46, 199
  sejunctum, 198
  sulphurescens, 199
  sulphureum, 46, 199
  titans, 172
Tricholomopsis
  decora, 46, 200
  formosa, 46, 200
  rutilans, 200
  sulfureoides, 200
Trogia crispa, 99
Tylopilus
  alboater, 56, 239
  atratus, 240
  atronicotianus, 240
  ballouii, 56, 240
  conicus var. *conicus*, 56, 240
  conicus var. *reticulatus*, 240
  felleus, 214, 242, 243
  ferrugineus, 243
  gracilis, 207
  griseocarneus, 240
  indecisus, 243
  peralbidus, 209
  plumbeoviolaceus, 56, 241
  rhoadsiae, 56, 241

  rhodoconius, 242
  rubrobrunneus, 241
  subflavidus, 208
  tabacinus var. *amarus*, 57, 242
  tabacinus var. *tabacinus*, 57, 242
  variobrunneus, 57, 243
  violatinctus, 57, 243

Underwoodia columnaris, 75, 294
Urnula
  craterium, 78, 302
  geaster, 312
Ustulina deusta, 87, 325

Vascellum
  depressum, 322
  pratense, 86, 321
Verpa conica, 283
Volvariella
  bombycina, 201
  gloiocephala, 46, 200
  speciosa, 201
  volvacea, 201

Wolfina aurantiopsis, 78, 302

Xanthoconium
  affine var. *affine*, 244
  affine var. *maculosus*, 57, 244
  separans, 57, 245
  stramineum, 57, 245
Xerocomus
  griseus, 216
  pulverulentus, 223
Xeromphalia
  campanella, 46, 201
  kauffmanii, 201
Xylaria
  cubensis, 294
  magnoliae, 87, 325
  oxyacanthae, 88, 325
  persicaria, 88, 325
  polymorpha, 88, 326
  tentaculata, 88, 326
Xylobolus frustulatus, 89, 330
Xylocoremium flabelliforme, 75, 294

Zelleromyces
  cinnabarinus, 86, 322
  ravenelii, 322

PHOTO CREDITS

Except for the following, all photographs were taken by Alan E. Bessette, William C. Roody, Arleen R. Bessette, and Dail L. Dunaway.

Brian Akers: *Copelandia westii, Leucoagaricus hortensis, Psilocybe cubensis*

Ulla Benny: *Coprinus floridanus*

Meredith Blackwell: *Laetiporus persicinus*

Robert Chapman: *Bolbitius vitellinus, Chalciporus pseudorubinellus, Cudonia circinans*

Tim Geho: *Aseroe rubra* (A, B)

Don Gray: *Blumenavia angolensis, Lysurus periphragmoides, Mutinus elegans* (B)

Dan Guravich: *Amanita chlorinosma, Anellaria sepulchralis* (B), *Austroboletus subflavidus, Boletellus ananas, Boletus curtisii, Boletus flammans, Boletus floridanus, Boletus luridellus, Boletus rubricitrinus, Cantharellus cinnabarinus, Cantharellus confluens, Coprinus americanus, Cordyceps melolonthae, Craterellus odoratus, Gymnopilus fulvosquamulosus, Gyromitra caroliniana* (A, B), *Gyroporus subalbellus, Hygrophorus roseibrunneus, Lactarius agglutinatus, Lactarius yazooensis, Leccinum nigrescens, Lentinus crinitis* (A), *Lentinus strigosus, Lepiota besseyi, Leucocoprinus birnbaumii* (B), *Leucocoprinus lilacinogranulosus, Lysurus periphragmoides, Marasmius cohaerens, Marasmius fulvoferrugineus, Omphalotus olearius* (A), *Peziza ammophila, Phallus hadriani, Pholiota polychroa, Tylopilus tabacinus, Tylopilus tabacinus* var. *amarus*

Richard Homola: *Geastrum fornicatum, Hypoxylon fragiforme, Hypoxylon rubiginosum* (A, B), *Leucocoprinus birnbaumii* (A), *Psathyrella hydrophila*

David Lewis: *Hebeloma syriense*

Joe Liggio: *Pulveroboletus ravenelii, Underwoodia columnaris*

Owen McConnell: *Boletus dupainii, Xanthoconium stramineum*

Van Metzler: *Cortinarius lewisii, Lysurus gardneri, Macrocybe titans* (A)

Jim Murray: *Chorioactis geaster* (A, B, C)

Robert Williams: *Amanita mutabilis* (B), *Ganoderma zonatum, Macrocybe titans* (B), *Pholiota polychroa* (A), *Stropharia bilamellata*

David Work: *Flammulina velutipes, Ganoderma tsugae, Syzygospora mycetophila*

Alan E. Bessette is a professional mycologist and professor of biology at Utica College of Syracuse University. He has published numerous professional papers in the field of mycology and authored eighteen books, including *Mushrooms of Northeastern North America*, *Mushrooms of the Adirondacks*, *Mushrooms: A Quick Reference Guide to Mushrooms of North America*, and *Edible Wild Mushrooms of North America*. Alan has presented numerous mycological programs, is the scientific advisor to the Mid-York Mycological Society, and serves as a consultant for the New York State Poison Control Center. He has been the principal mycologist at national and regional forays. He received the 1987 Northeast Mycological Foray Service Award and the 1992 North American Mycological Association Award for Contributions to Amateur Mycology.

William C. Roody is a mycologist and field biologist with the Wildlife Diversity Program at the West Virginia Division of Natural Resources. Author of the *Preliminary Checklist of Macrofungi and Myxomycetes of West Virginia*, *North American Boletes — A Guide to the Fleshy Pored Mushrooms*, and *Mushrooms of West Virginia and the Central Appalachians*, he frequently lectures on various aspects of mycology and has taught many mushroom-identification workshops. He has won numerous awards in the North American Mycological Association's annual photo competition, including top honors in both the documentary and pictorial divisions. In 2000, he received the North American Mycological Association's Award for Contributions to Amateur Mycology.

Arleen R. Bessette is a mycologist and botanical photographer who has been collecting and studying wild mushrooms for many years. She has authored ten books, including *Taming the Wild Mushroom: A Culinary Guide to Market Foraging*, *Mushrooms of North America in Color*, *North American Boletes — A Guide to the Fleshy Pored Mushrooms*, *The Rainbow Beneath My Feet: A Mushroom Dyer's Field Guide*, and *Mushrooms of Cape Cod and the National Seashore*. She has won several national awards for her photography and teaches introductory courses in mycology, dyeing with mushrooms, and mycophagy for the North American Mycological Association and other organizations.

Dail L. Dunaway is a mycologist and botanical photographer who has been studying fungi for many years. A native of Mississippi, Dail has extensive knowledge of the mushroom flora of the southeastern United States and has collected and described many new species from this region. His excellent photographs have been included in several mycological publications.